Fluvial Processes – 2nd Edition

IAHR Monograph

Series editors

Peter A. Davies
Department of Civil Engineering,
The University of Dundee,
Dundee,
United Kingdom

Robert Ettema
Department of Civil and Environmental Engineering,
Colorado State University,
Fort Collins,
USA

The International Association for Hydro-Environment Engineering and Research (IAHR), founded in 1935, is a worldwide independent organisation of engineers and water specialists working in fields related to hydraulics and its practical application. Activities range from river and maritime hydraulics to water resources development and eco-hydraulics, through to ice engineering, hydroinformatics and continuing education and training. IAHR stimulates and promotes both research and its application, and, by doing so, strives to contribute to sustainable development, the optimisation of world water resources management and industrial flow processes. IAHR accomplishes its goals by a wide variety of member activities including: the establishment of working groups, congresses, specialty conferences, workshops, short courses; the commissioning and publication of journals, monographs and edited conference proceedings; involvement in international programmes such as UNESCO, WMO, IDNDR, GWP, ICSU, The World Water Forum; and by co-operation with other water-related (inter)national organisations. www.iahr.org

IAHR AIRH

Supported by
Spain Water
and IWHR, China

Fluvial Processes

2nd Edition

Ana Maria Ferreira da Silva & M. Selim Yalin

Queen's University, Kingston, Canada

CRC Press
Taylor & Francis Group
Boca Raton London New York

CRC Press is an imprint of the
Taylor & Francis Group, an **informa** business
A BALKEMA BOOK

Published by: CRC Press/Balkema
P.O. Box 447, 2300 AK Leiden, The Netherlands
e-mail: Pub.NL@taylorandfrancis.com
www.crcpress.com – www.taylorandfrancis.com

First issued in paperback 2020

ISBN-13: 978-0-367-57319-5 (pbk)
ISBN-13: 978-1-138-00138-1 (hbk)

Typeset by MPS Limited, Chennai, India

Library of Congress Cataloging-in-Publication Data

About the IAHR Book Series

An important function of any large international organisation representing the research, educational and practical components of its wide and varied membership is to disseminate the best elements of its discipline through learned works, specialised research publications and timely reviews. IAHR is particularly well-served in this regard by its flagship journals and by the extensive and wide body of substantive historical and reflective books that have been published through its auspices over the years. The IAHR Book Series is an initiative of IAHR, in partnership with CRC Press/Balkema – Taylor & Francis Group, aimed at presenting the state-of-the-art in themes relating to all areas of hydro-environment engineering and research.

The Book Series will assist researchers and professionals working in research and practice by bridging the knowledge gap and by improving knowledge transfer among groups involved in research, education and development. This Book Series includes Design Manuals and Monographs. The Design Manuals contain practical works, theory applied to practice based on multi-authors' work; the Monographs cover reference works, theoretical and state of the art works.

The first and one of the most successful IAHR publications was the influential book "*Turbulence Models and their Application in Hydraulics*" by W. Rodi, first published in 1984 by Balkema. I. Nezu's book "*Turbulence in Open Channel Flows*", also published by Balkema (in 1993), had an important impact on the field and, during the period 2000–2010, further authoritative texts (published directly by IAHR) included *Fluvial Processes* by S. Yalin and A. da Silva and *Hydraulicians in Europe* by W. Hager. All of these publications continue to strengthen the reach of IAHR and to serve as important intellectual reference points for the Association.

Since 2011, the Book Series is once again a partnership between CRC Press/Balkema – Taylor & Francis Group and the Technical Committees of IAHR and I look forward to helping bring to the global hydro-environment engineering and research community an exciting set of reference books that showcase the expertise within IAHR.

Peter A. Davies
University of Dundee, UK
(Series Editor)

Table of contents

Preface to the 2nd edition

Much research has been carried out on fluvial processes and river morphodynamics since the publication of the 1st edition of this book. In the preparation of this version, I tried as much as possible to incorporate the results of such research, while adhering to the central theme of the book as stated in the 'Preface to the 1st edition', and without losing its coherence and logical sequence.

An effort was made to incorporate also comments on the book kindly provided throughout the years by several colleagues and students.

All chapters were updated and revised. The book consists of seven chapters. Just like in the 1st edition, the first chapter is devoted to the basics of turbulent flow and sediment transport. Chapter 2 deals with the origin, existence regions and geometric properties of bed forms (ripples, dunes, alternate and multiple bars). Chapter 3 concerns the flow over undulated beds. Chapters 2 and 3 are a significantly expanded version of the original Chapter 2. The extension enabled the inclusion of: 1- a more comprehensive presentation and discussion of works focusing on the characteristic scales and dynamics of large-scale turbulence coherent structures; 2- a new section on the geometric properties of alternate bars; and 3- a new section on the internal structure of flow over dunes. The remaining chapters follow the same order as in the 1st edition. Accordingly, Chapter 4 concerns the regime concept, and focuses on the reason for the formation of regime channels and also the computation of their pertinent geometric characteristics. In Chapter 5 we provide an analysis on how regime processes take place, with the time-development of meandering and braiding streams being discussed in the light of the regime-trend. The formation of deltas is also briefly considered. Chapter 6 deals with the geometry and mechanics of meandering streams. This chapter too is considerably expanded in comparison to its counterpart in the 1st edition, incorporating a more detailed description of the mechanics of meandering flows and their interaction with the deformable bed, as derived from recent laboratory and numerical experiments. Finally, Chapter 7 deals with the computation of bed deformation and planimetric evolution of meandering streams. All chapters, with the exception of Chapter 2, include a related set of problems. However, the problems at the end of Chapter 3, which concern primarily the determination of the resistance factor and sediment transport rate of flow past undulated beds, rely on methods and equations regarding the geometric properties of bed forms introduced in Chapter 2 – and as such provide ample opportunity to practice the application to real problems of the methods introduced in Chapter 2. The supplementary material, consisting of the solutions manual for the sets of problems in the book and the computer programs

RFACTOR and BHS-STABLE (both in Fortran and MATLAB), can be downloaded from the book site on the CRC Press website.

This book is published on the 10th anniversary of Prof. Yalin's death. Even if he could not factually contribute with his writing to the preparation of this edition, he has contributed by remaining a great source of inspiration, and more directly through our work together and many pertinent discussions up to 2007.

The book includes several methods due to the authors whose development and verification rest on experimental data. An effort was made to include in the work also data from the most recent data sources. All data sources are listed in Appendices A to E. Several of the explanations in the book are guided by the authors' views and convictions. It should, however, be pointed out that some of the topics dealt with remain a matter of debate (e.g. the origin of bed forms), while in other cases we have at present only incomplete descriptions of the phenomenon (e.g. large-scale turbulence). A special effort was made to identify such topics, and wherever appropriate, suggest subjects for future research.

My sincere thanks go to Larry Harris, from the Creative Services Unit at Queen's University. I would also like to express my thanks to my former Ph.D. student Arash Kanani, for his invaluable assistance with the typesetting software at the early stages of preparation of this manuscript; and my present Ph.D. student Yunshuo Cheng, for converting the Fortran programs RFACTOR and BHS-STABLE into MATLAB. Finally, but not least, I would like to thank Janjaap Blom and Lukas Goosen and their teams at CRC Press, Taylor & Francis Group, for their excellent work and support in the preparation of this book.

Ana Maria Ferreira da Silva
Kingston, Ontario
June, 2017

Preface to the 1st edition

A stream flowing in alluvium deforms its bed surface so as to form ripples, dunes, bars, etc., and, in many instances, it deforms its channel as a whole so as to create, in plan view, meandering or braiding patterns. One can say that, in general, an alluvial stream and its deformable boundary undergo a variety of *fluvial processes* which lead to the emergence of a variety of *alluvial forms*.

This monograph concerns the understanding and quantitative formulation of fluvial processes and the associated alluvial forms. It is designed for researchers and graduate students of hydraulic engineering, water resources, and the related branches of earth sciences. However, it may have some appeal to practicing professionals as well.

The central theme of the book is that the *initiation* of periodic (in flow direction) large-scale alluvial forms is due to large-scale *turbulence*, the subsequent *time-development* of the so-initiated alluvial forms being guided by the *regime trend*.

The text is of a deductive nature: the content of any chapter presupposes the knowledge of the contents of preceding chapters; hence, the text might not appear comprehensible if it is not read in sequence. Dimensional methods are extensively used, and the importance of the "agreement with experiment" is stressed. The stability-approach is not a part of the research field of the authors, and it is not used in this text. (The reader interested in the study of fluvial processes by stability is referred to the excellent works of G. Seminara, G. Parker, M. Tubino, T. Hayashi).

It was gratifying for the senior author to learn that some of his previous books are used in graduate courses in some universities. Considering this, the authors have tried to make this text "student-oriented". Thus each chapter is supplemented by a set of related problems, and, wherever appropriate, some topics for future research are suggested. The FORTRAN programs for the computation of resistance factor and regime channels are also included.

The representation of an experimental point-pattern by an appropriate equation which can be used for computational purposes (in short, by a "comp-eq.") is always desirable. Hence a number of "comp-eqs." is suggested in this text. These equations have no claim other than that their graphs pass through the midst of the respective point-patterns.

The first chapter of this book presents the basics of turbulent flow and sediment transport; the second deals with the bed forms and flow resistance. These two (classical) chapters are but the updated and revised extensions of their counterparts in the book "River Mechanics" (M.S. Yalin, Pergamon Press, 1992). Chapter 3 concerns the regime concept and its thermodynamic formulation. The development of meandering and braiding streams in the light of the regime trend is discussed in Chapter 4, where the formation of deltas is also considered. Much of the ongoing research in fluvial hydraulics is related to meandering streams, and Chapter 5 is devoted exclusively to the study of the geometry and mechanics of these streams. The computation of flow, bed deformation, and migration-expansion of meandering streams forms the topic of Chapter 6.

The authors are grateful to Prof. H. Scheuerlein (Chairman Division III – IAHR) for his continual and enthusiastic help and encouragement. They feel also indebted to Prof. G. Di Silvio and Prof. M. Jaeggi, for reviewing this manuscript and making a number of valuable suggestions. For any imperfection that may still remain, it is solely the authors who are to be blamed.

The authors would like to express their thanks to Dr. T. J. Harris, Dean of the Faculty of Applied Science, Queen's University, and to Dr. D. Turcke, Head of the Department of Civil Engineering, Queen's University, for their generous financial support. Their thanks go also to Larry Harris, Graphic Design Unit, Queen's University.

M. S. Yalin and A. M. Ferreira da Silva
Kingston, Ontario
August, 2000

Scope of the monograph

In this text the term *alluvium* is used as an abbreviation for *cohesionless* granular material or medium, *alluvial stream* being the flow whose channel, or only the channel bed, is alluvial.

The natural alluvial streams are usually "wide": their aspect ratio (width-to-depth ratio) usually is larger than ≈ 10, say. The aspect ratio increases with the flow rate, and in large natural rivers this ratio is often a three-digit number. The (turbulent) flow in such rivers is almost always sub-critical.

In physical sciences, a natural process is studied with the aid of its idealized counterpart, where the "natural arbitrariness" is removed, and the process is brought into a form suitable for mathematical treatment; a fluvial process is no exception.

The present text "aims" at large alluvial streams (rivers). Hence, it is assumed throughout this text that the idealized alluvial stream (ideal river) is *wide*, and that its flow is *turbulent* and *sub-critical*.

The studies of regime channels, and of meandering and braiding streams are carried out nowadays for a constant (representative) flow rate Q – and so it is done in the following. (The authors' views on the representative Q are summarized in Section 4.7).

List of symbols

1. General

f_A	dimensional function determining a quantity A
$\Phi_A, \phi_A, \Psi_A, \psi_A$	dimensionless functions determining a quantity A
$\alpha, \alpha', \beta, \beta'$	dimensionless coefficients (not necessarily constants) in the expression of a quantity
$\approx$	approximately equal to, comparable with
$\sim$	proportional to (proportionality factor may not be a constant)
∇	nabla-operator ("del")

2. Average values

$\overline{A}$	vertically-averaged value of a quantity A
A_m	cross-sectional average value of a quantity A
A_{av}	channel average value of a quantity A

(see Sub-section 6.3.1 for definitions of average values stated)

3. Subscripts

a, O	mark the values of a quantity at the apex- and crossover-sections of a meandering stream, respectively
b	marks the value of a quantity at the bed, or related to the bed
cr	marks the value of a quantity corresponding to the initiation of sediment transport (to the "critical stage")
max	marks the maximum value of a quantity
min	marks the minimum value of a quantity
R	marks the regime value of a quantity
0	marks, as a rule, the value of a quantity at $t = 0$; exceptions are τ_0, c_{M0}, and θ_0

4. Coordinates

t	time
x	direction of rectilinear flow; also general direction of meandering flow
y	direction horizontally perpendicular to x
z	vertical direction in general (also elevation of a point – see "5. Pertinent quantities")
z^+	dimensionless "wall" coordinate ($= \upsilon_* z/\nu$)

l	longitudinal coordinate of a meandering flow
l_c	longitudinal coordinate along the centreline of a meandering flow
n	radial coordinate of a meandering flow; $n=0$ at the flow centreline
n_s	direction horizontally normal to the streamline s; radial natural coordinate
r	radial coordinate of a meandering flow; $r=0$ at the centre of channel curvature
s	streamline; longitudinal natural coordinate
ϕ	angular polar coordinate
ξ_c	dimensionless counterpart of l_c ($\xi_c = l_c/L$)
η	dimensionless counterpart of n ($\eta = n/B$)
ζ	dimensionless counterpart of z ($\zeta = z/h_{av}$)

5. Pertinent quantities

A	mechanical quantity in general; also area
a_1, a_2	consecutive apex-sections (in Chapter 4)
a_i, a_{i+1}	consecutive apex-sections (in Chapters 6 and 7)
A_*	energy-related property of flow (subjected to minimization during the regime channel formation); also coefficient in the bed-load formula of H.A. Einstein (Problem 1.11)
B	flow width at the free surface
B_c	width of the central region of the stream cross-section
B_s	roughness function
c	total dimensionless (Chézy) resistance factor
c_f	pure friction component of c
c_Δ	bed form component of c
c_M	total local dimensionless resistance factor of a meandering flow
c_{M0}	local bed resistance component of c_M (bed form effect included)
C	local dimensionless volumetric concentration of suspended-load
C_ϵ	the value of C at $z=\epsilon$
CV	control volume
D	typical grain size (usually D_{50})
e_i	internal energy per unit fluid mass at a space point m
e_k	kinetic energy per unit fluid mass at a space point m
e_p	pressure work per unit fluid mass at a space point m
e	total energy per unit fluid mass at a space point m ($e = e_i + e_k + e_p$)
E_i	internal energy of the fluid in Sys or CV
E	total energy of the fluid in Sys or CV
e_V, e_H	eddy-burst forming eddies of vertical and horizontal turbulence, respectively
$\mathcal{F}_i$	energy transfer per unit time through a cross-section A_i (energy flux plus the displacement work)
g	acceleration due to gravity
h	flow depth
$\mathbf{i}_\alpha$	unit vector in the direction of α
J	longitudinal free surface slope
$\mathcal{J}$	radial free surface slope

$J_0(\theta_0)$	Bessel function of first kind and zero-th order (of θ_0)
k_s	granular roughness of bed surface
K_s	total effective bed roughness: granular+bed form roughness
L	meander length (measured along l_c)
L_V, L_H	length scale of large-scale turbulence vertical and horizontal coherent structures, respectively
n	number of rows of horizontal coherent structures (in plan view)
O_1, O_2	consecutive crossover-sections in Chapter 4; also locations along the flow direction where large-scale coherent structures originate (Chapter 2)
O_i, O_{i+1}	consecutive crossover-sections in Chapters 6 and 7
p	porosity of granular material
Q	flow rate
Q_{bf}	bankfull flow rate
Q_s	volumetric sediment transport rate (through the whole cross-section of flow)
$\dot{Q}_*$	net heat exchange time rate between surroundings and *Sys* or *CV*
q	specific flow rate ($q = Q/B$)
q_{sb}	specific volumetric bed-load rate (within ϵ)
q_{ss}	specific volumetric suspended-load rate (within $h - \epsilon$)
q_s	total specific volumetric transport rate ($q_s = q_{sb} + q_{ss}$)
R	curvature radius of the centreline of a meandering flow
$\mathcal{R}$	hydraulic radius
r_s	curvature radius of a streamline s
S	bed slope
S_c	bed slope along the centreline of a meandering flow
S_v	valley slope
s_*	specific entropy
S_*	entropy of *Sys* or *CV*
Sys	fluid system (coincidental with *CV*)
$\mathbf{T}$	local "side stress+bed shear stress"-resultant vector
T°	absolute temperature (degree Kelvin)
T_b	development duration of the bed of a meandering stream
$\hat{T}_0$	development duration of the flow width
T_i	development duration of a characteristic i of flow or its (moveable) boundary
T_V, T_H	time scale of large-scale turbulence vertical and horizontal coherent structures, respectively
T_R	development duration of the regime channel
T_Δ	development duration of bed forms (sand waves)
$(T_\Delta)_i$	development duration of bed form i ($i = a$ (alternate bars); $= d$ (dunes); $= r$ (ripples); etc.)
$\mathbf{U}$	local flow velocity vector
U	magnitude of $\mathbf{U}$
$\mathbf{U}_b$	flow velocity vector at the bed
u_b	scalar projection of $\mathbf{U}_b$ in longitudinal direction; also velocity at the bed of a two-dimensional flow in a straight channel

u, υ, w	scalar projections of **U** in longitudinal, transversal and vertical directions, respectively
V	volume
υ_*	shear velocity ($\upsilon_* = \sqrt{\tau_0/\rho}$)
υ_β	translatory component of radial velocity of a meandering flow
υ_Γ	cross-circulatory component of radial velocity of a meandering flow
W	local displacement velocity of the flow boundary surface (in the direction normal to this surface)
W_1', W_2'	local radial displacement velocities of the inner bank 1 and outer bank 2, respectively (in horizontal plan)
W'	local radial displacement velocity of the centreline
W_a'	value of W' at the apex-section
$\mathcal{W}_x$	migration velocity of a meandering channel (in x-direction)
W_x, W_l	migration velocity of bed forms along x and l, respectively
$\dot{W}_*$	net work exchange time rate between surroundings and *Sys* or *CV*
w_s	terminal (settling) velocity of grains
z	elevation of a point (also vertical direction in general – see "4. Coordinates")
z_b	bed elevation at any time t
$(z_b)_0$	bed elevation at $t = 0$
$(z_b)_T$	bed elevation at $t = T_b$
z_c	elevation of the centroid
z_f	elevation of the free surface at any time t ($z_f = z_b + h$)
z'	positive or negative increment of z_b at the time t ($z' = z_b - (z_b)_0$)
z_T'	positive or negative increment of z_b at the time T_b ($z_T' = (z_b)_T - (z_b)_0$)
γ	specific weight of fluid; also coefficient in Eq. (2.5)
γ_s	specific weight of grains in fluid
Γ	cross-circulation
Δ, Λ, δ	developed bed form height, bed form length, and bed form steepness ($\delta = \Delta/\Lambda$) respectively, in general; δ stands also for boundary layer thickness (Sub-section 2.2.1)
$\Delta_i, \Lambda_i, \delta_i$	developed bed form height, bed form length, and bed form steepness corresponding to the bed form i ($i = a$ (alternate bars); $= d$ (dunes); $= n$ (n-row bars); $= r$ (ripples))
ϵ	thickness of bed-load region
θ	deflection angle of a meandering flow at a section l_c
θ_0	deflection angle at $l_c = 0$
Θ	dimensionless time (or stage) of the channel development
κ	von Kármán constant (≈ 0.4)
λ_c	ratio of the resistance factor c to the friction factor c_f ($\lambda_c = c/c_f$); also distance (normalized by L) measured along the centreline from the upstream crossover of a meander loop (with positive R) to the upstream end of the erosion-deposition zone partially or fully contained in that loop and exhibiting erosion at the left bank (Chapter 6)
Λ_M	meander wavelength
ν	fluid kinematic viscosity
ν_t	kinematic eddy viscosity

ρ	fluid density
ρ_s	grain density
σ	sinuosity of a meandering flow ($\sigma = L/\Lambda_M$); also area of the side surface of an imaginary vertical prism of height h and base area A (Chapter 1)
τ	shear stress
τ_0	magnitude of the bed shear stress vector $\vec{\tau}_0$
χ	magnitude of the local "side stress" resultant vector $\vec{\chi}$
ϕ_r	angle of repose
ω	deviation angle (angle between the streamlines s and coordinate lines l of a meandering flow)
ω_c	value of ω along the centreline of a meandering flow
ω_z	vertical component of vorticity (Chapter 2)

6. Dimensionless combinations

Fr	flow Froude number ($Fr = \mathcal{V}^2/(gh)$, where $\mathcal{V}$ is a typical flow velocity ($\mathcal{V} = \bar{u}$, u_m, u_{av}, ..., etc.))
Re	flow Reynolds number ($Re = \mathcal{V}h/\nu$, where $\mathcal{V}$ is as above)
Re_*	roughness Reynolds number ($Re_* = \upsilon_* k_s/\nu$)
X	grain size Reynolds number ($X = \upsilon_* D/\nu$)
Y	mobility number ($Y = \rho\upsilon_*^2/(\gamma_s D)$)
Z	relative depth ($Z = h/D$)
W	density ratio ($W = \rho_s/\rho$)
Ξ^3	material number ($\Xi^3 = X^2/Y = \gamma_s D^3/(\rho\nu^2)$)
η_*	relative flow intensity ($\eta_* = Y/Y_{cr}$)
ϕ	Einstein's dimensionless transport rate ($\phi = \rho^{1/2} q_s/(\gamma_s^{1/2} D^{3/2})$)
N	dimensionless specific flow rate ($N = Q/(BD\upsilon_{*cr})$)

7. Abbreviations

[CD]	*L*/2-long convergence-divergence flow region
CS	turbulence coherent structure; also control surface (Chapter 4)
LSHCS	large-scale horizontal coherent structure (occasionally referred to as HCS)
LSVCS	large-scale vertical coherent structure (occasionally referred to as VCS)

Chapter 1

Fundamentals

Any flow-induced deformation of an alluvial channel (or only of its bed) is by means of the grain motion *en mass*, i.e. by means of *sediment transport*. Hence we start this text by considering some aspects of sediment transport and related topics.

1.1 SEDIMENT TRANSPORT

The grains forming the boundary of an alluvial stream have a finite weight (in fluid) and a finite coefficient of friction. Consequently they cannot be brought into motion by the flow if the shear stress τ_0 acting at a point on the flow boundary (Figure 1.1) is less than a certain "critical" value $(\tau_0)_{cr}$ (corresponding to that point). This means that the sediment can be transported only over that part $A'B'$ of the flow boundary $AA'B'B$ where

$$\eta_* = \frac{\tau_0}{(\tau_0)cr} > 1. \tag{1.1}$$

In this text the ratio η_* will be referred to as *relative tractive force* or *relative flow intensity*. Only the grains forming the uppermost grain layer of the flow boundary can be detached and transported by flow (in the flow direction x). The detachment is due to τ_0; the transport, to the longitudinal flow velocities u.

If η_* is smaller than a certain value, η_{*1} say $(1 < \eta_* < \eta_{*1})$, then the grains are transported by the deterministic "jumps" P_b in the neighbourhood of the bed ($k_s < z < \epsilon$ in Figure 1.2). This mode of grain transport is referred to as *bed-load*.

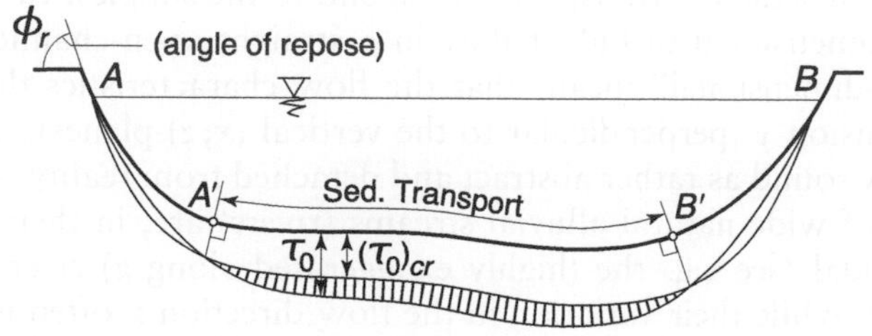

Figure 1.1 Schematic representation of cross-sectional distribution of shear stress acting on the flow boundary.

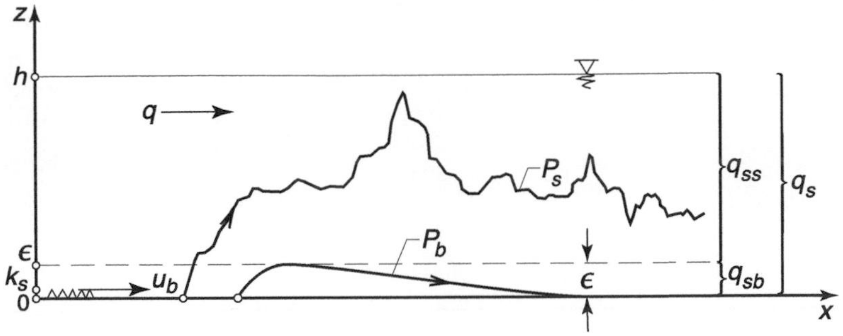

Figure 1.2 Definition sketch: modes of sediment transport.

If η_* is "large" ($\eta_* > \eta_{*1}$), then some of the transported grains "diffuse", by turbulence, into the remaining flow region $\epsilon < z < h$, while some others are still transported in the form of bed-load, i.e. within $k_s < z < \epsilon$. The transport of grains within $\epsilon < z < h$, along the probabilistic paths P_s, is referred to as *suspended-load*. It should thus be clear that suspended-load is not a "substitute for", but an "addition to" the bed-load.

It appears that η_{*1} must be a certain function of the grain size Reynolds number $v_* D/\nu$ – which will be denoted in this text by X (see Section 1.3 later on). It has not yet been settled what form this function must possess; for contemporary views on the topic see, e.g., Cheng and Chiew (1999).

The total volume of grains passing through a flow cross-section per unit flow width and per unit time is referred to as the specific and volumetric *total transport rate* q_s. This is the sum of the *bed-load rate* q_{sb} and the *suspended-load rate* q_{ss} (each of which is, of course, also specific and volumetric):

$$q_s = q_{sb} + q_{ss} \qquad (\text{with } q_s \geq q_{sb};\ q_s > q_{ss}). \tag{1.2}$$

The dimension of any q_{si} is

$$[q_{si}] = [\text{length}] \cdot [\text{velocity}]. \tag{1.3}$$

1.2 TURBULENT FLOW

1.2.1 General

The basic laws of sediment transport correspond to the simplest case of a steady and uniform two-dimensional turbulent flow in a straight open-channel. In the present context, "two-dimensional" means that the flow characteristics do not vary along the *third* dimension y (perpendicular to the vertical $(x; z)$-planes). The simplest case mentioned may sound as rather abstract and detached from reality. This is not so; the cross-sections of wide natural alluvial streams (rivers) are, in their straight reaches, nearly trapezoidal (see e.g. the (highly exaggerated along z) river cross-sections in Figures 1.3a,b), while their variation in the flow direction x often is insignificant. In such reaches, the flow in the (practically unaffected by the bank friction) "central region" B_c (Figure 1.3a) approximates closely to a uniform two-dimensional flow.

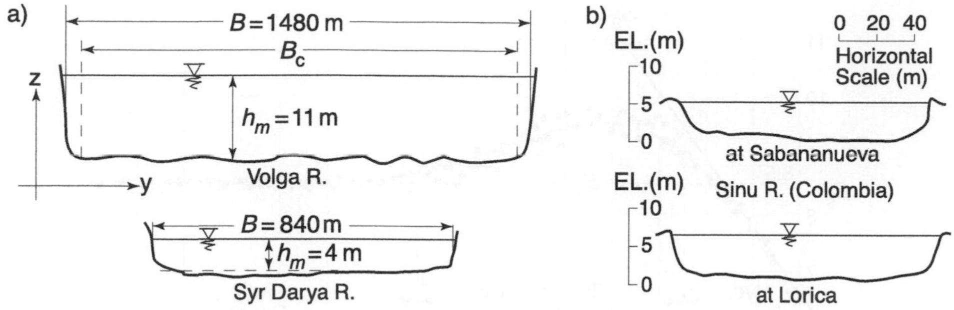

Figure 1.3 Examples of river cross-sections in straight reaches: (a) Volga and Syr Darya Rivers (adapted from Velikanov 1995); (b) Sinu River (adapted from Monsalve and Silva 1983).

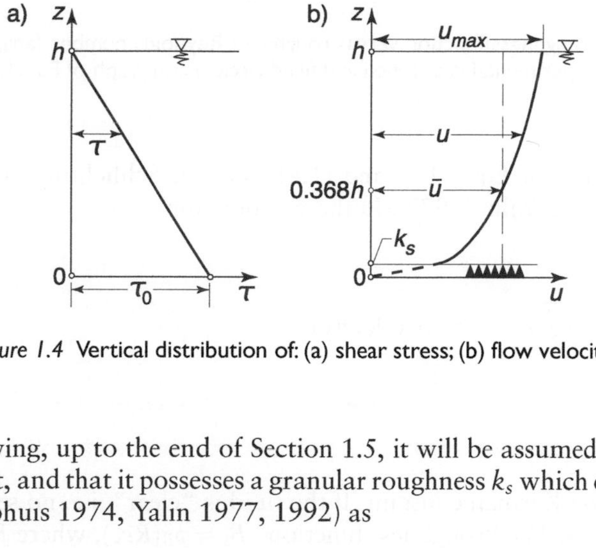

Figure 1.4 Vertical distribution of: (a) shear stress; (b) flow velocity.

In the following, up to the end of Section 1.5, it will be assumed that the mobile bed surface is flat, and that it possesses a granular roughness k_s which can be evaluated (following Kamphuis 1974, Yalin 1977, 1992) as

$$k_s \approx 2D. \tag{1.4}$$

1.2.2 Vertical distributions of shear stress and flow velocity

The distribution of the shear stresses τ $(=\tau_{zx})$ along z in the above described two-dimensional flow is given by the linear form (see Figure 1.4)

$$\tau = \tau_0 \left(1 - \frac{z}{h}\right), \tag{1.5}$$

while for the distribution of (time-averaged) flow velocities u, we have the logarithmic form

$$\frac{u}{v_*} = \frac{1}{\kappa} \ln \frac{z}{k_s} + B_s = \frac{1}{\kappa} \ln \left(A_s \frac{z}{k_s}\right) \qquad (\text{with } \kappa \approx 0.4). \tag{1.6}$$

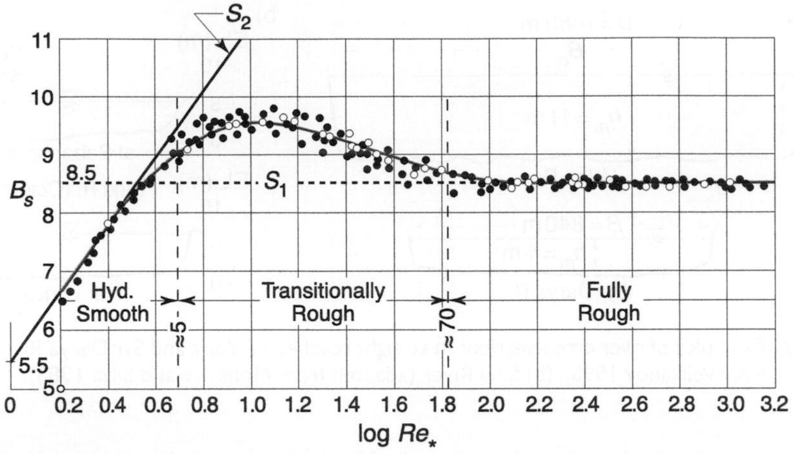

Figure 1.5 Plot of roughness function versus roughness Reynolds number (adapted from Schlichting 1968): experimental data (open and filled circles) and graph of Eq. (1.10) (solid red line).

For the derivation of Eqs. (1.5) and (1.6), see e.g. Schlichting (1968), Monin and Yaglom (1971) and Yalin (1977). In these expressions

$$\tau_0 = \gamma S h, \tag{1.7}$$

$$\upsilon_* = \sqrt{\frac{\tau_0}{\rho}} = \sqrt{gSh} \quad \text{(shear velocity)}, \tag{1.8}$$

and[1]

$$A_s = e^{\kappa B_s}. \tag{1.9}$$

Here κ is the von Kármán constant. If the fluid is "clear", i.e. no sediment in suspension, then $\kappa \approx 0.4$. The "roughness function" $B_s = \phi_B(Re_*)$, where $Re_* = \upsilon_* k_s/\nu$ is the *roughness Reynolds number*, is determined by the experimental curve in Figure 1.5. This curve is well represented, throughout its existence-region ($0.2 < \log Re_* < 3.2$), by the comp-eq.

$$B_s = (2.5 \ln Re_* + 5.5)e^{-0.0705(\ln Re_*)^{2.55}} + 8.5\left[1 - e^{-0.0594(\ln Re_*)^{2.55}}\right] \tag{1.10}$$

introduced in da Silva and Bolisetti (2000). The solid line in Figure 1.5 is the graph of Eq. (1.10).

Note that:

1 If $Re_* <\approx 5$, then the turbulent flow is said to be in the *hydraulically smooth regime*; if $\approx 5 < Re_* <\approx 70$, then it is in the *transitionally rough regime*; and if

[1] If the two-dimensional flow is not uniform, then its free surface slope J would differ from the bed slope S, and the value of τ_0 would be given by $\tau_0 = \gamma J h$ ($\neq \gamma S h$).

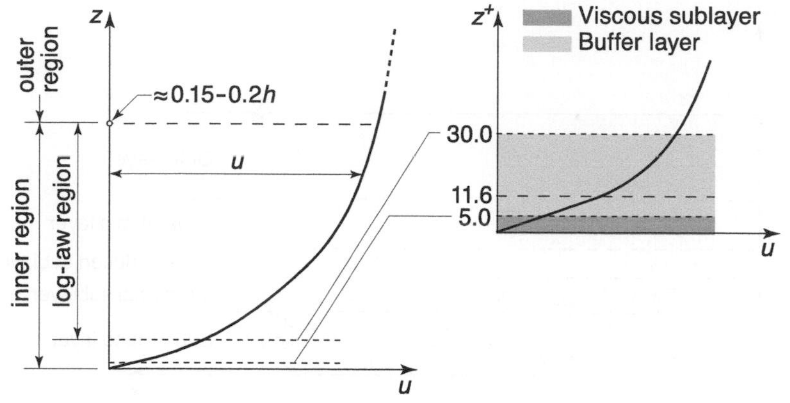

Figure 1.6 Schematic representation of the velocity profile in a hydraulically smooth flow.

$Re_* > \approx 70$, then it is in the *fully rough regime* (or it is a fully rough turbulent flow).

2 For practical purposes, it is usually assumed that Eq. (1.6) represents the velocity profile throughout the region $z = z_{min}$ to h, where z_{min} is either equal to $11.6\nu/\upsilon_*$ or k_s, whichever is the larger. This approach is adopted throughout this book.

With regard to point 2 above, it should, however, be noted that in reality the vertical velocity profile is more complex than implied by Eq. (1.6). Indeed, even though generally regarded as a good approximation to the entire velocity profile (above z_{min}), in fact Eq. (1.6) is strictly valid within the inner flow region ($z/h < \approx 0.15$ to 0.2). The deviations of the velocity profile from the logarithmic law in the outer region ($z/h > \approx 0.15$ to 0.2) can be adequately taken into account by adding to Eq. (1.6) an additional term, known as "wake function" $\phi_w(z/h)$. The most commonly invoked wake function is that by Coles (1956), which can be expressed as $\phi_w(z/h) = (2\Pi/\kappa)\cdot \sin^2(\Pi z/2h)$ (see also Hinze 1975). Here Π is the Coles's wake strength parameter. However, other formulae for the wake function, or modified versions of Coles' original function, are available in the literature (see e.g. Granville 1976, Krogstad et al. 1992, and Guo et al. 2005).[2]

The conditions near the bed in the case of hydraulically smooth flows and the effect of relative submergence in the case of rough beds deserve also some further discussion.

Under hydraulically smooth conditions, the flow in a thin region near the bed consists of two layers, namely the viscous sublayer ($z^+ < \approx 5$) and the buffer layer ($\approx 5 < z^+ < \approx 30$) (see Figure 1.6). Here $z^+ = \upsilon_* z/\nu$. Within the viscous sublayer, the u-distribution is given by the linear form $u/\upsilon_* = \upsilon_* z/\nu$. In the buffer layer, neither this linear form nor the logarithmic law (Eq. (1.6)) hold. While direct expressions for the u-distribution within the buffer layer are not available (Nezu and Nakagawa 1993,

[2]Since the sediment is transported mainly in the lower fluid layers, the function $\phi_w(z/h)$ will be disregarded in this text.

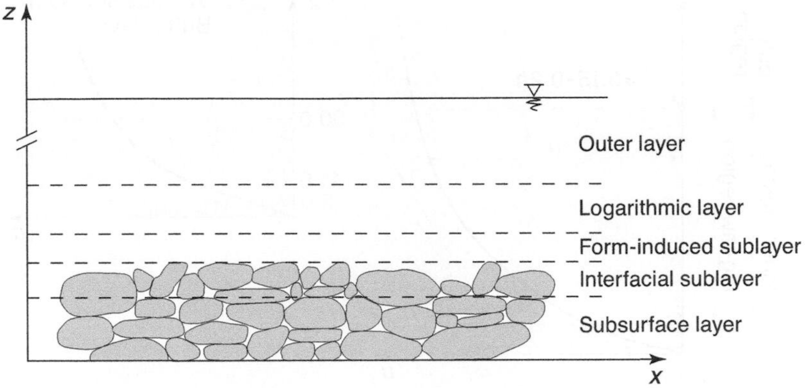

Figure 1.7 Flow layers in a fully rough flow (after Nikora et al. 2001).

Nezu 2005), the just mentioned linear form provides a reasonable approximation to the velocity profile up to $z^+ \approx 11.6$; while, on the other hand, the logarithmic law reasonably approximates the velocity profile within $\approx 11.6 < z^+ < \approx 30$ (see e.g. Monin and Yaglom 1971). This explains the reason for the statement "... where z_{min} is equal to $11.6\nu/\upsilon_*$ ..." in the point 2 above.

For the case of fully rough flows, it has since long been established that, as long as $h \gg k_s$, the vertical velocity profile exhibits properties similar to those of hydraulically smooth flows – in the sense that the profile follows the logarithmic law within the inner region down to some distance from the bed (see e.g. Raupach et al. 1991, Nezu and Nakagawa 1993, Jiménez 2004). Open-channel flows with small relative submergence (small h/k_s), such as gravel rivers where the roughness elements can protrude the flow up to the free surface, have also been the focus of considerable research especially in more recent years (see e.g. Dittrich and Koll 1997, Nikora et al. 2001, Koll 2006, McSherry et al. 2016, among others). It follows that fully rough flows can most appropriately be sub-divided into five layers (see Figure 1.7): outer layer, logarithmic layer, form-induced sublayer (just above the roughness crests where the flow is influenced by individual roughness elements and the velocity u acquires a linear profile), interfacial sublayer (between roughness crests and throughs), and in the case of a permeable bed, also a sub-surface layer (Nikora et al. 2001, Koll 2006). Accordingly, the flow can further be sub-divided into three types: 1- flow with high relative submergence ($h \gg k_s$), exhibiting all the above layers (and in which the outer layer is similar to the outer layer of hydraulically smooth flows, in the sense of Nezu and Nakagawa 1993); 2- flow with small relative submergence ($k_s \leq h < 2$ to $5k_s$), with the form-induced sublayer as the upper flow region (and where therefore the logarithmic layer is suppressed); and 3- flow with a partially inundated bed ($h < k_s$), with the interfacial sublayer as the upper flow region. Clearly, Eq. (1.6) applies only to the first of these three types – in which case for practical purposes, and as follows from point 2 above, Eq. (1.6) is assumed to provide a reasonable approximation of the velocity profile from $z_{min} = k_s$ to h. Figure 1.4b illustrates this case.

1.2.3 Average flow velocity

The average velocity of a two-dimensional flow, that is, its vertically-averaged velocity $\overline{u}$, is equal to u at the relative level $z/h = e^{-1} = 0.368$ – for any regime of the turbulent flow (see e.g. Yalin 1977). i.e.

$$\frac{\overline{u}}{v_*} = \frac{1}{\kappa}\ln\left(0.368\frac{h}{k_s}\right) + B_s = \frac{1}{\kappa}\ln\left(0.368A_s\frac{h}{k_s}\right) \tag{1.11}$$

which, in the case of a fully rough turbulent flow ($Re_* > \approx 70$), where $B_s = 8.5$, reduces to

$$\frac{\overline{u}}{v_*} = \frac{1}{\kappa}\ln\left(11\frac{h}{k_s}\right). \tag{1.12}$$

This relation is sometimes approximated by the power form (see e.g. Raudkivi 1990, Chen 1991, Yen 2002, Cheng 2007)

$$\frac{\overline{u}}{v_*} \approx 7.66\left(\frac{h}{k_s}\right)^{1/6}. \tag{1.13}$$

The ratio $\overline{u}/v_*$ is called the (dimensionless Chézy) *friction factor*. Henceforth this will be denoted by c_f:

$$c_f = \frac{\overline{u}}{v_*}. \tag{1.14}$$

Note that Eq. (1.11), which gives the $\overline{u}$-value of a uniform two-dimensional flow past a flat bed having a granular roughness $k_s \approx 2D$, gives, at the same time, the value of the friction factor of that flow:

$$c_f = \frac{1}{\kappa}\ln\left(0.368\frac{h}{k_s}\right) + B_s. \tag{1.15}$$

Using Eq. (1.8), one can express Eq. (1.14) in the form of the (Chézy's) *friction equation*

$$\overline{u} = c_f\sqrt{gSh}. \tag{1.16}$$

1.2.4 Resistance equation

If the surface of a mobile bed is "undulated" (i.e. if it is covered by ripples and/or dunes or any other kind of periodic along x bed forms), then its effective roughness K_s is larger than k_s, while its (total) *resistance factor* c is smaller than the friction factor c_f. In the case of an undulated bed the friction equation (1.16) is to be generalized into the *resistance equation*

$$\overline{u} = c\sqrt{gSh}. \tag{1.17}$$

(The evaluation of K_s and c will be carried out in Chapter 3).

The resistance equation (1.17) can be expressed in dimensionless form as

$$Fr = c^2 S, \tag{1.18}$$

where

$$Fr = \frac{\bar{u}^2}{gh} \tag{1.19}$$

is the Froude number. The (very important) relation (1.18) will often be used in this text. It should be noted that the Froude number reflects the energy-structure of flow (it is an "energy-related" flow characteristic). Indeed, the kinetic energy of the unit fluid volume can be characterized by $e_k = (1/2)\rho\bar{u}^2$, while its cross-sectional potential energy by $e_p = \rho gh$; hence the Froude number is a "measure" of the ratio e_k/e_p.[3] It should also be noted that in this and the next two chapters, we will be dealing only with two-dimensional flows, or with flows which can be treated as such (trapezoidal cross-section, "large" B/h, negligible bank friction). In the latter case, any expression given in these three chapters in terms of the vertically-averaged flow velocity $\bar{u}$ is to be viewed as that given in terms of the cross-sectional average flow velocity u_m (e.g. $Fr = \bar{u}^2/(gh)$ is to be considered as $Fr = u_m^2/gh$).

1.3 TWO-DIMENSIONAL TWO-PHASE MOTION

(i) The simultaneous motion of the transporting fluid and the transported sediment forms an inseparable mechanical totality which is referred to as the *two-phase motion*. A steady and uniform rectilinear two-dimensional two-phase motion can be defined by the following seven characteristic parameters (Yalin 1965, 1971, 1977)

$$\rho,\ \nu,\ \rho_s,\ D,\ \gamma_s,\ h,\ v_* \tag{1.20}$$

(see List of Symbols), and any quantity (A', say) related to this motion can thus be expressed as

$$A' = f_{A'}(\rho, \nu, \rho_s, D, \gamma_s, h, v_*). \tag{1.21}$$

However, in this text we will be dealing only with that subset $\{A\}$ of $\{A'\}$ which corresponds to the grain motion *en mass*, and which can thus not be dependent on the parameter ρ_s determining the motion of individual grains only (Yalin 1965, 1971, 1977). Excluding ρ_s, one obtains for A

$$A = f_A(\rho, \nu, D, \gamma_s, h, v_*). \tag{1.22}$$

The (generally dimensional) relation (1.22) can be expressed in dimensionless form as

$$\Pi_A = \phi_A(X, Y, Z), \tag{1.23}$$

[3] In the literature on open-channel flows the Froude number is, as a rule, identified with $\bar{u}/\sqrt{gh}$ (see e.g. Chow 1959). In this book we deviate from this convention, merely in order to have the expression of the Froude number acting as an obvious reminder of the fact that Fr is a measure of e_k/e_p. The matter is related to the content of Chapters 4 and 5, relying on an analysis of the energy structure of flow. The identification of Fr with $\bar{u}^2/(gh)$ instead of $\bar{u}/\sqrt{gh}$ is totally inconsequential, as $Fr = 1$ marks the transition from subcritical to supercritical flows, no matter which definition is adopted.

where the dimensionless variables X, Y and Z are given by

$$X = \frac{\upsilon_* D}{\nu}, \quad Y = \frac{\rho \upsilon_*^2}{\gamma_s D}, \quad Z = \frac{h}{D}, \tag{1.24}$$

the dimensionless counterpart Π_A of A being

$$\Pi_A = \rho^{x_A} \upsilon_*^{y_A} D^{z_A} A. \tag{1.25}$$

Here x_A, y_A and z_A must be determined so that Π_A becomes dimensionless.

(ii) Suppose that A stands for the sediment transport rate q_s. From Eqs. (1.23) and (1.25), it follows that

$$\Pi_{q_s} = \frac{q_s}{\upsilon_* D} = \overline{\phi}_{q_s}(X, Y, Z). \tag{1.26}$$

Multiplying both sides of this relation by $\sqrt{\rho \upsilon_*^2 / \gamma_s D} = \sqrt{Y}$, we obtain

$$\phi = \frac{\rho^{1/2} q_s}{\gamma_s^{1/2} D^{3/2}} = \phi_{q_s}(X, Y, Z) \qquad (\text{with } \phi_{q_s} = \sqrt{Y} \cdot \overline{\phi}_{q_s}). \tag{1.27}$$

Here, the dimensionless combination on the left is the "Einstein's ϕ" (Einstein 1942, 1950). This has the advantage of being related to q_s by the proportionality factor $\rho^{1/2}/(\gamma_s^{1/2} D^{3/2})$ which does not vary when the flow varies (for it does not contain υ_* and/or h).

At the inception of sediment transport (i.e. at the "critical stage" of a mobile bed), q_s is (just) zero and $Z = h/D$ is not a determining parameter. Indeed, the critical stage is determined solely by

$$(\tau_0)_{cr} = \gamma (hS)_{cr} \qquad (= \rho \upsilon_{*cr}^2), \tag{1.28}$$

which can be realized for *any* h (by an appropriate adjustment of S). Substituting $q_s \sim \phi = 0$, excluding Z and marking X and Y with the subscript cr, one obtains from Eq. (1.27)

$$0 = \phi_{q_s}(X_{cr}, Y_{cr}) \quad \text{i.e.} \quad Y_{cr} = \Phi(X_{cr}), \tag{1.29}$$

which is the Shields' *sediment transport inception function.* Here

$$X_{cr} = \frac{\upsilon_{*cr} D}{\nu} \quad \text{and} \quad Y_{cr} = \frac{\rho \upsilon_{*cr}^2}{\gamma_s D} = \frac{(\tau_0)_{cr}}{\gamma_s D}. \tag{1.30}$$

The experimental curve representing the function (1.29) is shown in Figure 1.8.

As is well known (Sedov 1960, Yalin 1971, 1977, 1992), a dimensionless variable can always be replaced by a function of that variable and some (or all) of the remaining variables. Consider e.g. the following combination of X and Y

$$\Xi^3 = \frac{X^2}{Y} = \frac{\gamma_s D^3}{\rho \nu^2}. \tag{1.31}$$

This combination reflects the influence of the solid (γ_s, D) and liquid (ρ, ν) phases involved and its value does not vary depending on the stage of the flow (for it does not

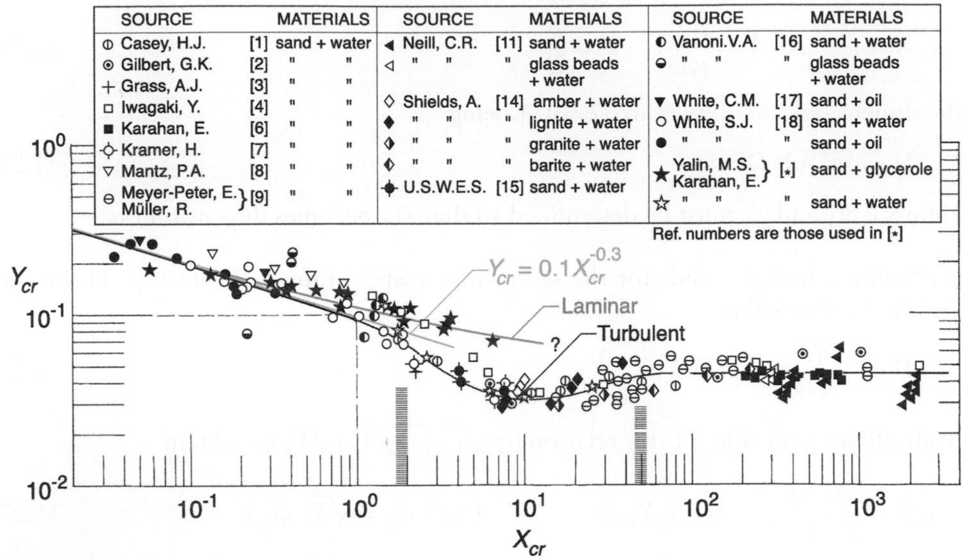

Figure 1.8 Shields' sediment transport inception function (adapted from Yalin and Karahan 1979).

involve υ_* and/or h). But then, Ξ^3 can equally well be expressed in terms of X and Y corresponding to the critical stage:

$$\Xi^3 = \frac{X_{cr}^2}{Y_{cr}} \quad \text{i.e.} \quad X_{cr} = \sqrt{\Xi^3\, Y_{cr}}. \tag{1.32}$$

Eliminating X_{cr} between Eqs. (1.29) and (1.32), one determines

$$Y_{cr} = \Phi(\sqrt{\Xi^3\, Y_{cr}}) \quad \text{i.e.} \quad Y_{cr} = \Psi(\Xi), \tag{1.33}$$

which is a *modified sediment transport inception function* with the advantage of supplying Y_{cr} and thus υ_{*cr} without "trial and error". The graph of the function $Y_{cr} = \Psi(\Xi)$ is shown in Figure 1.9. This function can be adequately represented, throughout the extent of Ξ-values in Figure 1.9, by the following comp-eq. (due to da Silva and Bolisetti 2000)

$$Y_{cr} = 0.13\, \Xi^{-0.392} e^{-0.015\Xi^2} + 0.045\left[1 - e^{-0.068\Xi}\right]. \tag{1.34}$$

Observe that the relative flow intensity $\eta_* = \tau_0/(\tau_0)_{cr}$ can be expressed, with the aid of Eq. (1.33), as

$$\eta_* = \frac{\tau_0}{(\tau_0)_{cr}} = \frac{Y}{Y_{cr}} = \frac{Y}{\Psi(\Xi)}. \tag{1.35}$$

Solving X and Y from Eqs. (1.31) and (1.35), one obtains

$$X = \sqrt{\Xi^3\, \eta_*\, \Psi(\Xi)} \quad \text{and} \quad Y = \eta_*\, \Psi(\Xi). \tag{1.36}$$

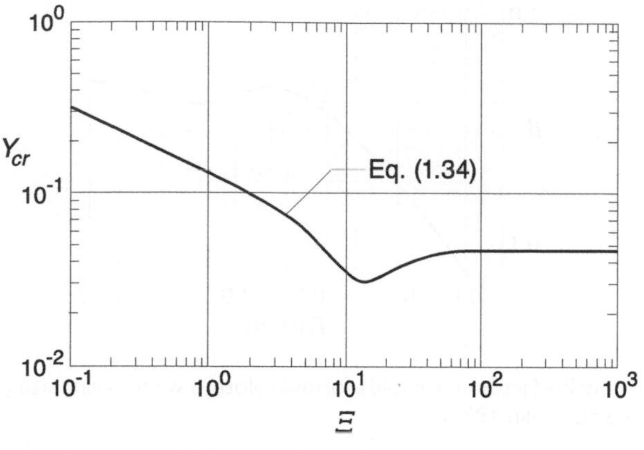

Figure 1.9 Modified sediment transport inception function (adapted from da Silva and Bolisetti 2000).

Using these expressions for X and Y in the basic form (1.23), one determines

$$\Pi_A = \phi_A\left(\sqrt{\Xi^3}\,\eta_*\,\Psi(\Xi),\ \eta_*\Psi(\Xi),\ Z\right) \tag{1.37}$$

i.e.

$$\Pi_A = \overline{\phi}_A(\Xi, \eta_*, Z), \tag{1.38}$$

which indicates that if Π_A is a function of X and Y, then it can equally well be regarded and treated as a function of Ξ and η_*. The form (1.38) will often be used in this text.

1.4 BED-LOAD RATE; BAGNOLD'S FORMULA

Numerous equations are available for the quantification of the rate q_{sb} of the bed-load transported in the neighbourhood ϵ (Figure 1.2) of a flat bed (for reviews, see e.g. Yalin 1977, Garde and Raju 1985, Raudkivi 1990, Yang 1996, Chien and Wan 1999, García 2008, Dey 2015). However, for the purposes of discussion and calculations, throughout this text q_{sb} will be evaluated with the aid of Bagnold's formula

$$q_{sb} = \beta u_b(\tau_0 - (\tau_0)_{cr})/\gamma_s, \tag{1.39}$$

where u_b is the flow velocity at the bed (vertically-averaged u within $k_s < z < \epsilon$), and β is a function of Ξ ($\beta = \phi_\beta(\Xi)$). In the case of a fully rough turbulent flow, this function reduces to $\beta = 0.5$. If the flow is not fully rough, then β can be determined from Figure 1.10. Since this figure is for the case of sand and water, the grain size D used as abscissa is to be viewed as a surrogate of Ξ ($= 25296D$ if $\rho = 1000kg/m^3$, $\nu = 10^{-6}m^2/s$ and $\gamma_s = 16186.5N/m^3$).

Bagnold's formula is preferred, because it is simple, it is as accurate as any, and it reflects clearly the meaning of the bed-load rate. Indeed, q_{sb} can be viewed as the product of the grain-volume v_s moving over the unit area of the bed, with the downstream displacement velocity u_s of that volume: $q_{sb} = v_s u_s$. The relation (1.39) clearly conveys

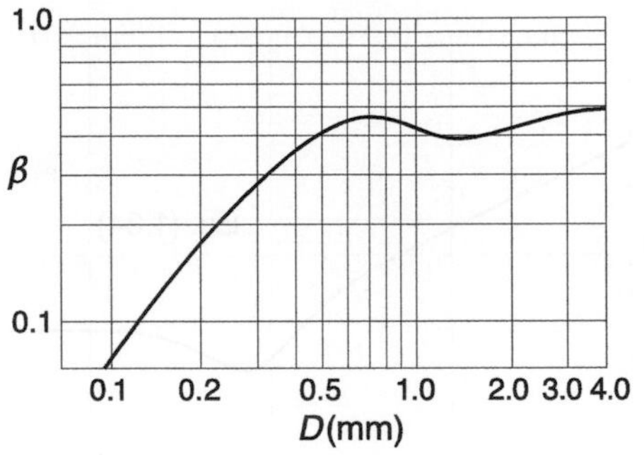

Figure 1.10 Proportionality factor in Bagnold's formula plotted versus grain size (adapted from Bagnold 1956; see also Yalin 1977).

that $u_s \sim u_b$, while $v_s \sim (\tau_0 - (\tau_0)_{cr})/\gamma_s$. Both of these proportionalities are realistic and dimensionally correct.

The flow velocity at the bed u_b can be typified by the average flow velocity in the ϵ-layer,

$$u_b \approx \frac{u_{k_s} + u_\epsilon}{2}, \tag{1.40}$$

where u_{k_s} and u_ϵ, which correspond to $z = k_s \approx 2D$ and $z = \epsilon$, are given by Eq. (1.6) as

$$u_{k_s} = v_* B_s \quad \text{and} \quad u_\epsilon \approx v_* \left(\frac{1}{\kappa} \ln \left(\frac{\epsilon}{2D} \right) + B_s \right). \tag{1.41}$$

If η_* is just above unity and the grain trajectories are thus very flat (ϵ amounts only to two or four grain-sizes), then u_ϵ is very close to u_{k_s} and one can adopt

$$u_b \approx u_{k_s} \quad (= v_* B_s), \tag{1.42}$$

(as done in Bagnold's original derivation; see Bagnold 1956). In this case Eq. (1.39), which can then be expressed in dimensionless form e.g. as

$$\phi = (B_s \beta) Y^{1/2} (Y - Y_{cr}) \quad \text{or} \quad \phi = [(B_s \beta) Y_{cr}^{3/2}] \eta_*^{1/2} (\eta_* - 1), \tag{1.43}$$

does not depend on $h \sim Z$. And the value of $q_{sb} \sim \phi$ (corresponding to a given granular material and fluid) is determined solely by τ_0 ($\sim v_*^2 \sim Y$), but not by the average flow velocity $\bar{u}$ – which is determined by τ_0 as well as by $h \sim Z$ (see Eq. (1.11), noting that it contains $v_* = \sqrt{\tau_0/\rho}$ and $h/k_s \approx Z/2$).

However, more often than not (and especially when the bed is not exactly flat but covered by ripples or dunes, while Z is "large"), the thickness of the bed-load region ϵ, though only a small fraction of h, can be much larger than D ($\approx k_s/2$). [e.g. one can have $h = 0.5m$, $\epsilon = 0.5cm$, $D = 0.2mm$, and thus $Z = 2500 \gg \epsilon/h = 0.01$; $\epsilon/D = 25$.] In such cases u_ϵ, and consequently u_b and q_{sb} are certainly affected by Z and thus by $\bar{u}$. Indeed, from Eqs. (1.41) and (1.11) it follows that

$$\frac{u_\epsilon}{v_*} = \frac{1}{\kappa} \ln \left(\frac{\epsilon}{k_s} \right) + B_s = \frac{\bar{u}}{v_*} - \frac{1}{\kappa} \ln \left(0.368 \frac{h}{\epsilon} \right). \tag{1.44}$$

Owing to this reason, the consideration of Eq. (1.39), which contains u_b, is often replaced by the form

$$q_{sb} = \beta' \bar{u}(\tau_0 - (\tau_0)_{cr})/\gamma_s, \tag{1.45}$$

which contains (the unambiguous) $\bar{u}$. Here, as can be inferred from Eqs. (1.11), (1.40) and (1.41),

$$\frac{\beta'}{\beta} = \frac{1 + \frac{1}{2\kappa B_s} \ln\left(\frac{\epsilon}{2D}\right)}{1 + \frac{1}{\kappa B_s} \ln\left(0.368 \frac{h}{2D}\right)} \qquad \left(= \frac{u_b}{\bar{u}}\right). \tag{1.46}$$

In most of the practical cases the flow characteristics (h, S, τ_0, $\bar{u}$, etc.) vary with time only gradually (i.e. non-impulsively), and in this text we will be dealing with such cases only. Consequently, the steady-state relations above will be considered as applicable even if the flow characteristics (gradually) vary in time. Thus if e.g. $(\bar{u}_1, \tau_{01})$ at the time t_1 change "slowly" into $(\bar{u}_2, \tau_{02})$ at t_2, then the corresponding $(q_{sb})_1$ and $(q_{sb})_2$ will be assumed to be determinable from the *same* expression (Eq. (1.45), say) – simply by evaluating it by $(\bar{u}_1, \tau_{01})$ and $(\bar{u}_2, \tau_{02})$, respectively. Considering this, the adjective "steady" will be omitted in the following.

1.5 VECTOR FORMS OF THE BED-LOAD RATE

Consider now some non-uniform open-channel flows; we continue to assume that the mobile bed surface is flat. The direction and magnitude of flow velocity varies from one space point $P(x; y; z)$ to another. Hence the flow velocity is to be treated as a vector

$$\mathbf{U} = \mathbf{i}_x u + \mathbf{i}_y v + \mathbf{i}_z w, \tag{1.47}$$

where $\mathbf{i}_x$, $\mathbf{i}_y$, $\mathbf{i}_z$ are the unit vectors along x, y, z, respectively; and u, v, w are the scalar components of $\mathbf{U}$. In general, u, v and w vary as functions of x, y, z.

In this text we will be dealing predominantly with the conditions corresponding to the central region of a wide channel. Consequently, we will assume that w is either zero or negligible, and that $\mathbf{U}$ is, in fact, determined as

$$\mathbf{U} = u\mathbf{i}_x + v\mathbf{i}_y \tag{1.48}$$

where, in general, $u = \phi_u(x, y, z)$ and $v = \phi_v(x, y, z)$. (Note that $w = 0$ does not mean that $\mathbf{U}$ does not vary with z).

In the present case of a flat bed, the vectors $\mathbf{U}_b = \mathbf{i}_b U_b$, $\vec{\tau}_0 = \mathbf{i}_\tau \tau_0$ and $\mathbf{q}_{sb} = \mathbf{i}_{q_{sb}} q_{sb}$ corresponding to the same point $P(x; y)$ of the bed surface have *always* the same direction – i.e. they are always collinear:

$$\mathbf{i}_b = \mathbf{i}_\tau = \mathbf{i}_{q_{sb}}. \tag{1.49}$$

Multiplying both sides of Eq. (1.39) with $\mathbf{i}_{q_{sb}} = \mathbf{i}_b$ (and bearing in mind that u_b means U_b) we determine

$$\mathbf{q}_{sb} = \mathbf{U}_b[\beta(\tau_0 - (\tau_0)_{cr})/\gamma_s], \tag{1.50}$$

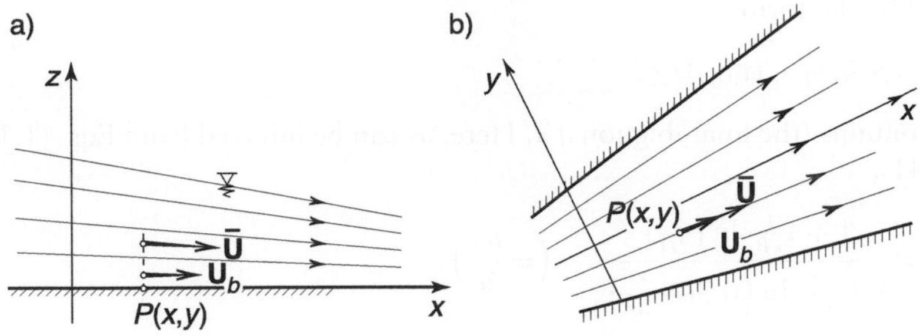

Figure 1.11 Examples of straight, gradually varying flows. (a) Side view of a converging flow; (b) top view of a diverging flow.

which is the vector form of the Bagnold's bed-load formula. This vector form can be used for all open-channel flows having a flat mobile bed.

The expression of $\mathbf{q}_{sb}$ in terms of $\overline{\mathbf{U}}$ is more involved:

1 If the time-average streamlines of a gradually varying non-uniform flow are straight (such as e.g. those in Figure 1.11), then $\overline{\mathbf{U}}$ and $\mathbf{U}_b$, and thus $\overline{\mathbf{U}}$ and $\mathbf{q}_{sb}$ (all at the same $P(x;y)$), can be taken as collinear:

$$\mathbf{i}_{q_{sb}} = \mathbf{i}_{\overline{\mathrm{U}}}. \tag{1.51}$$

Multiplying Eq. (1.45) with $\mathbf{i}_{q_{sb}} = \mathbf{i}_{\overline{\mathrm{U}}}$, we obtain

$$\mathbf{q}_{sb} = \overline{\mathbf{U}}[\beta'(\tau_0 - (\tau_0)_{cr})/\gamma_s]. \tag{1.52}$$

Hence the vector $\mathbf{q}_{sb}$ of a rectilinear flow can be described by both, Eqs. (1.50) and (1.52).

2 If the time-average streamlines are curved (flow in a meandering channel, or in any curved channel in general), then in addition to the vertically-averaged longitudinal fluid motion, we have also a cross-circulatory motion Γ (Figure 1.12). However, by definition the vertically-averaged value of Γ is zero (Nelson and Smith 1989, Yalin 1992, Geyer 1993), and therefore Γ cannot affect the vertically-averaged flow, and thus its average velocity $\overline{\mathbf{U}}$. Yet, Γ affects the flow at the bed, for it gives rise to the velocity $\mathbf{U}_b = (\mathbf{U}_b)_s + (\mathbf{U}_b)_\Gamma$ at the bed. Clearly, the directions of $\overline{\mathbf{U}}$ and $\mathbf{U}_b$, and thus of $\overline{\mathbf{U}}$ and $\mathbf{q}_{sb}$ (which is caused by $\mathbf{U}_b$), are different. But this means that Eq. (1.52) is no longer valid (although Eq. (1.50) is still valid, for the directions of $\mathbf{U}_b$ and $\mathbf{q}_{sb}$ are still in coincidence). It will be shown in Chapter 6 (see Sub-section 6.3.3) that for a given relative curvature (B/R), the prominence of Γ progressively decreases with the increment of the width-to-depth ratio B/h_m of a meandering stream. Hence, in large natural meandering streams (where, as a rule, $B/h_m > 100$, say) Γ can be ignored, and Eq. (1.52) can be used.

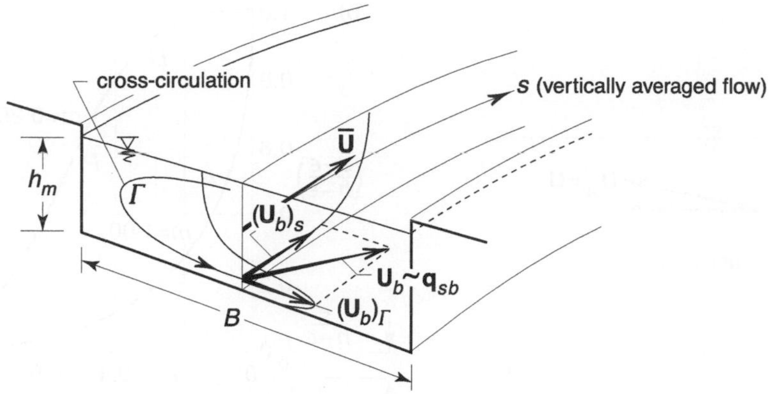

Figure 1.12 Vertically-averaged and near bed flow velocity vectors in a curved channel.

1.6 SUSPENDED-LOAD RATE

(i) Let C be the dimensionless volumetric concentration of suspended load particles (grains) at a location of flow,[4] and $\mathbf{U}_{ss}$ be the migration velocity of the "cloud" of suspended particles at that location. If the grains are sufficiently small, then $\mathbf{U}_{ss}$ can be identified, and usually *is* identified, with the local time-average flow velocity $\mathbf{U}$:

$$\mathbf{U}_{ss} = \mathbf{U}. \tag{1.53}$$

Consider now a (imaginary) vertical unit area at a space point P; the unit normal vector $\mathbf{n}$ of this area is, of course, identical to $\mathbf{i}_x$ (Figure 1.13a). The volumetric suspended-load rate q^*_{ss} passing through this unit area can be expressed (considering Eq. (1.53)) as

$$q^*_{ss} = \mathbf{n}\mathbf{U}_{ss}C = \mathbf{n}\mathbf{U}C = \mathbf{i}_x\mathbf{U}C = uC, \tag{1.54}$$

which yields for q_{ss} (corresponding to the case of a steady and uniform two-dimensional two-phase motion)

$$q_{ss} = \int_\epsilon^h q^*_{ss} dz = \int_\epsilon^h Cudz\,. \tag{1.55}$$

In the case of a "naturally acquired"[5] suspended-load, the value of C usually does not exceed ≈ 0.02, say (see e.g. Soo 1967), and u can be evaluated with the aid of Eq. (1.6).

The relation (1.55) can be expressed in vector form as

$$\mathbf{q}_{ss} = \int_\epsilon^h \mathbf{q}^*_{ss} dz = \int_\epsilon^h C\mathbf{U}dz, \tag{1.56}$$

where $\mathbf{U}$ is in the sense of Eq. (1.48).

[4] If V_{sed} is the volume of sediment contained in a water-sediment mixture of the volume V_{mix}, then $C = V_{sed}/V_{mix}$. Clearly, C is a dimensionless fraction: $0 \le C < 1$.
[5] i.e. by the dynamic action of flow on its bed – no "external input".

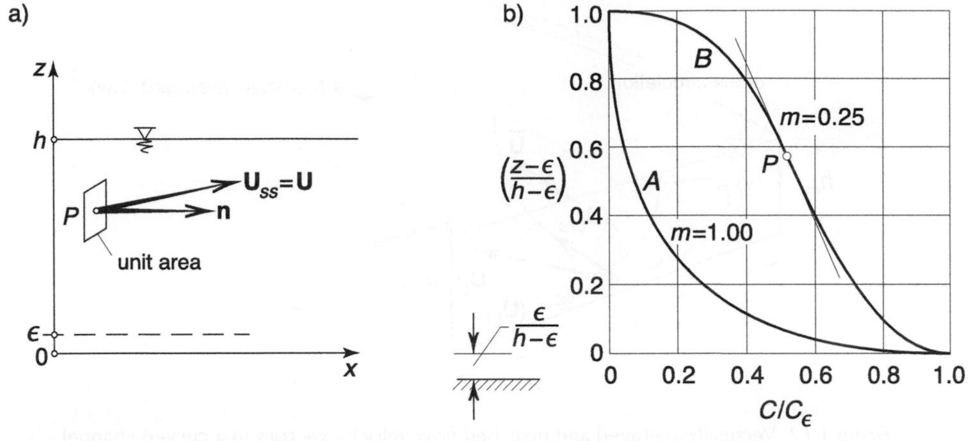

Figure 1.13 (a) Imaginary vertical unit area for definition of suspended-load rate; (b) vertical distribution of suspended-load concentration.

Consider the vectors U at the points of the interval $\epsilon < z < h$ of the z-axis (passing through a point $P(x; y)$ of the flow plan). If these vectors are collinear, then the unit vector $\mathbf{i}_s$ of the flow direction is the same for any $z \in [\epsilon, h]$ and we can write Eq. (1.56) as

$$\mathbf{q}_{ss} = \mathbf{i}_s \int_{\epsilon}^{h} CUdz. \tag{1.57}$$

Using the average value theorem of integral calculus (see e.g. Smirnov 1964), viz

$$\int_a^b \psi(z) f(z) dz = \psi(z_m) \int_a^b f(z) dz \quad (\text{with } a < z_m \ (= const) < b), \tag{1.58}$$

for $\psi(z) = C$ and $f(z) = U$, we determine

$$\int_{\epsilon}^{h} CUdz = \overline{\overline{C}} \int_{\epsilon}^{h} Udz = \overline{\overline{C}} h \overline{U}, \tag{1.59}$$

where the average value $\overline{\overline{C}} = \psi(z_m)$ is very close to (but may not be exactly the same as) the "usual" vertically-averaged $\overline{C}$. Taking into account this fact by introducing a near-to-unity coefficient α_c $(= \overline{\overline{C}}/\overline{C})$, and considering that $\overline{\mathbf{U}} = \mathbf{i}_s \overline{U}$, one can express Eq. (1.57) as

$$\mathbf{q}_{ss} = \alpha_c h \overline{\mathbf{U}}\, \overline{C}. \tag{1.60}$$

(ii) Consider the distribution of concentration (of a naturally acquired suspended-load) C along the flow depth in the interval $\epsilon < z < h$. Several expressions have been produced to date for this purpose, and among them the Rouse form (Rouse 1937; see also e.g. Bogardi 1974, Yalin 1977, Chang 1988, Chiu et al. 2000)

$$C = C_{\epsilon} \left[\frac{\epsilon}{z} \cdot \frac{h - z}{h - \epsilon} \right]^m \qquad (\text{with } m = 2.5 w_s / v_*) \tag{1.61}$$

is the most popular.[6] Here C_ϵ is the concentration at $z=\epsilon$, and w_s is the settling (terminal) velocity of grains. The curves in Figure 1.13b give an idea on the nature of C-distribution along z.

For C_ϵ and ϵ the following relations of van Rijn (1984, 1985), expressed here for a flat bed, can be adopted:

$$C_\epsilon = 0.05\, \Xi^{-1}\, (\eta_* - 1), \tag{1.62}$$

$$\frac{\epsilon}{D} = 0.3\, \Xi^{0.7} (\eta_* - 1)^{0.5}. \tag{1.63}$$

1.7 SEDIMENT TRANSPORT CONTINUITY EQUATION

(i) Consider a non-steady and non-uniform two-phase motion – the non-steadiness and non-uniformity are gradual. We select as the control volume (CV) a fixed in space imaginary vertical prism: height h, base area A (Figure 1.14). The volume of CV will be denoted by V, the area of its side-surface, by σ. The outward unit normal $\mathbf{n}$ will be used: the possible variations of h and ϵ ($\ll h$) within A will be assumed to be negligible.

The net bed-load rate $\mathcal{F}_b$ passing at an instant t through (the lower part ϵ of) σ is given by

$$\mathcal{F}_b = \oint_l \mathbf{n}\mathbf{q}_{sb} dl = \int_A \nabla \mathbf{q}_{sb}\, dA, \tag{1.64}$$

where l is the perimeter of A (Figure 1.14); the net suspended-load rate $\mathcal{F}_s$ passing through (the upper part $h-\epsilon$ of) σ being

$$\mathcal{F}_s = \oint_l \mathbf{n} \left(\int_\epsilon^h \mathbf{q}^*_{ss} dz \right) dl = \oint_l \mathbf{n}\mathbf{q}_{ss} dl = \int_A \nabla \mathbf{q}_{ss} dA, \tag{1.65}$$

where the second step is by the use of Eq. (1.56).

The volume of granular material accumulated in CV per unit time because of q_{sb}, viz $-\mathcal{F}_b$, causes solely the rise of the bed level (z_b) in CV per unit time. Denoting the average velocity of the increment of the bed level of CV (due to q_{sb}) by (the positive) W_b, one can express $-\mathcal{F}_b$ as

$$-\mathcal{F}_b = (1-p) W_b\, A, \tag{1.66}$$

where p (<1) is the porosity of granular material.

The material accumulated in CV because of q_{ss}, viz $-\mathcal{F}_s$, partly causes the rise of the bed, with the averaged over A (positive) velocity W_s, and partly the increment $\mathcal{J}_s$ of the volume of suspended solids (per unit time) in CV:

$$-\mathcal{F}_s = (1-p) W_s\, A + \mathcal{J}_s. \tag{1.67}$$

[6] For the purposes of estimating the C-distribution, even more realistic results (than those produed by the Rouse form) may be obtained using some of the recent C-formulae, including those proposed by Chiu et al. (2000) and Wright and Parker (2004). Such expressions are best viewed as modified versions of Eq. (1.61), as they differ from it only because they are derived by using a different (from log-distribution) velocity variation formula, and/or a different expression of the diffusion coefficient, etc.

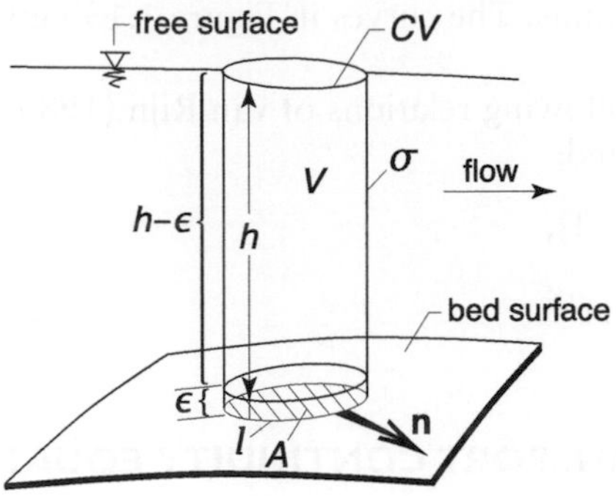

Figure 1.14 Control volume used in the derivation of the sediment transport continuity equation.

Here, $\mathcal{J}_s$ is determined as

$$\mathcal{J}_s = \frac{\partial}{\partial t}\int_V C dV = \frac{\partial}{\partial t}\int_A \left(\int_\epsilon^h C dz\right) dA$$

$$= \frac{\partial}{\partial t}\int_A (h\overline{C}) dA = \int_A \frac{\partial (h\overline{C})}{\partial t} dA \qquad (1.68)$$

where $\overline{C}$ is the vertically-averaged concentration; the actual product $(h-\epsilon)\overline{C}$ is replaced by $h\overline{C}$ (for $\epsilon \ll h$). [The vertical displacement velocity W_s at a point $P(x;y)$ of the bed surface (due to q_{ss}) is equal to the difference between the volume of particles $(w_s C)$ *landing* on the unit area at $P(x;y)$ and the volume of particles $(\nu_t(\partial C/\partial z)_{z=\epsilon})$ *detaching* from that area – both per unit time.]

Since the displacement velocity at a point $P(x;y)$ is given by $W = \partial z_b/\partial t$, the average vertical displacement velocity W_A of the area A can be expressed as

$$W_A = \frac{1}{A}\int_A W dA = \frac{1}{A}\int_A \frac{\partial z_b}{\partial t} dA. \qquad (1.69)$$

Any change of the bed elevation of CV is solely due to $\mathbf{q}_{sb}$ and $\mathbf{q}_{ss}$, and therefore (the scalar) W_A is but the algebraic sum of W_b and W_s. Considering this, and taking into account that W_b and W_s are determined by Eqs. (1.66) and (1.67), we obtain

$$(1-p)W_A = (1-p)(W_b + W_s) = -\frac{1}{A}(\mathcal{F}_b + \mathcal{F}_s + \mathcal{J}_s). \qquad (1.70)$$

Substituting here for W_A, $\mathcal{F}_b$, $\mathcal{F}_s$, and $\mathcal{J}_s$ their expressions (1.69), (1.64), (1.65) and (1.68), we determine

$$\int_A \left((1-p)\frac{\partial z_b}{\partial t} + \nabla \mathbf{q}_{sb} + \nabla \mathbf{q}_{ss} + \frac{\partial (h\overline{C})}{\partial t}\right) dA = 0, \qquad (1.71)$$

and since the area A is arbitrary,

$$(1-p)\frac{\partial z_b}{\partial t} + \nabla \mathbf{q}_{sb} + \nabla \mathbf{q}_{ss} + \frac{\partial}{\partial t}(h\overline{C}) = 0, \qquad (1.72)$$

which is the sediment transport continuity equation in its general form.

In this text we will be dealing mainly with those cases where $\partial(h\overline{C})/\partial t$ can be neglected (steady-state or negligible $\overline{C}$), and therefore Eq. (1.72) will be used mostly in its reduced form

$$(1-p)W = (1-p)\frac{\partial z_b}{\partial t} = -\nabla \mathbf{q}_s, \tag{1.73}$$

where $\mathbf{q}_s = \mathbf{q}_{sb} + \mathbf{q}_{ss}$.

Moreover, the vectors $\overline{\mathbf{U}}$, $\mathbf{U}_b$, $\mathbf{q}_{sb}$ and $\mathbf{q}_{ss}$ will be treated mostly as collinear. i.e. it will be often assumed that they can be expressed as $\overline{\mathbf{U}} = \mathbf{i}_s\overline{U}$, $\overline{\mathbf{U}}_b = \mathbf{i}_s U_b$, $\mathbf{q}_{sb} = \mathbf{i}_s q_{sb}$, and $\mathbf{q}_{ss} = \mathbf{i}_s q_{ss}$ (where $\mathbf{i}_s$ is the unit vector of the vertically-averaged streamlines (s) at the point $P(x;y)$ under study). The considerations below are developed accordingly.

Since q_s is a strongly increasing function of $\overline{U}$, the positive (or negative) value of $\partial\overline{U}/\partial s$ is invariably associated with the positive (or negative) value of $\partial q_s/\partial s$ and, consequently, of $\nabla \mathbf{q}_s$.[7] Thus Eq. (1.73) indicates that:

1. If the flow at a location $P(x;y)$ is convectively *accelerated*, i.e. if $\partial\overline{U}/\partial s > 0$ and thus $\nabla \mathbf{q}_s > 0$, then $\partial z_b/\partial t < 0$ (*erosion* takes place);
2. If the flow (at $P(x;y)$) is convectively *decelerated*, i.e. if $\partial\overline{U}/\partial s < 0$ and thus $\nabla \mathbf{q}_s < 0$, then $\partial z_b/\partial t > 0$ (*deposition* takes place);
3. No vertical displacement of the bed surface can take place if $\partial\overline{U}/\partial s = 0$, i.e. if the flow is uniform.

(ii) It is intended now to express $\mathbf{q}_s$ in Eq. (1.73) in terms of the quantities determining it. We have, according to Eqs. (1.50) and (1.60),

$$\mathbf{q}_s = \mathbf{q}_{sb} + \mathbf{q}_{ss} = \mathbf{U}_b\beta(\tau_0 - (\tau_0)_{cr})/\gamma_s + \overline{\mathbf{U}}\alpha_c h\overline{C}. \tag{1.74}$$

But since $\mathbf{U}_b$ and $\overline{\mathbf{U}}$ are treated as collinear, i.e. since

$$\mathbf{U}_b = \mathbf{i}_b U_b = (\mathbf{i}_s\overline{U})\frac{\beta'}{\beta} = \overline{\mathbf{U}}\frac{\beta'}{\beta}, \tag{1.75}$$

one can express Eq. (1.74) as

$$\mathbf{q}_s = (h\overline{\mathbf{U}})\,\psi_q \tag{1.76}$$

where the dimensionless function ψ_q, implying the ratio of the specific transport rate q_s to the specific flow rate q $(= h\overline{U})$, i.e.

$$\psi_q = \frac{q_s}{q}, \tag{1.77}$$

[7]See Problem 1.17 at the end of the chapter; it contains the expression

$$\nabla \mathbf{q}_s = \frac{\partial q_s}{\partial s} - \frac{q_s}{q}\frac{\partial(\overline{U}h)}{\partial s}.$$

Since the variation of q_s $(\sim \overline{U}\tau_0 \sim \overline{U}^3)$ with $\overline{U}$ is much stronger than that of $q = \overline{U}h$, and since $q_s/q \ll 1$, the magnitude of the first term on the right is larger than that of the second – at least in the (natural) transport phenomena we will be dealing with in this text.

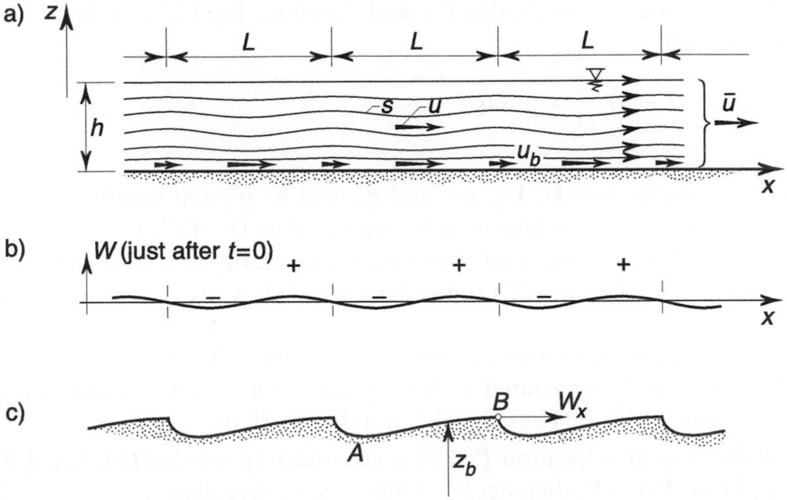

Figure 1.15 Schematic representation of flow exhibiting a periodic internal non-uniformity. (a) Wave-like streamlines; (b) diagram of bed vertical displacement velocity; (c) resulting bed forms.

is given by

$$\psi_q = \beta' \left(\frac{\tau_0 - (\tau_0)_{cr}}{\gamma_s h} \right) + \alpha_c \overline{C}. \tag{1.78}$$

Differentiating Eq. (1.76), one obtains

$$\nabla \mathbf{q}_s = (h\overline{\mathbf{U}})\, \nabla \psi_q + \psi_q\, \nabla (h\overline{\mathbf{U}}) = (h\overline{\mathbf{U}}) \nabla \psi_q, \tag{1.79}$$

for $\nabla(h\overline{\mathbf{U}}) = \nabla \mathbf{q} = 0$ (by virtue of continuity). Observe from Eq. (1.79) that if at each point $P(x;y)$ the vectors $\overline{\mathbf{U}}$ and $\nabla\psi_q$ are perpendicular to each other, then $\nabla \mathbf{q}_s \equiv 0$, and thus $W = \partial z_b/\partial t \equiv 0$. Hence the deformation of the bed surface terminates (its equilibrium state is achieved) when it acquires such a "topography" as to render the $\psi_q = const$ lines to be congruent with the streamlines (s).

(iii) Consider a two-dimensional turbulent straight open-channel flow past a flat initial mobile bed (at $t = 0$). Suppose that this flow, which is assumed to be basically uniform, has acquired a periodic along x "internal non-uniformity": its (initially straight) time-average streamlines s become deformed into the wave-like ones (Figure 1.15a): the wave-length is L, the wave-amplitude is "small". How such a deformation can originate will be explained in Section 2.2, and here we will be concerned only with how it can affect the bed surface.

The velocities u (including u_b) of the above described flow must periodically vary along x – and so must do q_s, $\nabla \mathbf{q}_s$ $(= \partial q_s/\partial x)$, and W: the velocity W and $\partial q_s/\partial x$ vary around zero average value (Figure 1.15b). But this means that the flat (initial) bed *must* with the passage of time deform into an "undulated" one, i.e. it must become covered by *bed forms* or *sand waves* of the length L (Figure 1.15c).

The bed forms emerging (just after $t = 0$) in this way, affect in turn the structure of flow, and thus its action on the bed, ..., and so on. This "reciprocal adjustment" continues until the state of equilibrium is reached at a time $t = T_\Delta$, say. The duration T_Δ will be referred to as the "duration of development" of bed forms. No characteristic of the developed bed (and flow) is supposed to vary for $t > T_\Delta$.

(iv) Consider the two-dimensional developed bed forms ($t > T_\Delta$) and suppose that $q_s = q_{sb}$. We have, on the basis of Eq. (1.73),

$$(1-p)\frac{\partial z_b}{\partial t} = -\frac{\partial q_{sb}}{\partial x} \quad \text{i.e.} \quad \frac{\partial q_{sb}}{\partial x} + (1-p)\frac{\partial z_b}{\partial t} = 0 \tag{1.80}$$

which is the familiar Exner-Polya equation.

We assume that the bed forms are periodic and that they are migrating along x (without changing their size and shape) with a constant velocity W_x (Figure 1.15c). Then

$$\frac{\partial z_b}{\partial t} = -\frac{\partial z_b}{\partial x}\frac{\partial x}{\partial t} = -\frac{\partial z_b}{\partial x}W_x, \tag{1.81}$$

and Eq. (1.80) becomes

$$\frac{\partial}{\partial x}\left(q_{sb} - (1-p)W_x z_b\right) = 0, \tag{1.82}$$

which yields (with regard to an observer moving with the bed forms)

$$q_{sb} - (1-p)W_x z_b = const. \tag{1.83}$$

Since the flow past the bed forms is converging along their upstream face AB, the largest q_{sb} occurs at the bed form crest B, where z_b is also the largest. i.e.

$$(q_{sb})_{max} - (1-p)W_x (z_b)_{max} = const. \tag{1.84}$$

Subtracting Eq. (1.83) from Eq. (1.84), one obtains

$$\frac{(q_{sb})_{max} - q_{sb}}{(z_b)_{max} - z_b} = (1-p)W_x, \tag{1.85}$$

where the right-hand side is a constant. Eq. (1.85) indicates that the shape of a migrating bed form is strongly associated with the bed-load distribution past it: the q_{sb}-deficit is related to the z_b-deficit by a constant proportion.

1.8 ADDITIONAL REMARKS

(i) The relation $k_s/D \approx 2$ (Eq. (1.4)) is adequate for commonly encountered conditions in open-channel and river flows. However, it is known that when Y is "large", then k_s/D increases with Y. The analysis in Yalin (1992) suggests that Eq. (1.4) can be used for $Y <\approx 1$ (i.e. up to $\eta_* \approx 20$); while for $Y >\approx 1$, k_s/D is reasonably well approximated by the following relations:

$$\frac{k_s}{D} = \begin{cases} Y + (Y-4)^2(0.043Y^3 - 0.289Y^2 - 0.203Y + 0.125), & \text{if } 1 < Y < 4 \\ 5Y, & \text{if } Y > 4 \end{cases} \tag{1.86}$$

Equations for k_s/D expressing a dependency on Y (for sufficiently large values of Y) have been proposed also by other authors (for reviews, see e.g. Camenen et al. 2006 and Dey 2015). Yet, the more recent analyses by Camenen et al. (2006) and Matoušek and Krupička (2009) suggest that for "large" values of Y, k_s/D may, in addition to Y, depend also on other parameters – even if there is no agreement yet on what exactly are those additional parameters.

(ii) Clearly, Eq. (1.6) implies that the maximum flow velocity u_{max} occurs at the free surface ($z = h$). However, it should be kept in mind that Eq. (1.6) is strictly valid as long as the conditions are as stated in Sub-section 1.2.1. If the flow width-to-depth ratio is small ($B/h < 5$ to 7, say), then the velocity profile deviates from that implied by Eq. (1.6), with the maximum flow velocity in reality occurring somewhat below the free surface even in the midst of the flow cross-sections – an occurrence known as the "velocity dip phenomenon". The purpose of this paragraph is merely to raise awareness of the phenomenon: for more information on the topic, the readers are referred to the recent reviews and treatments of the phenomenon by e.g. Guo et al. (2005) and Absi (2011).

(iii) The material in real alluvial rivers and streams is often a mixture of grains of different sizes – while the classical expressions for sediment transport (Bagnold, Meyer-Peter, etc.) assume that the material is uniform, or can be viewed as such. Given the practical significance of non-uniform sediment transport, substantial research has been devoted to it for many years, and several methods are available in the literature to estimate non-uniform sediment transport. While the matter is not dealt with in this book, detailed information on the topic can be gained from Wu et al. (2000), Wu et al. (2003), Wilcock and Crowe (2003), and Parker (2008), among several others.

PROBLEMS

To solve the problems below, take $\gamma_s = 16186.5N/m^3$, $\rho = 1000kg/m^3$, $\nu = 10^{-6}m^2/s$ (which correspond to sand or gravel and water).

1.1 Prove that in the case of a two-dimensional turbulent flow, the dimensionless velocity deficit

$$\frac{u_{max} - u}{v_*}$$

(universal velocity distribution law) is a function of the dimensionless position z/h only: assume that the logarithmic u-distribution is valid throughout the flow depth.

1.2 Prove that in the case of a uniform turbulent flow, the velocity deficit

$$\frac{u_{max} - \overline{u}}{v_*}$$

has the same constant value for any flow regime (hydraulically smooth, transitionally rough and fully rough).

a) What is this constant value in the case of an open-channel flow?
b) What is this constant value in the case of a flow in a circular pipe?

1.3 Consider a two-dimensional fully rough turbulent open-channel flow. Integrating the logarithmic u-distribution (1.6) between k_s and h and assuming that $k_s \ll h$, derive the expression (1.11).

1.4 In addition to Chézy's resistance equation (1.17), we have for the turbulent open-channel flows also Manning's and Darcy-Weisbach's resistance equations. For the case of a wide flow ($\mathcal{R} \approx h$, in which $\mathcal{R}$ is hydraulic radius) these can be expressed as

$$\overline{u} = \frac{1}{n} h^{2/3} S^{1/2}$$

and

$$\overline{u} = \sqrt{\frac{8}{f}} \sqrt{ghS},$$

respectively. Determine the interrelations between c and n, c and f, and n and f.

1.5 Consider the log-log $(X; Y)$-plane in Figure 1.8, where X and Y are given by the relations (1.24). Let m_1 and m_2 be two points on this plane which represent the same granular material and fluid for two different flow stages $(v_*)_1$ and $(v_*)_2$. What is the inclination of the straight line m_1m_2? What physical meaning can be attached to this straight line?

1.6 Note from Figure 1.8 that the turbulent sediment transport initiation curve merges to the straight line $Y_{cr} = 0.1X_{cr}^{-0.3}$ when the values of X_{cr} are small ($X_{cr} < 1$, say). Knowing this, determine (analytically) the expression of the corresponding straight line in the plot in Figure 1.9.

1.7 The slope of a two-dimensional open-channel flow is $S = 0.15 \times 10^{-4}$, the typical grain size of the bed is $D_{50} = 0.18mm$. What is the value of the flow depth which corresponds to the initiation of sediment transport?

1.8 It is intended to study the initiation of sediment transport of the open-channel flow of Problem 1.7 with the aid of a physical model where the movable bed will be formed by polystyrene ($\gamma_s/\gamma = 0.05$). What must be the grain size D' of the model bed material and the slope S' of the model flow if the (model) flow depth is $h' = 10cm$?
(Hint: model and prototype must have identical values of X_{cr} and Y_{cr} for dynamic similarity).

1.9 Consider the steady and uniform two-dimensional flow in a central part of a wide river. The flow depth and slope are $h = 2m$ and $S = 0.00067$. The flow bed is a uniform sand of the grain size $D = 2mm$: treat the mobile bed surface as flat. Determine the specific volumetric sediment transport rate q_{sb} using Bagnold's bed-load formula.

1.10 The Pembina River (in Alberta) has a rather large width-to-depth ratio, and therefore in the central part of its cross-section, the flow can be treated as two-dimensional.

In a certain region of this river, the bed consists of a cohesionless sand having the average grain size $D=0.4mm$. The flow depth and slope are (at a certain time of the year) $h=5m$ and $S=0.00025$. Treat the bed surface as flat.
a) Determine the friction factor c_f.
b) Determine the flow rate Q (assuming that the average river width is $B=100m$).
c) Determine the specific volumetric bed-load rate q_{sb} using Bagnold's formula.

1.11 Consider the bed-load formula of H.A. Einstein, viz

$$1-\frac{1}{\sqrt{\pi}}\int_{-(B_*\Psi+1/\eta_0)}^{(B_*\Psi-1/\eta_0)} e^{-\xi^2}\,d\xi=\frac{A_*\phi}{1+A_*\phi}$$

where Ψ is the reciprocal of the Y-number ($\Psi=1/Y$), and $A_*=43.50$, $B_*=0.143$ and $\eta_0=0.5$; and consider also the bed-load formula of M.S. Yalin, viz

$$\phi=0.635s\sqrt{Y}\left[1-\frac{1}{as}\ln(1+as)\right]$$

where

$$s=\frac{Y}{Y_{cr}}-1 \quad \text{and} \quad a=2.45\frac{\sqrt{Y_{cr}}}{(\rho_s/\rho)^{0.4}}.$$

For which value(s) of Y do these formulae give the same value of the bed-load rate if $Y_{cr}=0.043$? (Take $\rho_s/\rho=2.65$).

1.12 Prove that Eq. (1.61), viz

$$C=C_\epsilon\left(\frac{h/z-1}{h/\epsilon-1}\right)^m \qquad (m=2.5w_s/v_*),$$

implies

$$C=\phi_C(X,Y,Z,z/D).$$

1.13 The shape of the C-distribution curve can either be like A in Figure 1.13b or like B with a point of inflection P. This depends on the numerical value of $m=2.5w_s/v_*$. Determine the range of values of m that yields the shape B.

1.14 In a certain river, the suspended-load concentration C at the level $z/h=0.75$ is half the concentration at the level $z/h=0.25$. Is the C-distribution curve like A or B in Figure 1.13b?

1.15 Consider a two-dimensional flow in a stream having a bed that consists of a cohesionless sand having the average grain size $D=0.30mm$. The flow depth and slope are $h=0.80m$ and $S=0.0002$. Treat the bed surface as flat.
a) Determine the bed-load rate q_{sb} using Bagnold's formula.
b) Determine the thickness ϵ of the bed-load region.
c) Determine the concentration C_ϵ at $z=\epsilon$.
d) Determine the specific volumetric suspended-load rate q_{ss}. (Take $w_s=0.03m/s$).

1.16 At a certain stage of the Mississippi River (at St. Louis) the flow depth and maximum velocity are $h = 12m$ and $u_{max} = 1.5m/s$. In the interval $0.5 < z/h < 0.6$, the concentration distribution curve can be approximated by the straight line $C = 0.0002(1 - z/h)$. Adopting for the velocity distribution the expression $u = u_{max}(z/h)^{1/7}$, determine the specific volumetric suspended-load rate q_{ss} passing through the interval $0.5 < z/h < 0.6$.

1.17 Prove that $\partial(\overline{U}h)/\partial s$, $\partial q_s/\partial s$ and $\nabla \mathbf{q}_s$ are interrelated as follows

$$\nabla \mathbf{q}_s = \frac{\partial q_s}{\partial s} - \frac{q_s}{q}\frac{\partial(\overline{U}h)}{\partial s}.$$

Hint: Use the open forms of $\nabla \mathbf{q}_s$ and $\nabla(h\overline{\mathbf{U}}) = 0$ (continuity).

1.18 Prove that Eq. (1.79) (viz $\nabla \mathbf{q}_s = (h\overline{\mathbf{U}})\nabla\psi_q$, where $\psi_q = q_s/q$) and the expression of $\nabla \mathbf{q}_s$ in Problem 1.17 are identical to each other.

REFERENCES

Absi, R. (2011). An ordinary differential equation for velocity distribution and dip-phenomenon in open channel flows. *Journal of Hydraulic Research*, 49(1), 82-89.

Bagnold, R.A. (1956). The flow of cohesionless grains in fluids. *Philosophical Transactions of the Royal Society of London, Series A, Mathematical and Physical Sciences*, Vol. 249, Issue 964, 235-262.

Bogardi, J. (1974). *Sediment transport in alluvial streams*. Akadémiai Kiadó, Budapest.

Camenen, B., Bayram, A. and Larson, M. (2006). Equivalent roughness height for plane bed under steady flow. *Journal of Hydraulic Engineering*, 132(11), 1146-1158.

Chang, H.H. (1988). *Fluvial processes in river engineering*. John Wiley and Sons, Inc.

Chen, C.L. (1991). Unified theory on power laws for flow resistance. *Journal of Hydraulic Engineering*, 117(3), 371-389.

Cheng, N.-S. (2007). Power-law index for velocity profiles in open channel flows. *Advances in Water Resources*, 30(8), 1775-1784.

Cheng, N.-S. and Chiew, Y.M. (1999). Analysis of initiation of sediment suspension from bed load. *Journal of Hydraulic Engineering*, 125(8), 855-861.

Chien, N. and Wan, Z. (1999). *Mechanics of sediment transport*. ASCE Press, Virginia, USA.

Chiu, C.-L., Jin, W. and Chen, Y.-C. (2000). Mathematical models of distribution of sediment concentration. *Journal of Hydraulic Engineering*, 126(1), 16-23.

Chow, V.T. (1959). *Open-channel hydraulics*. McGraw-Hill, New York.

Coles, D. (1956). The law of the wake in the turbulent boundary layer. *Journal of Fluid Mechanics*, Vol. 1, 191-226.

da Silva, A.M.F. and Bolisetti, T. (2000). A method for the formulation of Reynolds number functions. *Canadian Journal of Civil Engineering*, 27(4), 829-833.

Dey, S. (2015). *Fluvial hydrodynamics: hydrodynamic and sediment transport phenomena*. GeoPlanet: Earth and Planetary Sciences Book Series, Springer.

Dittrich, A. and Koll, K. (1997). Velocity field and resistance of flow over rough surface with large and small relative submergence. *International Journal of Sediment Research*, 12(3), 21-33.

Einstein, H.A. (1950). *The bed-load function for sediment transportation in open channel flows*. U.S. Department of Agriculture, Soil Conservation Service Technical Bulletin, 1026.

Einstein, H.A. (1942). Formulae for the transportation of bed load. *Transactions of the American Society of Civil Engineers*, 107, 561-577.

García, M.H. (2008). Sediment transport and morphodynamics. Chapter 2, in *Sedimentation engineering: processes, measurements, modeling and practice*, edited by M.H. García, MOP 110, ASCE, 21-163.

Garde, R.J. and Ranga Raju, K.G. (1985). *Mechanics of sediment transportation and alluvial stream problems*. 2nd edition, Wiley Eastern Ltd., New Delhi.

Geyer, W. R. (1993). Three dimensional tidal flow around headlands. *Journal of Geophysical Research*, 98(C1), 955-966.

Granville, P.S. (1976). A modified law of the wake for turbulent shear layers. *Journal of Fluids Engineering*, 98(3), 578-580.

Guo, J., Julien, P.Y. and Meroney, R.N. (2005). Modified log-wake law for zero-pressure-gradient turbulent boundary layers. *Journal of Hydraulic Research*, 43(4), 421-430.

Hinze, J.O. (1975). *Turbulence*. 2nd edition, McGraw-Hill, New York.

Jiménez, J. (2004). Turbulent flows over rough walls. *Annual Review of Fluid Mechanics*, Vol. 36, 173-96.

Kamphuis, J.W. (1974). Determination of sand roughness for fixed beds. *Journal of Hydraulic Research*, 12(2), 193-203.

Koll, K. (2006). Parameterisation of the vertical velocity profile in the wall region over rough surfaces. *Proceedings of River Flow 2006, 3rd International Conference on Fluvial Hydraulics*, edited by Rui M.L. Ferreira, Elsa C.T.L. Alves, João G.A.B. Leal and António H. Cardoso, CRC Press, Taylor & Francis Group, 163-171.

Krogstad, P.-A., Antonia, R. and Browne, L. (1992). Comparison between rough-and-smooth-walled turbulent boundary layers. *Journal of Fluid Mechanics*, Vol. 245, 599-617.

Matoušek, V. and Krupička, J. (2009). On equivalent roughness of mobile bed at high shear stress. *Journal of Hydrology and Hydromechanics*, 57(3), 191-199.

McSherry, R., Chua, K., Stoesser, T. and Falconer, R.A. (2016). Large Eddy Simulations of rough bed open channel flow with low submergence and free surface tracking. *Proceedings of River Flow 2016, 8th International Conference on Fluvial Hydraulics*, edited by George Constantinescu, Marcelo García and Dan Hanes, CRC Press, Taylor & Francis Group, 85-90.

Monin, A.S. and Yaglom, A.M. (1971). *Statistical fluid mechanics: mechanics of turbulence*. Vol. 1, MIT Press, Cambridge, Mass. Originally published in 1965 by Nauka Press, Moscow (In Russian).

Monsalve, G.C. and Silva, E.F. (1983). Characteristics of a natural meandering river in Colombia: Sinu River. In *River Meandering: Proceedings of Rivers'83*, edited by Charles M. Elliott, ASCE, 77-88.

Nelson, J.M. and Smith, J.D. (1989). Evolution and stability of erodible channel beds. In *River Meandering: Water Resources Monograph*, 12, edited by S. Ikeda and G. Parker, American Geophysical Union, 321-378.

Nezu, I. (2005). Open-channel flow turbulence and its research prospect in the 21st century. *Journal of Hydraulic Engineering*, 131(4), 229-246.

Nezu, I. and Nakagawa, H. (1993). *Turbulence in open-channel flows*. IAHR Monograph, A.A. Balkema, Rotterdam, The Netherlands.

Nikora, V., Goring, D., McEwan, I. and Griffiths, G. (2001). Spatially averaged open-channel flow over rough bed. *Journal of Hydraulic Engineering*, 127(2), 123-133.

Parker, G. (2008). Transport of gravel and sediment mixtures. Chapter 3, in *Sedimentation engineering: processes, measurements, modeling and practice*, edited by M.H. García, MOP 110, ASCE, 165-251.

Raudkivi, A.J. (1990). *Loose boundary hydraulics*. 3rd edition, Pergamon Press, Oxford.

Raupach, M.R., Antonia, R.A. and Rajagopalan, S. (1991). Rough wall turbulent boundary layers. *Applied Mechanics Reviews*, 44(1), 1-25.

Rouse, H. (1937). Modern conceptions of the mechanics of fluid turbulence. *Transactions of the American Society of Civil Engineers*, 102, 463-543.

Schlichting, H. (1968). *Boundary layer theory*. 6th edition, McGraw-Hill Book Co. Inc., Verlag G. Braun.

Sedov, L.I. (1960). *Similarity and dimensional methods in mechanics*. Academic Press Inc., New York.

Smirnov, V.I. (1964). *A course of higher mathematics*. Vol. 1, Addison-Wesley Pub. Co., Reading, Mass.

Soo, S.L. (1967). *Fluid dynamics of multiphase systems*. Blaisdall Publishing Co., Waltham, Massachussetts, Toronto, London.

van Rijn, L.C. (1985). *Sediment transport*. Delft Hydraulic Laboratory, Publication No. 334, Feb.

van Rijn, L.C. (1984). Sediment transport. Part II: suspended-load transport. *Journal of Hydraulic Engineering*, 110(11), 1613-1641.

Velikanov, M.A. (1995). *Dynamics of alluvial streams*. State Publishing House for Physico-Mathematical Literature, Moscow.

Wilcock, P.R. and Crowe, J.C. (2003). Surface-based transport model for mixed-size sediment. *Journal of Hydraulic Engineering*, 129(2), 120-128.

Wright, S. and Parker, G. (2004). Flow resistance and suspended load in sand-bed rivers: simplified stratification model. *Journal of Hydraulic Engineering*, 130(8), 796-805.

Wu, W., Wang, S.S.Y. and Jia, Y. (2000). Nonuniform sediment transport in alluvial rivers. *Journal of Hydraulic Research*, 38(6), 427-434.

Wu, B., Molinas, A. and Shu, A. (2003). Fractional transport of sediment mixtures. *International Journal of Sediment Research*, 18(3), 232-247.

Yalin, M.S. (1992). *River mechanics*. Pergamon Press, Oxford.

Yalin, M.S. and Karahan, E. (1979). Inception of sediment transport. *Journal of the Hydraulics Division*, ASCE, 105(11), 1433-1443.

Yalin, M.S. (1977). *Mechanics of sediment transport*. 2nd edition, Pergamon Press, Oxford.

Yalin, M.S. (1971). *Theory of hydraulic models*. MacMillan and Co., Ltd., London.
Yalin, M.S. (1965). *Similarity in sediment transport by currents*. Hydraulics Research Papers, No. 6, Hydraulics Research Station, London.
Yang, C.T. (1996). *Sediment transport: theory and practice*. McGraw Hill.
Yen, B.C. (2002). Open channel flow resistance. *Journal of Hydraulic Engineering*, 128(1), 20-39.

Chapter 2

Bed forms

2.1 GENERAL

The research on bed forms is not motivated by an academic interest only: the dimensions and geometry of bed forms determine as practically relevant quantities as the effective roughness of a mobile bed and the resistance factor. As mentioned in the "Scope of the Monograph", the present text concerns wide streams having small values of Fr; hence, antidunes will not be considered here. The longitudinal ridges, whose practical significance is only limited, will also not be discussed. We will be dealing in this chapter with the periodic (along the flow direction x) bed forms known as dunes, bars and ripples only, and we will assume that they develop starting from a *flat* mobile bed surface: the (wide) open-channel is straight.

Dunes and *bars* are large-scale bed forms, with the length of dunes being proportional to the flow depth h, and that of bars to the flow width B. Bars can have various configurations in plan view, the simplest configuration being that of *alternate bars* (also known as *one-row bars* or *single-row bars*). The length of ripples does not depend on the external dimensions of flow (h or B). Instead, ripple length is proportional to the grain size D. The longitudinal sections of ripples are, however, similar to those of dunes. Dunes, bars and ripples continuously migrate downstream. Photos of dunes, dunes with ripples superimposed on them, and alternate bars are shown in Figures 2.1a-c, respectively.

Throughout this book, the length and height of the developed bed forms of the type i ($i = d, r, a$ for dunes, ripples and alternate bars, respectively) will be denoted by Λ_i and Δ_i, respectively; the ratio $\Delta_i/\Lambda_i = \delta_i$ will be referred to as the "steepness" of the bed forms i.

2.2 ORIGIN OF BED FORMS

In Section 1.7 (iii) it has been shown that if a uniform flow has a periodic along x internal non-uniformity, then its initially flat mobile bed must turn into an undulated one (it must become covered by bed forms). It has not been explained, however, how does such an internal non-uniformity originate, and this forms the topic of the present section.

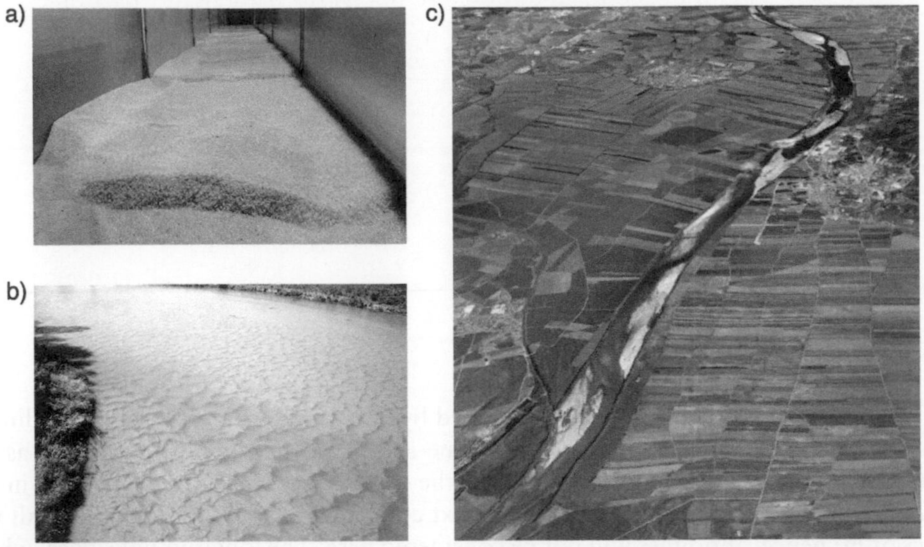

Figure 2.1 Examples of bed forms. (a) Dunes generated in the 0.76 m wide sediment transport flume of Queen's University; view looking upstream (Photo credit: Mr. Joshua Wiebe, M.Sc. Queen's 2006); (b) dunes with superimposed ripples migrating along the bed of the Niobrara River, U.S.A. (Image courtesy of the 'Bedforms in Unsteady Flows' project, funded by the U.K. Natural Environment Research Council (see www.bedform.co.uk); (c) alternate bars in the Tagus River, near Santarém, Portugal (from Google Earth; photo sources: Digital Globe and IGP/DGRF; imagery date: May 29, 2013).

2.2.1 Dunes and bars

Introductory considerations: large-scale turbulence and its coherent structures

Many prominent researchers dealing with fluvial processes were convinced since a long time that the large-scale bed forms, i.e. dunes and bars, are caused by large-scale turbulence (Matthes 1947, Velikanov 1955, Kondratiev et al. 1959, Grishnanin 1979, etc.). However, this view could not be properly elaborated before the discovery of organized motion (in turbulence) and of the related coherent structures by turbulence researchers dealing with different boundary-layer flows.[1]

In the wake of such a discovery, it has become clear that coherent structures (CS's) exist in open-channel and river flows also at a wide range of spatial and temporal scales, varying from near-wall (small-scale) to large-scale coherent structures extending throughout the body of fluid (Yalin 1992, Nezu and Nakagawa 1993, Pope 2000, Adrian and Marusic 2012). Following Hussain (1983), large-scale coherent structures can be defined as the largest conglomeration of eddies with a prevailing sense of rotation. These structures can be vertical or horizontal (Yokosi 1967, Utami and Ueno 1977, Yalin 1992, Yalin and da Silva 1992, da Silva 2009, Rodi et al. 2013): the

[1]For an interesting review of the history of research on coherent motions in turbulent flows, see Robinson (1991).

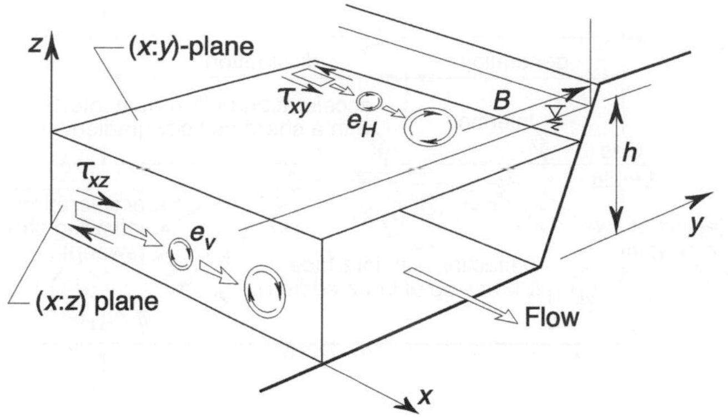

Figure 2.2 Vertical and horizontal planes of rotation of coherent structures (from Kanani and da Silva 2015, reprinted with permission from Springer).

former rotate in vertical $(x;z)$-planes around horizontal axes and scale with the flow depth, while the latter rotate in horizontal (x,y)-planes around vertical axes and scale with the flow width (see the schematic Figure 2.2).

In the following we consider the formation of dunes and bars in view of the present understanding on large-scale vertical and horizontal coherent structures, respectively. The turbulence processes we will be dealing with, which form a "traceable" and thus deterministic component of a turbulent flow, are described below in a schematical manner – all possible deviations and distortions due to the strong "random element" ever-present in any turbulent flow are disregarded in this description. The characteristics of vertical turbulence are marked in this text by the subscript V; those of horizontal turbulence, by the subscript H.

Vertical coherent structures; dunes

(i) Large-scale vertical coherent structures (LSVCS's) in wall-bounded flows in general, and open-channel flows in particular, have by now been the object of numerous works (for reviews, see e.g. Yalin 1992, Nezu and Nakagawa 1993, Franca 2005, Balakumar and Adrian 2007).[2] Even though several details of how exactly these structures originate and develop remain unclear, their life-cycle can, on the basis of Blackwelder (1978), Cantwell (1981), Hussain (1983), Gad-el-Hak and Hussain (1986), Rashidi and Banerjee (1988), Jiménez (1998), and several others, be briefly synthesized as illustrated in the conceptual Figure 2.3 (showing, in a stationary frame of reference, the life-cycle of a LSVCS in an open-channel flow).

A vertical coherent structure originates around a point P, at a location O_i, with the rolling-up at time $t=0$, say, of an eddy e_V ("tranverse vortex"). This is then ejected away from the bed, together with the fluid around it. This total fluid mass moves towards the free surface as it is conveyed by the flow downstream (*ejection*

[2]In the recent literature, these structures have also been referred to as 'very large-scale coherent structures' (VLSCS's; Franca and Lemmin 2015) and 'super structures' (Adrian and Marusic 2012).

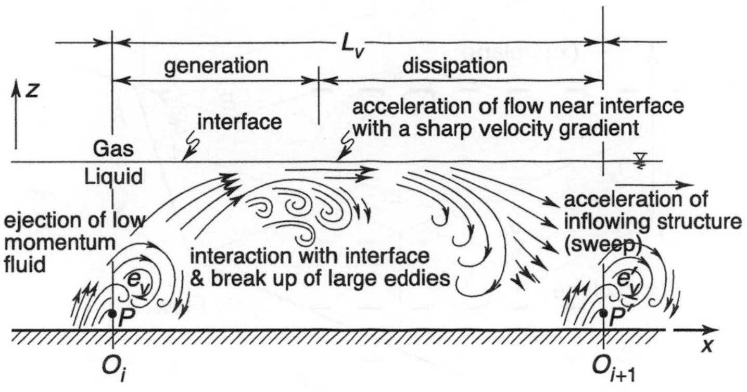

Figure 2.3 Conceptual representation of the life-cycle of a large-scale vertical coherent structure (LSVCS); longitudinal view (from da Silva 2006, reprinted with permission from IAHR).

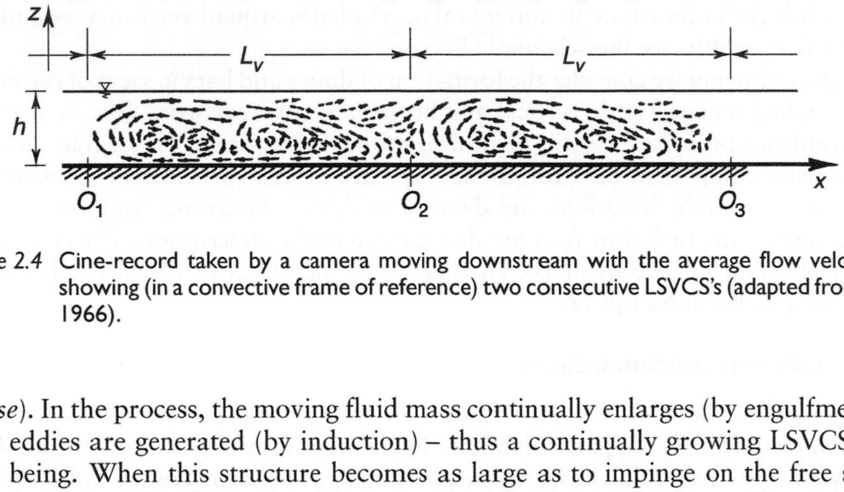

Figure 2.4 Cine-record taken by a camera moving downstream with the average flow velocity, and showing (in a convective frame of reference) two consecutive LSVCS's (adapted from Klaven 1966).

phase). In the process, the moving fluid mass continually enlarges (by engulfment) and new eddies are generated (by induction) – thus a continually growing LSVCS comes into being. When this structure becomes as large as to impinge on the free surface, it disintegrates (*break-up phase*) into a multitude of successively smaller and smaller eddies, until the size of the eddies is reduced to the last level of microturbulence (i.e. until the size of eddies become as small as the lower limit ν/υ_*), at which point their energy is dissipated by viscous action (as implied by the "Eddy-Cascade Theory"). In the process, the neutralized fluid mass moves downstream – towards the bed (*sweep stage*), with a substantially smaller velocity than that of ejection. The "break-up" of a coherent structure prompts the generation of the new eddy e'_V at O_{i+1}, which goes through the same cycle and eventually generate e''_V, ..., and so on. The fact that the (complete) disintegration of one coherent structure coincides (in space x, and time t) with the generation of the next, means that the coherent structures form a *sequence* along x. The cine-record in Figure 2.4, following from the work by Klaven (1966), shows (in a convective frame of reference) an instantaneous view of two consecutive LSVCS's in a longitudinal section $(x; z)$ of a turbulent open-channel flow (rigid bed, camera-speed is u_{av}).

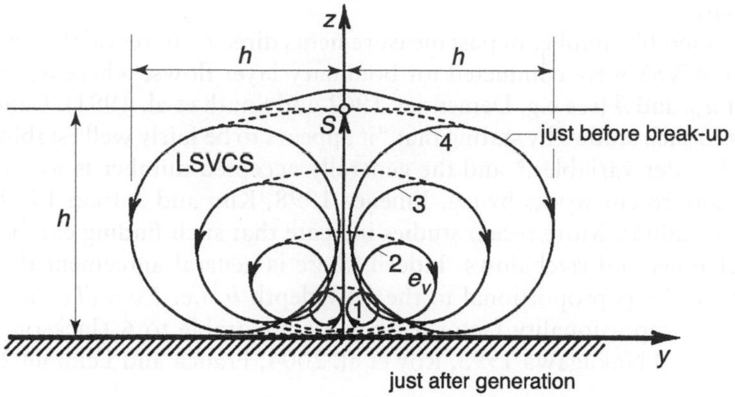

Figure 2.5 Growth of a LSVCS in cross-sectional planes.

The vertical coherent structures are, as a rule, three-dimensional. And it appears that in the cross-sectional $(y; z)$-planes they grow as indicated schematically in Figure 2.5 by 1, 2, 3, 4, while being conveyed downstream. The largest lateral extent of the LSVCS's is thus $\approx 2h$ (Jackson 1976, Yalin 1992, Nezu and Nakagawa 1993, Kadota and Nezu 1999, del Álamo and Jiménez 2003, etc.).[3] It follows that a (longitudinal) sequence of LSVCS's is thus confined to a "corridor" of the width $\approx 2h$ (and the height h).

(ii) In the following, and on the basis of e.g. del Álamo and Jiménez (2003), Franca and Lemmin (2008, 2015) and Kanani and da Silva (2015), among others, the term length scale will be used to designate the distance traveled in the streamwise direction by the (large-scale) coherent structures, from the location of their origin to that of their complete disintegration; and the term time scale, to designate the time taken by the structures to travel such a distance (i.e. to designate their "life-span") – with the symbols T_V and L_V being used to denote the time and length scales of the LSVCS's, respectively.[4] Assuming, with an accuracy sufficient for all practical purposes, that the

[3]The presence of the progressively growing while symmetrically counter-rotating vortices in the $(y; z)$-planes is in line with the contemporary conceptual ("horse-shoe" or "hairpin") models of the development of coherent structures. The existence of systematic cross-sectional vorticities in natural (turbulent) streams apparently has been detected as early as mid-forties by Matthes (1947) (see also Leliavsky 1959).

[4]In the authors' view, it would be more appropriate to refer to these as integral length and time scales (two terms which, on the other hand, may not be totally desirable, as the term integral time scale is invariably used in the literature to imply the area under the time correlation function); or perhaps, following Quadrio and Luchini (2003), as integral life length and time scales. However, for the sake of simplicity, the shorter forms length and time scales are adopted here.

It should also be noted that in this book, the terms length and time scales of the CS's are invariably used in the sense of the average of the length and time scales of the life-cycles of a (large) number of CS's. That is, for any given flow the symbols T_V and L_V (as well as T_H and L_H introduced later on) are to be interpreted as the average length and time scales of the CS's.

LSVCS's are conveyed downstream with the average flow velocity u_{av}, one can write $L_V = T_V \cdot u_{av}$.

A considerable number of past measurements directed to reveal the time and length scales of LSCVS's were conducted for boundary-layer flows, where u_∞ and δ appear instead of u_{av} and h (see e.g. Dementiev 1962 and Smith et al. 1991). Cantwell (1981) summarized such studies by stating that "it appears to be fairly well established that T_V scales with outer variables,[5] and the generally accepted number is $u_\infty T_V/\delta \approx 6$" (see also the more recent works by e.g. Jiménez 1998, Kim and Adrian 1999, del Álamo and Jiménez 2003). More recent studies indicate that such finding can be generalized to open-channel and river flows. Indeed, there is general agreement that the length scale of LSVCS's is proportional to the flow depth h, i.e. $L_V = (T_V \cdot u_{av}) = \alpha_V h$, the value of the proportionality factor α_V being comparable to 6 (Jackson 1976, Yalin 1992, Nezu and Nakagawa 1993, Roy et al. 2004, Franca and Lemmin 2008, 2015). That is,

$$L_V = \alpha_V h \approx 6h. \tag{2.1}$$

(iii) The occurrence of coherent structures is randomly distributed in space and time This means that under completely uniform conditions of flow and bed surface, there is an equal probability (or frequency) of occurrence of coherent structures for any region Δx and time interval Δt. Such a homogeneous or uniform distribution of coherent structures along the flow direction x cannot lead to an internal deformation of the flow that would 'imprint' itself as a periodic deformation of the bed surface (Yalin 1992). However, suppose that there exists in the flow a 'location of preference' associated with an increased frequency of occurrence of coherent structures (at that location). Since the break-up of one coherent structure (CS) triggers the generation of the next CS, such a location of preference leads also to the more frequent generation of sequences (along x) of coherent structures initiating from it. Such an increased frequency of occurrence of CS's at a location can be realized by means of a local discontinuity, d say (the section containing it thus becoming the preferential section $x = 0$, say). Under laboratory conditions, the discontinuity can be the beginning of mobile bed, an accidental ridge across the bed surface, etc. (see Yalin 1992).

It follows that in the presence of a discontinuity, the straight time-averaged initial flow becomes subjected to a perpetual action of sequences of CS's 'fired' from the (ideally speaking) same location (the discontinuity at $x = 0$). It is such action that must inevitably render the flow to acquire a sequence of periodic (along x and t) non-uniformities. Consequently, the time-averaged streamlines s – averaged over a multitude of T_V – must vary periodically along x only with a period T_V, i.e. they are as shown in Figure 1.15a. These wave-like streamlines deform, in turn, the bed surface (as explained in Section 1.7(iii)), so as to produce a sequence of bed forms, viz dunes, whose (average) length Λ_d is the same as L_V (Figure 2.6):

$$\Lambda_d = L_V \approx 6h. \tag{2.2}$$

[5] "Outer variables" are u_∞ and δ; "inner variables" are υ_*, ν and k_s (see e.g. Rao et al. 1971, Aubry et al. 1988).

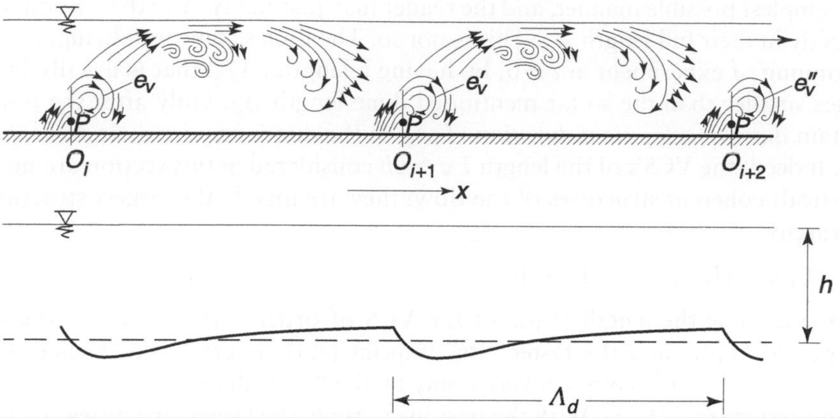

Figure 2.6 (a) Longitudinal sequence of LSVCS's in a flow past a flat bed; (b) longitudinal view of dunes.

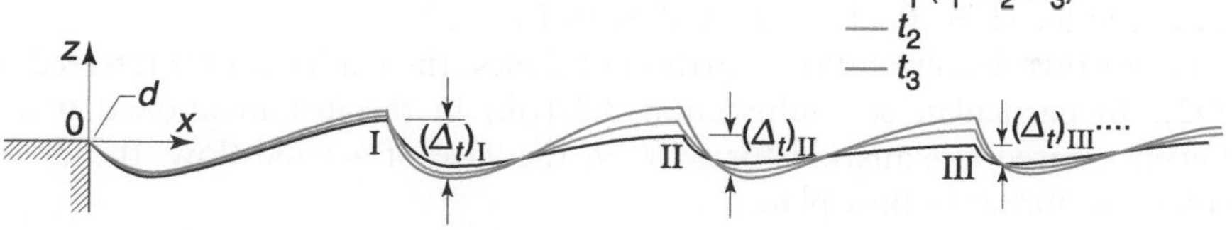

Figure 2.7 Schematic representation of dunes forming downstream of a discontinuity at different development times *t*.

As is known from research on sediment transport, the (average) length of dunes (produced by fully rough flows) can indeed be expressed as $\approx 6h$: e.g. $\Lambda_d = 5h$ (Yalin 1964, Allen 1985); $\Lambda_d = 7h$ (Hino 1968); $\Lambda_d \approx 2\pi h$ (Yalin 1977); $\Lambda_d = 7.3h$ (van Rijn 1984); $\Lambda_d = 6h$ (Yalin 1992).

It should be noted that the longitudinal sequences of CS's prompted by the upstream discontinuity must be expected to attenuate along x. That is, the activity started by the discontinuity should be sustained only within a confined region downstream of the discontinuity itself. In the absence of further discontinuities along x, there has to be a deterioration in the degree of organization of the internal fluid structure, with the conditions eventually reverting to those that would be observed in the absence of any discontinuities. The question thus arises of how can an 'attenuating' (along x) sequence of CS's produce dunes extending throughout the length of a long channel. The answer lies in the fact that the first emerging infinitesimal "ridges" (i.e. steps of the height Δ_t; see Figure 2.7) themselves act as new discontinuities, which perpetuate and propagate along x the activity started by d. (Note that, merely for the sake of simplicity, the fact that dunes grow in length with the passage of time is disregarded in the schematic Figure 2.7).

(**iv**) Owing to pedagogical reasons, the formation of dunes has been explained above in the simplest possible manner, and the reader may justifiably infer that the dunes emerge directly in their full length Λ_d. This is not so. The dunes come into being, just after the beginning of experiment at $t=0$, by having a length $(\Lambda_d)_0$ that is usually by multiple times smaller than the so far mentioned dune length Λ_d. Only after the passage of a certain *dune-development duration* $(T_\Delta)_d$ do the dunes acquire their *developed* length Λ_d. Indeed, the VCS's of the length $L_V \approx 6h$ considered in this section are not the only (vertical) coherent structures of the flow: they are merely the largest structures of the hierarchy

$$(L_V)_0 < (L_V)_1 < \ldots < L_V \approx 6h. \tag{2.3}$$

The smaller is the length $(L_V)_k$ of the VCS of order k ($k=0$, 1, ..., the shorter is its period $(T_V)_k$, and the faster "its" imprint $(\Lambda_d)_k$ is produced. Hence, soon after $t=0$, the bed first becomes covered only by the "mini-dunes" $(\Lambda_d)_0$, generated by the "mini-structures" $(L_V)_0$. With the passage of time, the larger structures $(L_V)_k$ begin to "catch up" with their bed form production. The emergence of the bed forms $(\Lambda_d)_k$ is associated with the elimination of the previous bed form $(\Lambda_d)_{k-1}$, the transition from $(\Lambda_d)_{k-1}$ to $(\Lambda_d)_k$ being, as is well known, by *coalescence*. This pseudo-growth of bed forms terminates (at $t=(T_\Delta)_d$) when the largest bed forms, viz the developed dunes Λ_d, are produced (by the largest VCS's with $L_V \approx 6h$).[6]

For further details on the formation of dunes, the reader is here referred to Yalin (1992). In particular, see Sub-section 3.2.1(iii) in the just mentioned work for a turbulence-based explanation for why, in the case of a wide flow, the dune crests become curvilinear in flow plan.

(**v**) As is well known (Yalin 1977, 1992), the multiplier ≈ 6 is applicable only to the length of dunes generated by fully rough initial flows (having $\upsilon_* k_s/\nu \approx 2X >\approx 70$). If the dune-generating turbulent flow is not fully rough, then the multiplier of h may deviate substantially from ≈ 6 (see Sub-section 2.3.2 later on). This opens the possibility that there is a systematic influence of viscous conditions at the bed (due to ν) on the value of the proportionality factor α_V. On the other hand, such a deviation may be a mere reflection of mobile beds having different $(\nu; k_s)$-conditions responding differently to the same action of LSVCS's.

(**vi**) In view of the content of the previous paragraph, it seems appropriate to end this sub-section by considering the possibility of α_V (in the relation $L_V = \alpha_V h$) being affected by the roughness Reynolds number Re_* $(=\upsilon_* k_s/\nu)$ and/or the relative roughness k_s/h – and indeed, as described below, some attempts to reveal the influence of these dimensionless variables have already been made.

A prominent example is provided in Figure 2.8, adapted from Figures 3 and 13 in Jackson (1976), and showing the plots of measured values of $T_V \cdot u_{av}/h = L_V/h$

[6]From this description, it should not be concluded that the height of dunes is directly related to the CS's, in the sense that the length is. The height ultimately reached by the dunes (as well as their shape) is likely the result of a combination of factors, related to sediment transport (see Section 1.7 (iii)) and overall energy of flow (see Section 5.4). That is, dune height is related to the CS's themselves only to the extent that these contribute to the establishment of sediment transport modes and rates.

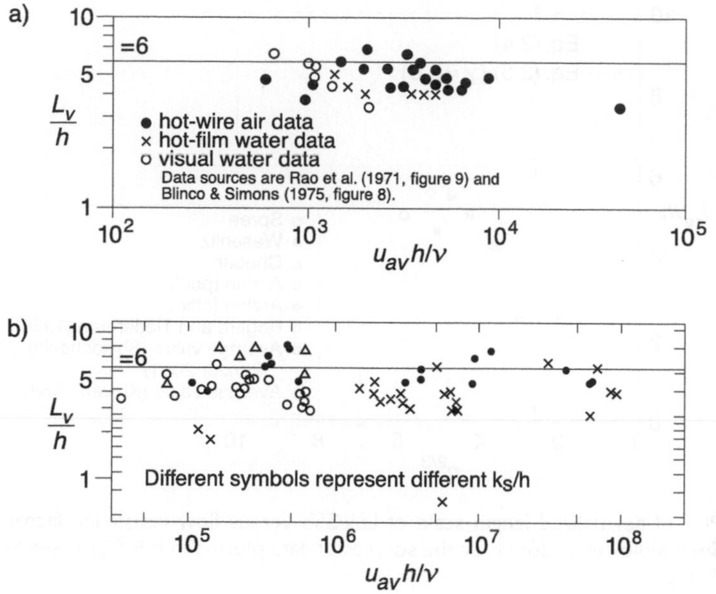

Figure 2.8 Plots of normalized length scale of LSVCS's versus flow Reynolds number (adapted from Jackson 1976), and including: (a) data from laboratory boundary-layer flows past a smooth bed (see Jackson 1976 for the references in this figure); (b) data from the Missouri River and various canals.

versus the flow Reynolds number $u_{av}h/\nu$. Figure 2.8a contains laboratory data of boundary-layer flows past a smooth bed, while Figure 2.8b contains field data from measurements carried out in the Missouri River and various canals.[7] Note that since $u_{av}h/\nu = c_f(h/k_s)Re_*$, where $c_f = \phi(Re_*, h/k_s)$ (see Eq. (1.15)), the consideration of $u_{av}h/\nu$ is equivalent to that of Re_* and k_s/h. No influence of $u_{av}h/\nu$ or k_s/h is detectable from Figures 2.8a,b, in which the gross scatter of experimental points around $L_V/h \approx 5$ to 6 appears random. However, as already pointed out by Yalin (1992), it would not be prudent to conclude from these plots alone that α_V is not affected by Re_* and/or k_s/h. Indeed, L_V/h may depend weakly on these variables, in which case it is possible for the data scatter to obfuscate the form of dependency on k_s/h and/or Re_*, if any. Additionally, the data plotted may not cover sufficiently small values of these parameters.

On the other hand, measurements carried out in laboratory flumes by Klaven (1966) and Klaven and Kopaliani (1973) suggest that α_V is affected by Re_* and/or k_s/h. On the basis of a theoretical analysis of turbulence, Grishanin (1979) (see also Sukhodolov et al. 2011) concluded that the influence of Re_* and/or k_s/h (on L_V/h) is

[7]The original coordinates of Figures 3 and 13 in Jackson (1976) are altered so as to correspond to the present text by identifying δ and u_∞ with h and u_{av}, respectively. The momentum thickness θ was evaluated as $(7/72)\delta \approx 0.097h \approx 0.1h$. The Eulerian integral time scale T_E was taken (in accordance with Jackson 1976) as $0.36T_V$.

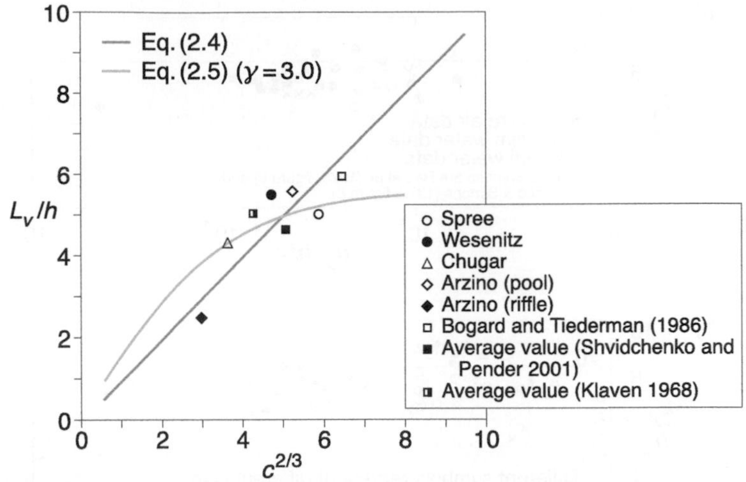

Figure 2.9 Plot of normalized length scale of LSVCS's versus flow resistance factor (adapted from Sukhodolov et al. 2011); for the sources of data plotted in this figure, see Sukhodolov et al. 2011.

by means of the resistance factor c ($=c_f$ if the bed is flat), which is a function of Re_* and k_s/h. This author proposed the following relation for L_V/h:

$$\frac{L_V}{h} = c^{2/3}. \tag{2.4}$$

Yalin (1992) argued that if L_V/h is a function of c, it should be an asymptotic function yielding $L_V/h = const \approx 6$ for sufficiently large values of c, instead of a monotonously increasing function of c as implied by Eq. (2.4). By analyzing the time-growth rate of an "eddy" along its path as it is conveyed downstream, Yalin (1992) arrived at the following expression for L_V/h:

$$\frac{L_V}{h} \approx 6\left[1 + \gamma\frac{1}{c}\right]^{-1}, \tag{2.5}$$

where the value of the coefficient γ is not yet known.

An attempt to verify Eq. (2.4) was recently made by Sukhodolov et al. (2011), using measurements carried out in a number of rivers in Europe (see their Figure 6). Figure 2.9 is a modified version of the just mentioned figure, in which in addition to the straight line representing Eq. (2.4), the curve representing Eq. (2.5) with $\gamma = 3.0$ is also plotted. As can be inferred from Figure 2.9, the latter is in reasonable agreement with the data. It should, however, be kept in mind that the dataset plotted in this figure is rather limited (too few data points) to arrive at definite conclusions.

The discussion above clearly points to the need for further research on the form of dependency of L_V/h on Re_* and k_s/h (or possibly c), if any. In conducting such research, it would be advisable to start by considering strictly the simplest case of flow past a flat bed. This will ensure that the value of c reflects only the influence of

Re_* and k_s/h, and is not affected by the presence of bed forms or other irregularities that may otherwise be present – and which introduce a complicating factor, likely to further obscure the situation. Furthermore, to be meaningful, such research should be carried out by considering the average length and time scales of the CS's (obtained by averaging the length and time scales of a sufficiently large number of coherent events).

Horizontal coherent structures; bars

(i) Where horizontal turbulence is concerned, considerable research efforts have already been devoted to the study of horizontal eddies and horizontal coherent structures originating at river irregularities, such as sudden changes in direction, confluences, bifurcations, etc. (see e.g. McLelland et al. 1996, Constantinescu et al. 2012, Konsoer and Rhoads 2014, among many others).

A related, but essentially distinct process, is the occurrence of large-scale horizontal coherent structures (LSHCS's) in rectilinear (or nearly rectilinear) channels without any such irregularities, and in which the LSHCS's are thus not induced by them. As summarized by Yalin (1992), these were brought to light through the earlier field and laboratory measurements reported by Yokosi (1967) and Grishanin (1979). Such LSHCS's have so far been the object of only a few isolated studies. In spite of this, the related observations and measurements suggest that in their life-cycle the LSHCS's follow, *mutatis mutandi*, a sequence of events as described for LSVCS's but instead occurring in a horizontal "flow ribbon". The difference appears to be mainly in the "length scale", with all "lengths" of the large-scale horizontal turbulence being proportional to the flow width B (instead of flow depth h, as is the case in large-scale vertical turbulence).

Let u be the time average flow velocity at a point of a straight open-channel flow. The velocity u is an increasing function of z, and it exhibits in each of the horizontal $z = const$ planes a certain distribution (variation along y). Clearly, the largest values of $\partial u/\partial y$, and thus the largest τ_{xy} (see Figure 2.2), are in the neighbourhood of the side walls (banks) at the free surface. For this reason, LSHCS's are likely to originate predominantly near the banks and the free surface (i.e. in the neighbourhood of the upper corners of the flow cross-section). Afterwards, they are conveyed downstream by the mean flow, while growing in size (see Figure 2.10a). As long as the width-to-depth ratio is not too "large", the LSHCS's will grow until their lateral extent becomes as large as B.[8] At this point, they interact with the banks and disintegrate into a cascade of successively smaller and smaller eddies. The neutralized fluid mass returns to its original bank so as to arrive there at $t = T_H$, the (average) length scale of the LSHCS's being given by $L_H = \alpha_H B$. There seems to be agreement that the coherent structures forming the large-scale horizontal turbulence of a wide open channel have the shape of horizontally positioned disks (Figure 2.10b) eventually extending (along z) throughout the flow thickness h (Grishanin 1979, Jirka and Uijttewaal 2004, Yalin 2006, da Silva and Ahmari 2009).

[8]The reason for the statement "As long as the width-to-depth ratio is not too large..." will be clarified in the paragraph (iii) below.

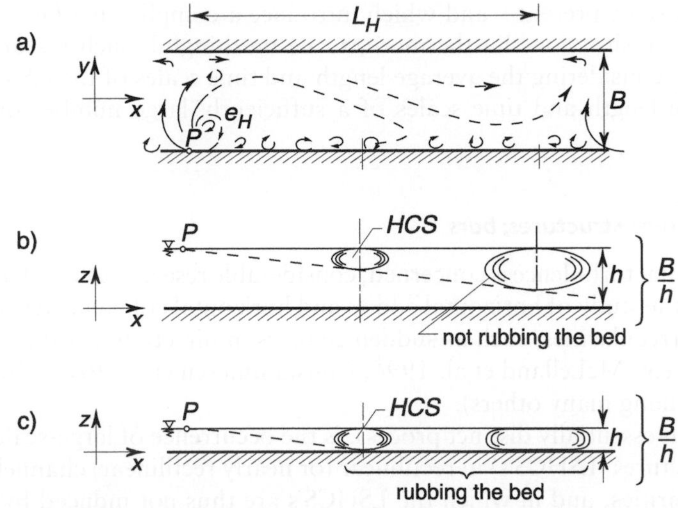

Figure 2.10 Conceptual representation of the life-cycle of a large-scale horizontal coherent structure – LSHCS (from da Silva 2006, reprinted with permission from IAHR). (a) Plan view; (b) side view in the case of "small" values of width-to-depth ratio; (c) side view in the case of comparatively larger values of width-to-depth ratio.

(ii) As pointed out by Yalin (1992), the growth rates of large-scale coherent structures in the flow direction (x) exhibit a remarkable similarity. But if so, then the value of α_H should be comparable to α_V, namely 6, i.e.

$$L_H = \alpha_H B \approx 6B. \tag{2.6}$$

As shown in paragraph (iv) below, this expectation is in agreement with existing measurements, however few.

On the other hand, it is known that the (average) length of alternate bars Λ_a is also comparable to $6B$ (Hayashi 1971, JSCE 1973, Jaeggi 1984, Ikeda 1984, Fujita and Muramoto 1985, etc.). This strongly suggests that, similarly to the case of LSVCS's and dunes, the length L_H of LSHCS's is 'imprinted' on the bed surface of a mobile bed as the alternate bar length:[9]

$$\Lambda_a = L_H \approx 6B. \tag{2.7}$$

Alternate bars are anti-symmetrical in plan view with respect to the x-axis (Figure 2.11c). But if so, then the sequences of LSHCS's issued from the right and left banks generating them, must also be anti-symmetrical in plan view – that is, they must be expected to be as shown in Figure 2.11a, where the LSHCS's are 'fired' from the right bank at O_1, O_2, etc.; and from the left bank, at O_1', O_2', etc. Here, O_1', O_2', ..., are located downstream of O_1, O_2, ..., by $L_H/2$.

[9]In conjunction with this point, see also Section 2.5.

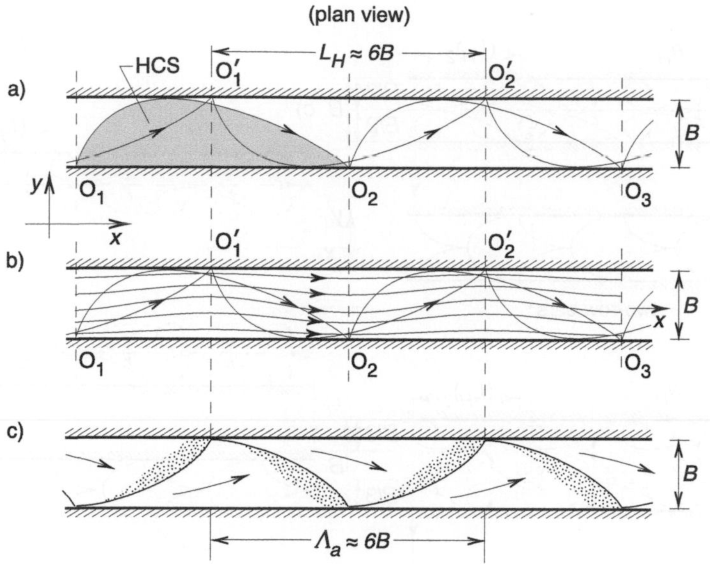

Figure 2.11 Plan views of: (a) longitudinal sequences of LSHCS's issued from the right and left banks, in a flow past a flat bed; (b) wave-like streamlines resulting from the superimposition of the sequences of LSHCS's on the mean flow; (c) alternate bars.

It should be noted here that alternate bars do not emerge directly in their full length. Just like dunes, they first come into being with a length much smaller than their final length $\Lambda_a \approx 6B$; and they grow step-by-step (by coalescence) during their development time. This is well illustrated e.g. in the experiments by Ikeda (1983) and Boraey (2014).

(iii) Consider now the influence of the B/h-ratio on the growth of the LSHCS's. If B/h is small, i.e. if it is smaller than a certain value $(B/h)_{0,1}$, say, then a LSHCS acquires its largest size B without rubbing the bed (Figure 2.10b). In this case, the LSHCS's can obviously not produce any bed forms. If, however, $B/h > (B/h)_{0,1}$, then the LSHCS's *are* rubbing the bed (Figure 2.10c) and they are thus capable of producing "their" bed forms, viz bars. In the latter case, different conditions can be encountered, as follows.

If B/h is larger than $(B/h)_{0,1}$, but smaller than a certain $(B/h)_{1,2}$, then the HCS's emitted from both banks exhibit a single-row configuration in plan view and produce the single-row bars or alternate bars (Figure 2.11). Suppose now that $B/h > (B/h)_{1,2}$. In this case, the relative flow width is too large, and the bed-rubbing HCS's are destroyed by the bed friction *before* their size can become as large as B. The HCS's emitted from both banks thus meet each other in the midst of the flow width; and instead of the *one-row* plan configuration of HCS's, we have *two-row* and consequently *two-row bars* (Figure 2.12a). The further increment of B/h leads to the *three-row*, ..., *n-row* configuration of the HCS's (Figures 2.12b,c) and thus to the *three-row*, ..., *n-row bars*

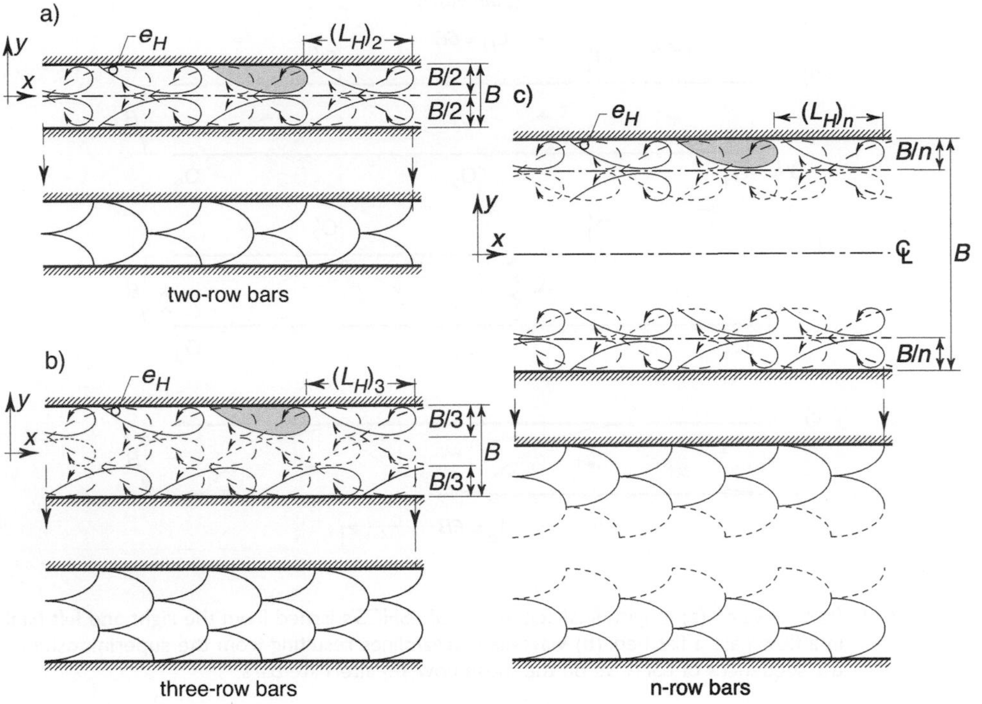

Figure 2.12 Different plan configurations of LSHCS's and bars in streams having large values of width-to-depth ratio: (a) two-row; (b) three-row; (c) n-row.

or *multiple bars*. The length of the n-row HCS's and bars ($(L_H)_n$ and Λ_n) is given by (Ikeda 1983, Fujita and Muramoto 1988, da Silva 1991, etc.)

$$(L_H)_n = \Lambda_n = \frac{6B}{n} \qquad \text{(where } \Lambda_1 \text{ implies } \Lambda_a\text{)}. \tag{2.8}$$

The structure of the large-scale horizontal turbulence of a fully rough flow in a wide trapezoidal open-channel (having a flat rigid bed covered by the granular roughness $k_s \approx 2D$) is determined solely by the ratios (da Silva 1991, Yalin 1992; see also Pope 2000)

$$\frac{B}{h} \quad \text{and} \quad \frac{h}{k_s} \quad \left(\text{or } \frac{h}{D} = Z\right). \tag{2.9}$$

Clearly, these variables are applicable also to the case of a mobile bed – as long as this bed is flat and the suspended-load is not extensive. Since the plan configuration of HCS's is an aspect of the structure of flow past the initial bed, n must be a certain (although so far unexplored) function of the variables (2.9):

$$n = \phi_n\left(\frac{B}{h}, \frac{h}{D}\right). \tag{2.10}$$

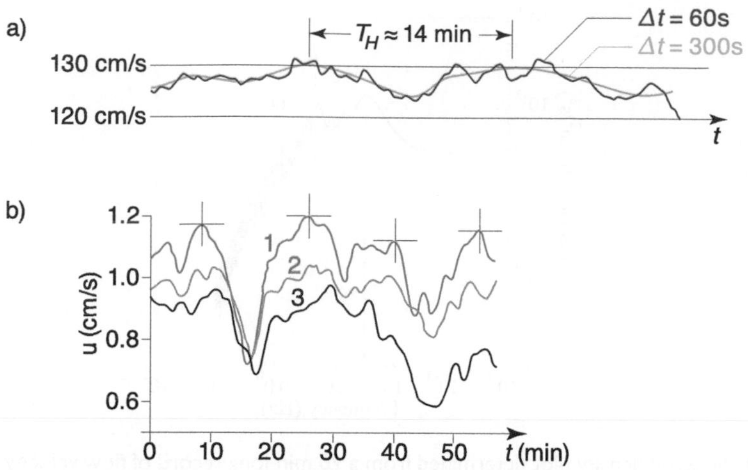

Figure 2.13 Smoothed oscillograms of longitudinal flow velocity resulting from measurements in: (a) the Uji River (adapted from Yokosi 1967); (b) the Syr-Darya River at Ak-Ajar (adapted from Grishanin 1979).

It should be noted that the emergence of bars merely regularizes and stabilizes the plan configuration of the HCS's – it does not alter it (Kondratiev et al. 1982, Yalin 1992).

(iv) As mentioned earlier, very few studies were carried out so far to investigate the LSHCS's under consideration. These invariably focused on the detection of the structures, and the establishment of their length and time scales. Such studies are limited to the earlier works by Yokosi (1967), Dementiev (1962) and Grishanin (1979); and the more recent works by da Silva and Ahmari (2009) and Kanani and da Silva (2015) (see also da Silva et al. 2012).

Yokosi (1967) conducted flow velocity measurements in the Uji River, Japan. This author then applied a simple filtering to the velocity records, by averaging them over consecutive time-intervals Δt. As is well known, by adopting such a procedure, the resulting smoothed velocity oscillograms contain only those velocity fluctuations whose "period" is larger than Δt. Thus, by selecting a sufficiently large Δt, it is possible to reveal the longest "periods", or lowest frequencies of the velocity fluctuations, caused by the passage of the largest coherent structures.

Figure 2.13a shows the smoothed oscillograms of longitudinal flow velocity resulting from the measurements by Yokosi (1967). These correspond to $\Delta t = 60s$ and $\Delta t = 300s$. The longest "period" T_H due to the passage of the largest HCS's is associated with the curve in green in Figure 2.13a ($\Delta t = 300s$). This yields $T_H = 14min$. Together with $B = 100m$, and $u_{av} = 1.10m/s$, one obtains $L_H/B\ (= u_{av}T_H/B) = 9.24 \approx 9$. [Note that Yokosi (1967) reported only the flow velocity at $z/h = 0.8$, namely $u = 1.28m/s$. For the present purposes, the average flow velocity was estimated by identifying u at $z/h = 0.8$ with u_{max}, and by taking into account that $(u_{max} - u_{av}) = 2.5v_*$ (see Problem 1.1). Here $v_* = \sqrt{gSh} = 0.07m/s$ ($h = 2m$, $S = 0.00026$). This yielded $u_{av} = 1.10m/s$.]

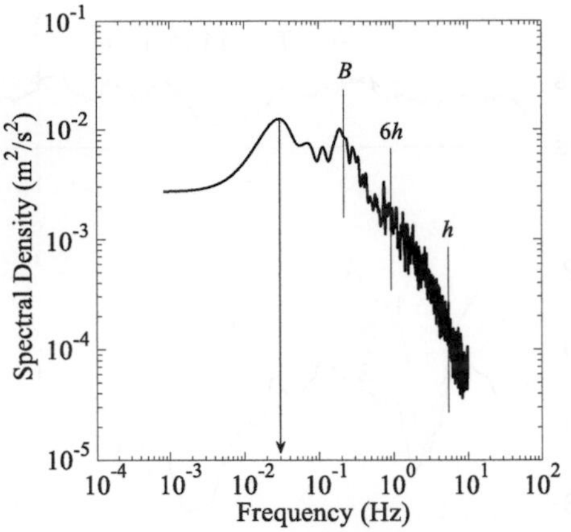

Figure 2.14 Spectral density plot determined from a 20 min long record of flow velocity (from Kanani and da Silva 2015, reprinted with permission from Springer).

Similar measurements were carried out by Dementiev (1962) and Grishanin (1979) in rivers of Central Asia. Figure 2.13b shows the only available velocity records, resulting from the measurements in the Syr-Darya River at Ak-Ajar. Here the curves 1-3 correspond to measurements at $z/h = 0.8$, 0.4 and 0.2, respectively; the oscillograms shown are smoothed (using $\Delta t = 120s$). The average interval between the peaks is $\approx 13min$ ($T_H \approx 13min$). Together with $B = 100m$ and $u_{av} = 0.9m/s$, one determines $L_H/B = 7.02 \approx 7$.

The values of α_H determined above, namely 9 and 7, even though larger than ≈ 6, are nonetheless comparable with it. When considering these values, it should be kept in mind that they are based on a limited number of "samples"; and thus cannot be expected to supply a reliable value of α_H.

da Silva and Ahmari (2009) and Kanani and da Silva (2015) analyzed an extensive series of velocity measurements conducted (by da Silva and Ahmari 2009) in a $21m$ long, $1m$ wide laboratory flume. The bed was flat; $D = 2.2mm$; $h = 0.04m$; $u_{av} = 0.225m/s$; $B/h = 25$. The velocity measurements were carried out at a total of 306 different locations (points) in the flow domain at the level $z/h = 0.75$. The sampling duration was $120s$; at a few locations, additional measurements were performed using a sampling duration of $20min$.

As an example, Figure 2.14 shows the spectral density plot determined from one of the $20min$-long velocity measurements. Following the terminology by Sukhodolov and Uijttewaal (2010), this is a composite spectrum, extended to the frequencies of horizontal coherent structures. The frequencies corresponding to the length scales of h, $6h$, and B are indicated by vertical lines. Note that the energy dissipation range (viscous subrange) was not captured by the measurements; and that the $-5/3$ inertial subrange (starting at roughly $6Hz$) was only partially captured.

The frequency content of the totality of the $120s$ velocity records was analyzed using both simple averaging (see Ahmari and da Silva 2009), as well as continuous wavelet transform (CWT) (see Figures 2.15a,b for examples of the results for two

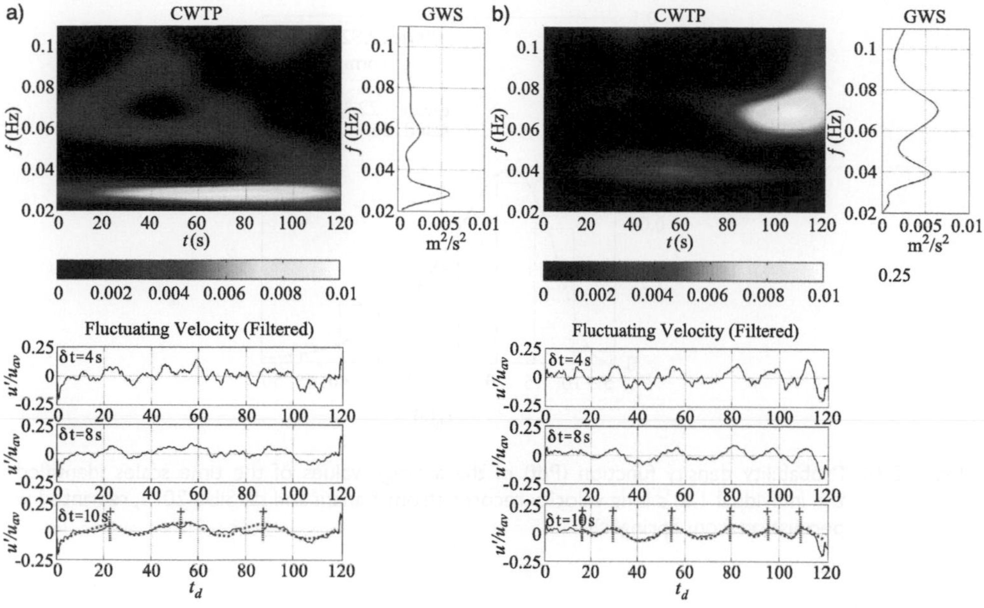

Figure 2.15 Continuous wavelet transform power (CWTP) contour-plots for frequency band 0.02-011 Hz, corresponding global wavelet spectrum (GWS), and filtered oscillograms of u' at two points located on a cross-section 9 m from the channel entrance (from Kanani and da Silva 2015, reprinted with permission from Springer). (a) At ≈ 0.18 m from the left bank; (b) at ≈ 0.35 m from the left bank.

distinct velocity records). Large-scale HCS's could be identified in all of the velocity records, and appeared as quasi-cyclic, sustained features in the flow. On the basis of the 306 velocity records, the average time scale of the LSHCS's was found to be equal to 22.3*s* (see Figure 2.16), implying a length scale of five times the flow width (i.e. $L_H/B = 5$). Again, this value is consistent with the considerations in paragraph (ii) above.

2.2.2 Ripples

Ripples are prominent when the turbulent flow generating them is viscous at the bed, i.e. when it is hydraulically smooth ($\upsilon_* k_s/\nu \approx 2X <\approx 5$). Hence it is highly unlikely that they are caused by the large-scale coherent structures whose length is proportional to h and/or B. [Apparently the viscosity at the bed (that is, the near bed flow affected by viscosity) shields the initial bed surface from the action of the large-scale CS's].

Experiment shows that vertical turbulence disturbs the viscous flow at the bed so as to produce in the $(x; y)$-plane at a level z near the bed surface a series of adjacent high- and low-speed zones, in short *viscous flow structures*, which vary (slowly) with the passage of time (Johansson and Alfredsson 1989, Rashidi and Banerjee 1990, Lu and Smith 1991, Smith et al. 1991, Nezu and Nakagawa 1993). A sample of such quasi-periodic (along x and y) zones is shown in Figure 2.17. Here they are depicted by instantaneous contour-lines of vertical component of vorticity ω_z (at $z^+ = 56$), for

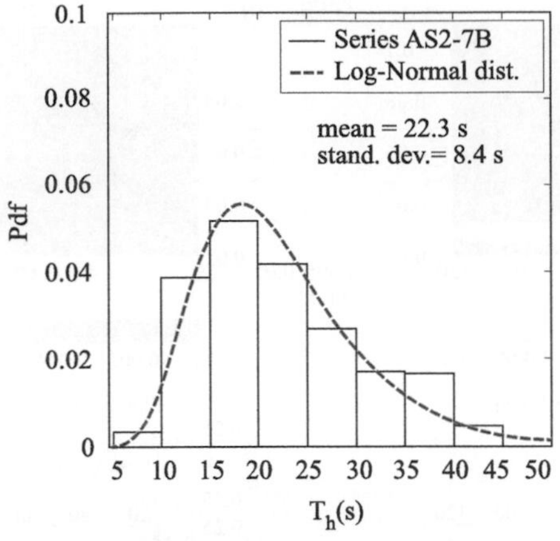

Figure 2.16 Probability density function (Pdf) of the average values of the time scales identified in the individual 120 s long velocity records (from Kanani and da Silva 2015, reprinted with permission from Springer).

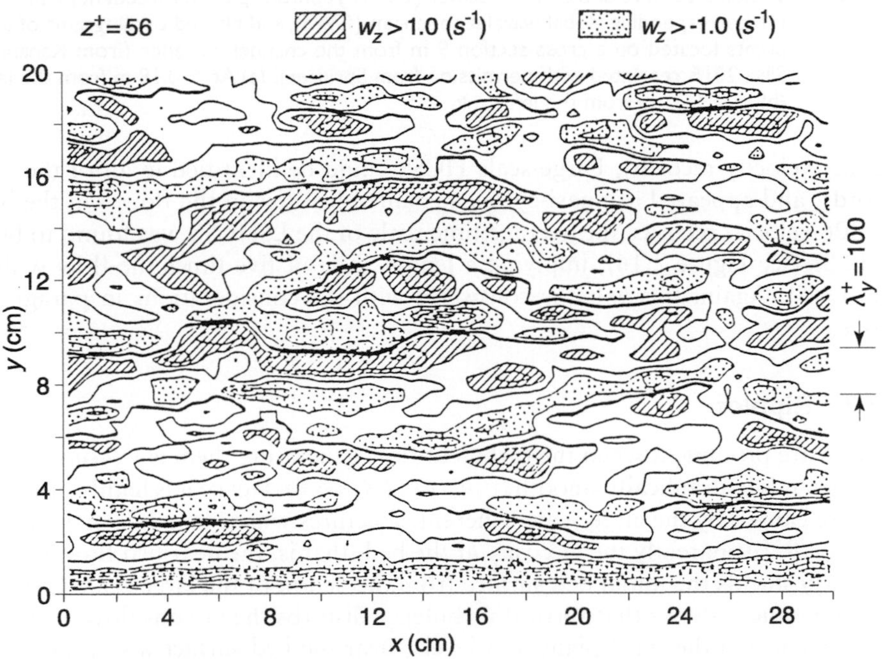

Figure 2.17 Distribution of vertical component of vorticity at the horizontal plane at $z^+ = 56$ obtained on the basis of flow visualization of tracer particles (from Nezu and Nakagawa 1993, reprinted with permission from A. Balkema).

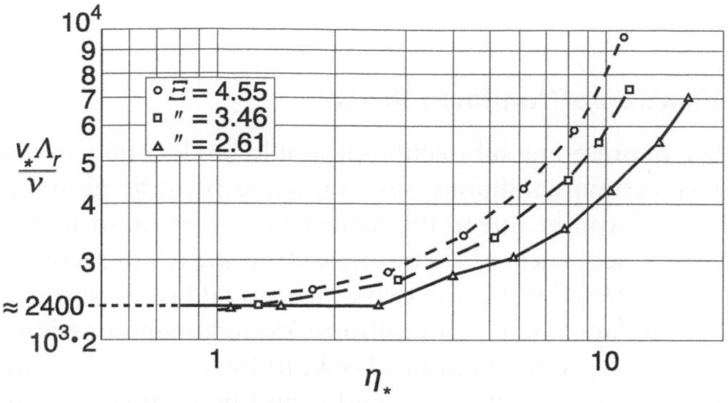

Figure 2.18 Plot of dimensionless ripple length versus relative flow intensity for three different values of the material number (for the sources of the data in this figure, see Yalin 1987).

a flow with $u_{av} = 6.5cm/s$, $\upsilon_* = 0.7cm/s$, $h = 4cm$, and $Re = 2600$. It has been found (Blackwelder 1978, Blackwelder and Eckelmann 1979, Cantwell 1981, Nezu and Nakagawa 1993; see also del Álamo and Jiménez 2003) that the average length of each zone in the x-direction is comparable with $\approx 1000\nu/\upsilon_*$,[10] the average longitudinal period, i.e. the average "wave length" λ_x, being thus given by

$$\frac{\upsilon_* \lambda_x}{\nu} \approx 2000. \tag{2.11}$$

λ_x is the only length which is inherently present at the bed in the structure of vertical turbulence of a hydraulically smooth flow. Hence we postulate (as it has been done in Yalin 1992) that Λ_r is but the "imprint" of the length λ_x of the aforementioned viscous flow structures. Note from Figure 2.18 that when η_* approaches unity, i.e. when the conditions of the two-phase motion approach to those of a clear fluid (for which the value ≈ 2000 was determined), then the values of $\upsilon_* \Lambda_r/\nu$, corresponding to various Ξ, tend to become ≈ 2400 (which is comparable with ≈ 2000).

In contrast to the case of LSVCS's and LSHCS's, there is no "hierarchy" of λ_x-values. On the other hand, the emergence of ripples renders the bed rougher. As a consequence, the plan dimensions of the viscous structures can be expected to change (increase) to some extent as the ripples develop (Yalin 1992). Such explanation is consistent with several experimental observations indicating that the length of ripples in general does not grow during their development as much as that of dunes, with ripples growing by a factor of 2 or 3 (Fok 1975, Baas 1994).

The viscous flow structures at the bed are not permanent. Yet, they may persist at an area for a sufficiently long duration to permit the initiation of ripples at that location. This would explain why a "local discontinuity" on the flow bed, though certainly facilitating the formation of ripples, is not absolutely necessary for their

[10] In Figure 2.17, the average length of each zone in the x-direction is about 10 cm, or $\approx 700\nu/\upsilon_*$, and thus roughly comparable with $\approx 1000\nu/\upsilon_*$. Note also from Figure 2.17 that the positive and negative regions of ω_z alternate also in the spanwise direction y, the mean spanwise spacing being $\lambda_y^+ \approx 100$.

occurrence – they can emerge spontaneously anywhere on the bed (see e.g. Schmid 1985).

2.2.3 Clarification of "uniform flows"

From the content of preceding sub-sections it should be clear that an externally uniform flow may, under certain conditions, contain some periodic along x "internal non-uniformities" (which can be due to the sequences of vertical or horizontal CS's (Sub-section 2.2.1), to the sequences of viscous flow structures (Sub-section 2.2.2), etc.). But this means that an externally uniform two-dimensional flow past the flat initial mobile bed (at $t = 0$) may, in fact, not be really uniform, even though its external characteristics u_m, h and S do not vary along x. In this book, following the convention, such a flow will still be referred to as "uniform" – and it will be treated as such. Yet, it will be invariably assumed tacitly that each of the internal characteristics, a say, of such a flow signifies its $(t;x)$-*average* value (time-space average value), and not the time-average value only. The periodicity can be present in the y-direction as well (multiple-row LSHCS's, viscous flow structures). In such cases any a will be assumed to signify the (t,x,y)-*average* value.

A flow past the periodic along x two-dimensional bed forms (of the length Λ, say) can also be considered as "uniform", in the sense that its non-uniformity along Λ does not vary with x. Following Yalin (1977, 1992) such flows will be referred to as "quasi-uniform".

2.2.4 Special aspects

(i) No aspects peculiar only to the open-channel flows were used in the present considerations regarding the origin of dunes and ripples. Hence, dunes and/or ripples can occur on the deformable bed of any turbulent flow having LSVCS's, and/or viscous bed structures. Indeed, the bed forms mentioned occur also in closed conduits and in the desert. In the latter case the length scale L_V (and thus Λ_d) is to be identified with $\approx 6\delta$, where δ is the boundary-layer thickness of the dominant air flow (dominant wind).

No side walls are present in the desert; hence no periodic (along x) horizontal coherent structures (HCS's) can be present there, and therefore no periodic "desert bars".

(ii) Let Λ_i and Λ_j be the lengths of bed forms i and j, respectively. If Λ_i and Λ_j are comparable, then the bed forms i and j are mutually exclusive, and only the bed forms which are due to the "stronger agent" materialize. If, however, Λ_i and Λ_j are substantially different (e.g. if $\Lambda_i \ll \Lambda_j$), then the bed forms i and j can co-exist – as mentioned later on in Sub-section 2.3.4 with regard to ripples superimposed on dunes. Consider the possibility of dunes being superimposed on alternate bars. A necessary (but not sufficient) condition for this to occur is the large value of the width-to-depth ratio, for $\Lambda_d \ll \Lambda_a$ implies $h \ll B$. And if $\Lambda_r \ll \Lambda_d \ll \Lambda_a$, then we can have ripples superimposed on dunes superimposed on alternate bars.

(iii) Since the large-scale horizontal turbulence "diffuses" from the banks towards the flow centerline, its intensity must progressively decrease when the relative distance to the banks (y/h) increases. But this means that multiple bars, which are generated by this turbulence, must become less and less prominent with the increment of y/h. Hence, in the midst of a flow having "very large" B/h, no large-scale horizontal turbulence (and thus no bars) can be expected to be present: we should have only the large-scale vertical turbulence (and thus ripples and/or dunes).

(iv) For a given value of the sediment transport rate, the duration of development $(T_\Delta)_i$ of a two-dimensional bed form i is proportional to the area $\Lambda_i\Delta_i \sim \Lambda_i^2$ of that bed form (see Eq. (3.22) in Yalin 1992); which means that for the bed forms i and j having comparable steepness, $(T_\Delta)_i/(T_\Delta)_j \approx (\Lambda_i/\Lambda_j)^2$. If e.g. $\Lambda_i \approx 20cm$ and $\Lambda_j \approx 100m$ (which are the values commonly encountered in desert ripples and dunes), and the bed form steepnesses are comparable, then $(T_\Delta)_i/(T_\Delta)_j \approx 4 \times 10^{-6}$ – which explains why the direction of dunes (forming by the dominant wind over several months) does not always coincide with the direction of ripples (forming in few hours by the daily changing winds). In the following, when referring to the duration T_Δ of the development of bed forms, we will refer automatically to the largest among the present $(T_\Delta)_i$'s.

(v) It is not known, and it may be worthwhile to explore, why desert ripples formed by air are very regular (two-dimensional), while those formed by water usually are not (three-dimensional).

2.2.5 Additional remarks: the ongoing debate on the origin of bed forms

Despite the considerations so far in this chapter, it should be emphatically pointed out here that the origin of dunes, bars and ripples remains a matter of controversy and debate. The main reasons for such controversy are outlined below.

(i) On one hand, recent years have been marked by the development of a much better understanding of the origin and fate of coherent structures in turbulent flows, including at the larger scales. The similarity between the (average) length scale of the LSVCS's of flows past flats beds, by now established independently by a large number of authors, and the (average) dune length is striking. The same applies to the length scale of the viscous structures near the bed of hydraulically smooth flows and the length of ripples. Even if scarce, existing studies on the length scale of LSHCS's, including of flows past flat beds, also very clearly indicate that this is comparable to the length of alternate bars. It is rather difficult to conceive that such length scale similarities are all a mere coincidence. As follows from the previous sub-sections, this appears as a strong indication that the large-scale coherent structures are responsible for the origin of the bed forms known as dunes and bars, while the viscous structures near the bed are responsible for the origin of ripples.

(ii) As pointed out by Yalin and da Silva (2001), if periodic (along x) bed forms were not to be found in laminar flows (as suggested by the earlier experiments by Tison 1949), then a hypothesis aiming to explain the origin of bed forms would have to necessarily rest on turbulence. Yet, a relatively recent development is that it has been found that bed forms occur also in laminar flows. This has been particularly well demonstrated by

Coleman and Eling (2000). Even though these authors refer to the bed forms resulting from their experiments as wavelets, they resembled ripples in their longitudinal profiles and scaling with the grain size. A photo of the bed forms in the just mentioned work is provided by Lajeunesse et al. (2010) (see their Figure 4B). Lajeunesse et al. (2010) subsequently argued that ripples, dunes as well as alternate and multiple bars can all occur in laminar flows. On the basis of the evidence provided, it is however difficult to assess the validity of such claim. Indeed, the bars in their Figure 6D are quite similar to rhomboid bars, which are a type of bed forms in their own right and not thought to be due to turbulence (see e.g. Ikeda 1983 and Devauchelle et al. 2010, providing examples of rhomboid bars produced under under both laminar and turbulent flows) – while the striations on the bed in their Figure 6C only vaguely resemble alternate bars.

It therefore appears that the question nowadays is no longer whether or not bed forms can occur in laminar flows, but rather what exactly are the bed forms produced in such flows. Are they bed forms in their own right, distinct from those in turbulent flows, or are they the same (even if somewhat distorted)? In this regard, it would seem that the only mechanisms capable of leading to bed forms in laminar flows are either instability of the (laminar) flow itself, or instability of the granular material.

In the authors' view, the points made above, more than anything point to two matters. The first is the existing scarcity of information on bed forms in laminar flows, highlighted by e.g. Valance and Langlois (2005). The second is the need to develop a proper understanding of the instability of laminar flows over granular beds. Directing research efforts to addressing both of these matters thus appears as particularly worthwhile.

(iii) On the other hand, as is well-known, stability analysis, even if with some limitations (see e.g. Zhou and Mendoza 2009), is capable of providing rather realistic descriptions of the growth of bed forms. Under such approach, bed forms emerge as the outcome of inherent instabilities of some equilibrium state of the system and in particular of the interface between the flowing mixture and the granular material (Seminara 1998). The basic state is a uniform (turbulent) flow in an infinitely long straight channel with a plane bed formed by cohesionless material. At present, the different explanations for the origin of bed forms (turbulence and stability analysis) appear to be viewed as mutually exclusive and thus almost invariably presented as completely different views of the same phenomenon (see e.g. Coleman and Nikora 2011). To the best knowledge of the authors, the only mention so far to the possibility of a connection between the "inherent instabilities" of a turbulent flow and organized motion in turbulence are by Pope (2000) and Yalin (2006).

2.3 GEOMETRY AND EXISTENCE REGIONS OF DUNES AND RIPPLES

2.3.1 Introductory considerations

(i) A straight steady and uniform[11] two-dimensional two-phase motion *en mass* is determined by three dimensionless variables (see Section 1.3)

$$X, Y, Z \quad \text{or} \quad \Xi, \eta_*, Z, \quad \ldots \quad \text{etc.} \tag{2.12}$$

[11]Bear in mind Sub-section 2.2.3, here and in the rest of the text.

In a factual alluvial channel, the flow width B is finite, and B/h is an additional variable. Hence (2.12) is to be augmented into

$$X, Y, Z, B/h \quad \text{or} \quad \Xi, \eta_*, Z, B/h, \quad \ldots \quad \text{etc,} \tag{2.13}$$

where the parameters forming the dimensionless variables are evaluated by their values corresponding to some specified locations, or, as it is usually done, by their cross-sectional average values. The four independent dimensionless variables above are necessary and sufficient to express *any* dimensionless characteristic of the two-phase motion, in *any* alluvial channel of a specified cross-sectional geometry. This, however, does not mean that each of these variables must necessarily be present in the expression of every characteristic (Sedov 1960, Yalin 1971, 1977, etc.). For example, experiment shows that the geometric characteristics of dunes and ripples in channels having sufficiently large values of B/h (larger than ≈ 5 to 7, say) are not affected by the value of this ratio. But this means that such characteristics can be considered as determined by three variables (2.12); i.e. as if they are caused by a two-dimensional flow. In contrast to this, the geometric characteristics of bars (i.e. of bed forms generated by horizontal turbulence) are determined *mainly* by B/h.

(ii) At this stage it may be mentioned that a particular phenomenon (e.g. a particular experimental run in a laboratory flume) is specified by some particular numerical values of its characteristic parameters: none of the parameters is supposed to vary during the progress of the phenomenon (in the course of a run). Yet, more often than not, a run in a laboratory flume is conducted for a constant flow rate Q and a constant slope S. In this case the growth of bed forms (starting from a flat initial bed) causes the increment of bed resistance and thus the increment of h. This would be perfectly acceptable if Q and S (but not h) were characteristic parameters. However, in the present case it is h and $v_* = \sqrt{ghS}$ (and not Q) which are selected as characteristic parameters, and therefore the run must be conducted so that h and S (and thus h and v_*) remain constant throughout the duration of a run. In this case the increment of bed resistance due to the growth of bed forms is to be accommodated by an appropriate continual adjustment (reduction) of Q. The variation of Q is permissible, for it is not a characteristic parameter in the present case. (More on this topic in Sub-section 3.1.4).

(iii) During the duration of development $(T_\Delta)_i$ of the bed forms i, the mobile bed deforms (by means of consecutive erosion-depositions, as shown schematically in Figure 2.19) "around" the flat initial bed; hence the space-averaged plane of the bed at any $t \in [0, (T_\Delta)_i]$ is the same as the plane of the initial bed. It follows that the flow depth h of an undulated mobile bed is thus to be measured, at any t, from this plane upwards.

2.3.2 Geometric characteristics of dunes

Dune length

Considering the aforementioned we ignore B/h and adopt (2.12). Moreover, we take into account that Λ_d does not depend on Y (Yalin 1977, 1992, etc.). Hence

$$\frac{\Lambda_d}{D} = \phi'_{\Lambda_d}(X, Z), \tag{2.14}$$

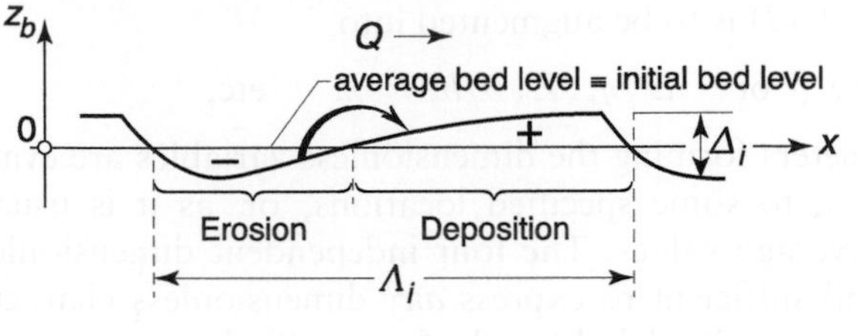

Figure 2.19 Consecutive erosion-deposition around the flat initial bed associated with a bed form.

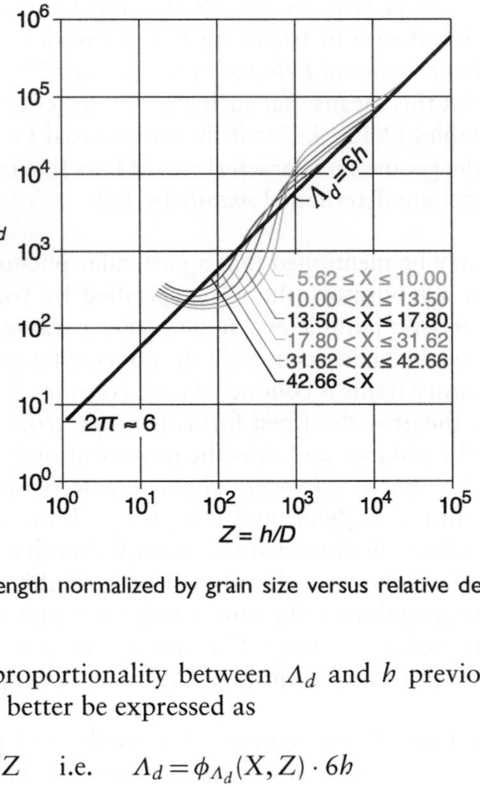

Figure 2.20 Plot of dune length normalized by grain size versus relative depth (adapted from Yalin 1977).

which, owing to the proportionality between Λ_d and h previously mentioned (see Sub-section 2.1.1), can better be expressed as

$$\frac{\Lambda_d}{D} = \phi_{\Lambda_d}(X, Z) \cdot 6Z \quad \text{i.e.} \quad \Lambda_d = \phi_{\Lambda_d}(X, Z) \cdot 6h \tag{2.15}$$

where

$$\phi_{\Lambda_d}(X, Z) \approx 1 \quad \text{when} \quad X > \approx 35. \tag{2.16}$$

The family of curves in Figure 2.20 is the graph of the function of two variables (2.15). This curve-family (which has been determined from the voluminous data reported in Yalin 1977) is well represented by the comp-eq.

$$\frac{\Lambda_d}{D} = 6Z \cdot \left[1 + 0.01 \frac{(Z - 40)(Z - 400)}{Z} e^{-m_\Lambda}\right] \tag{2.17}$$

$$(\text{with } m_\Lambda = 0.055\sqrt{Z} + 0.04X),$$

where the expression in square brackets implies $\phi_{\Lambda_d}(X, Z)$.

Dune steepness

Since dunes are to be treated as two-dimensional, the dune steepness $\Delta_d/\Lambda_d = \delta_d$ is, in general, determined by the dimensionless variables (2.12), i.e. as

$$\delta_d = \phi''_{\delta_d}(X, Y, Z) = \phi'_{\delta_d}(X, \eta_*, Z). \tag{2.18}$$

1- We consider first the case of a fully rough turbulent flow past the flat initial bed ($X >\approx 35$). In this case the Reynolds number X is no longer a variable and Eq. (2.18) reduces to

$$\delta_d = \phi_{\delta_d}(\eta_*, Z), \tag{2.19}$$

which can be expressed by a family of curves: η_*-abcissa, δ_d-ordinate, Z-curve-classifying parameter. As is well known, when η_* increases, then δ_d (corresponding to a specified $Z = cont$) first increases (starting from $\delta_d = 0$ when $\eta_* = 1$), then it reaches its maximum value ($\delta_d = (\delta_d)_{max}$ when $\eta_* = \hat{\eta}_{*d}$), and then it decreases as to yield the *flat bed at advanced stages* ($\delta_d \to 0$). This behaviour of δ_d is clearly conveyed by the experimental curves shown in Figure 2.21. From this figure (adapted from Yalin and Karahan 1979), one gets the impression that with the increment of Z, the δ_d-curves (i.e. the δ_d versus η_*-curves) corresponding to various $Z = (const)_i$ approach asymptotically to a "limiting δ_d-curve" (viz to the curve C_4 in Figure 2.21), which has the largest values of $(\delta_d)_{max}$ and $\hat{\eta}_{*d}$ (viz ≈ 0.06 and 16, respectively). However, it has become evident that this impression is not exactly realistic, but rather that it is due to the fact that most data-points in Figure 2.21 stem from laboratory measurements – only very few of them originate from rivers having large values of Z (Zhang 1999, da Silva and Zhang 1999). The additional river data corresponding to large Z, which became recently available to the authors, enabled them to extend the plot in Figure 2.21 in the Z-direction further. Thus the plots shown in Figures 2.22a,b were produced.[12] These graphs indicate that the largest $(\delta_d)_{max}$ is still ≈ 0.06. Yet, the curve C_4 is *not* a limiting curve, nor $\hat{\eta}_{*d} = 16$ is the largest of all $\hat{\eta}_{*d}$. Figures 2.22a,b (which contain large-river data) show that the increment of Z beyond ≈ 1000 alters the shape of the δ_d-curves so as to render $(\delta_d)_{max}$ to become smaller than ≈ 0.06, and $\hat{\eta}_{*d}$, larger than ≈ 16.

The experimentally determined curves representing the variation of $(\delta_d)_{max}$ and $\hat{\eta}_{*d}$ with Z are shown in Figure 2.23. These curves can be adequately reflected (throughout the range of Z-values in Figure 2.22) by the comp-eqs.

$$(\delta_d)_{max} = 0.00047Z^{1.2}e^{-0.17Z^{0.47}} + 0.04\left(1 - e^{-0.002Z}\right) \tag{2.20}$$

and

$$\hat{\eta}_{*d} = 35\left(1 - e^{-0.074Z^{0.4}}\right) - 5. \tag{2.21}$$

[12]The sources of the data-points in Figures 2.22a,b are given, among others, in Apendix A. These sources include
Laboratory data: [3a], [6a], [7a], [9a], [11a], [12a], [13a], [15a], [16a], [18a], [26a], [28a], [30a], [33a], [35a], [36a], [38a], [40a], [41a], [42a], [43a], [46a], [47a], [48a], and
Field data: [1a], [8a], [19a], [20a], [25a], [27a], [31a], [34a], [37a], [39a], [44a], [45a].

Symbol and Source	D_{50} (mm)	Z	Symbol and Source	D_{50} (mm)	Z
⊗ [6a] Bishop (flume)	0.54 to 1.10	100 to 240	∗ [40a] Simons, Richardson & Haushild (")	0.47	194 to 862
● [20a] Lane & Eden (Mississippi River)	0.27 to 0.56	$2.03 \cdot 10^4$ to $1.04 \cdot 10^5$	⊖ [42a] Singh (")	0.462	100 to 219
⊖ [39a] Shinohara & Tsubaki (flume)	1.26 to 1.46	100 to 508	ϕ [43a] Stein (")	0.40	312 to 777
⦶ [41a] Simons, Richardson & Albertson (")	0.45	162 to 608	⊕ [48a] Znamenskaya (")	0.80	91 to 783
⊙ [13a] Guy, Simons & Richardson (")	0.27 to 0.93	169 to 1043	★ Nordin (Amazon River)* (")	0.15 tp 0.50	$5.4 \cdot 10^4$ to
			*personal communication		

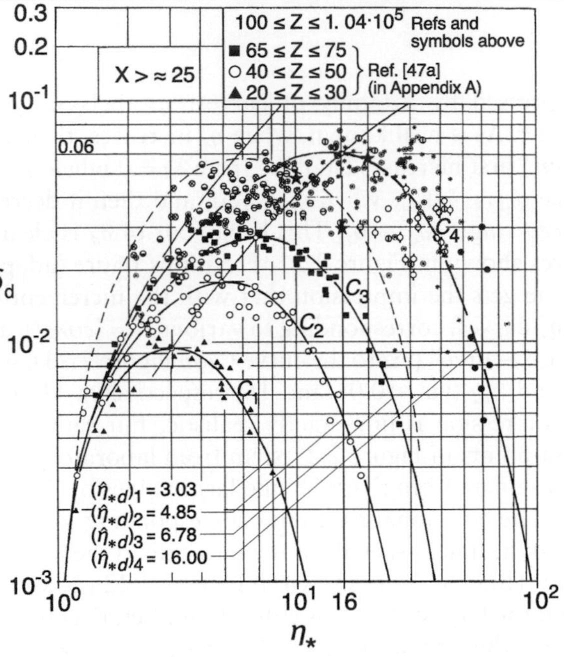

Figure 2.21 Plot of dune steepness versus relative flow intensity (adapted from Yalin and Karahan 1979).

The filled circles on the $(\delta_d)_{max}$-curve in Figure 2.23 are the average values of δ_d of the data-points forming the top of the δ_d-curves corresponding to various Z; the analogous is valid for the filled circles on the $\hat{\eta}_{*d}$-curve.

The extended family of the δ_d-curves themselves is plotted in Figure 2.24. This curve-family can be well approximated by the comp-eq.

$$\delta_d = (\delta_d)_{max}(\zeta_d e^{1-\zeta_d})^{m_\delta} \qquad (= \phi_{\delta_d}(\eta_*, Z)), \tag{2.22}$$

where

$$\zeta_d = \frac{\eta_* - 1}{\hat{\eta}_{*d} - 1}. \tag{2.23}$$

Here $(\delta_d)_{max}$ and $\hat{\eta}_{*d}$ are given by Eqs. (2.20) and (2.21) respectively, the value of m_δ being

$$m_\delta = 1 + 0.6e^{-0.1(5-\log Z)^{3.6}}. \tag{2.24}$$

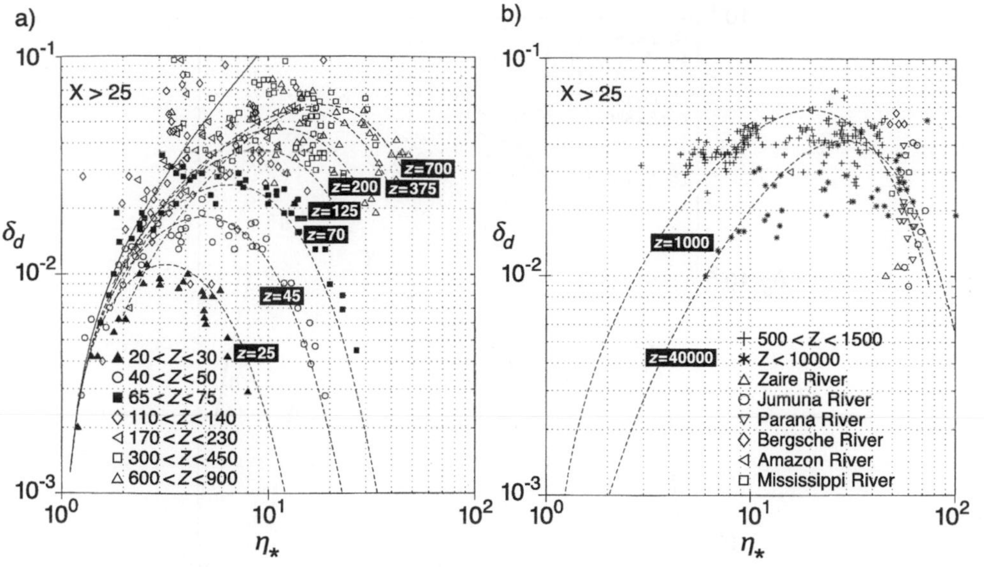

Figure 2.22 Plots of dune steepness versus relative flow intensity produced on the basis of both laboratory and field measurements, and including: (a) data with $Z < 900$; (b) data with $Z > 500$.

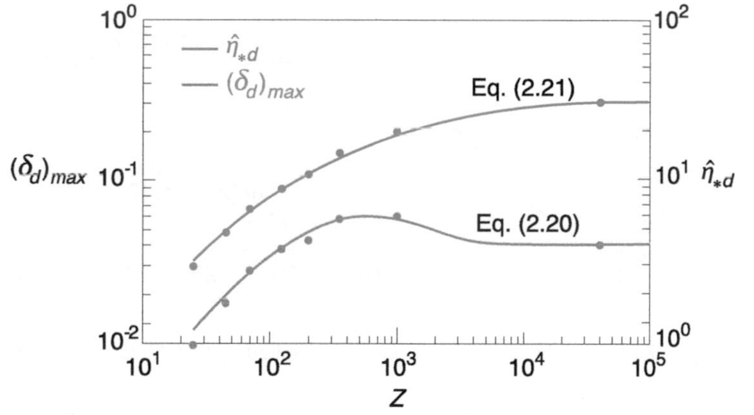

Figure 2.23 Plot of values of maximum dune steepness and corresponding values of relative flow intensity versus Z.

The comp-eqs. above were composed with the aid of the point-patterns corresponding to various intervals of Z.

2- It was assumed so far that the turbulent flow is fully rough ($X >\approx 35$). If, however, the dune-generating turbulent flow is transitionally rough ($\approx 2.5 < X <\approx 35$), then the dune steepness is smaller than that given by the expressions above. Indeed, for the same remaining conditions, δ_d progressively decreases with X in the interval $\approx 2.5 < X < \approx 35$

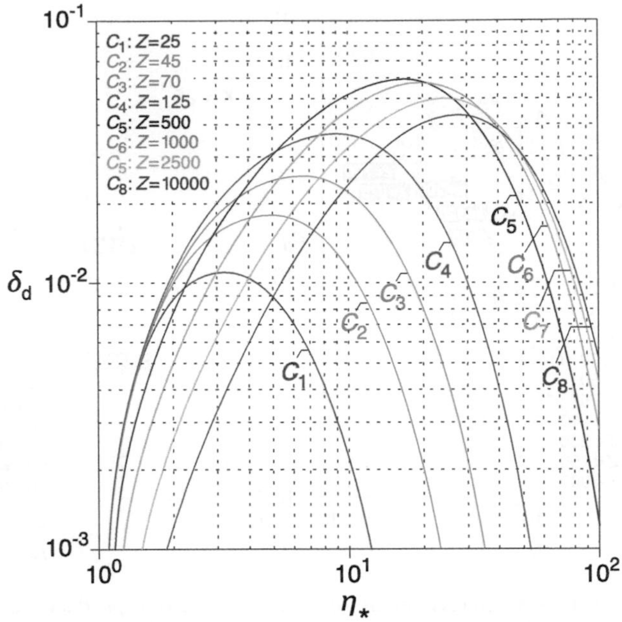

Figure 2.24 Graph of the dune steepness curve-family implied by Eq. (2.22).

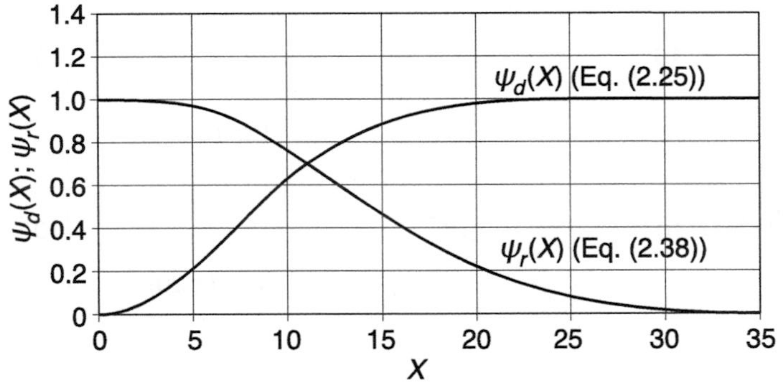

Figure 2.25 Graphs of the functions of X implied by Eqs. (2.25) and (2.38).

as to become $\delta_d \equiv 0$ for all $X <\approx 2.5$ (i.e. for the case of a hydraulically smooth regime of a turbulent flow). This fact can be taken into account by multiplying $(\delta_d)_{max}$, given by Eq. (2.20), with an appropriate function $\psi_d(X)$. This will automatically reduce the value of δ_d, given by Eq. (2.22), as well. The following form can be adopted (as comp-eq.) for $\psi_d(X)$

$$\psi_d(X) = 1 - e^{-(X/10)^2}. \tag{2.25}$$

The graph of this function, together with its counterpart $\psi_r(X)$ for ripples (which will be introduced in the next sub-section), is shown in Figure 2.25.

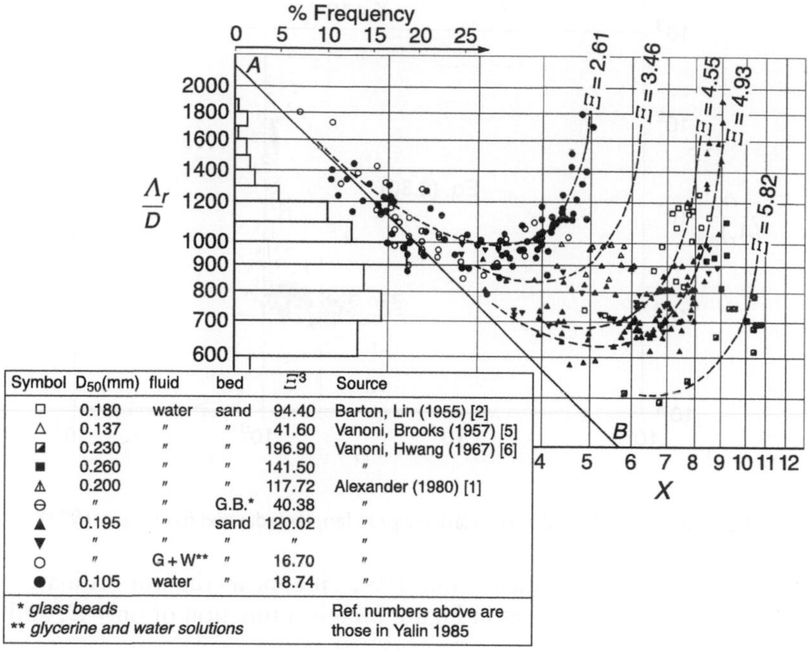

Figure 2.26 Variation of relative ripple length with independent variables (from Yalin 1985, reprinted with permission from ASCE).

The dune steepness δ_d was expressed in the point 1 above as a function $\delta_d = \phi_{\delta_d}(\eta_*, Z)$ (see Eq. (2.22)), while in point 2 the multiplier-function $\psi_d(X)$ was introduced. Hence, in total, δ_d is determined by the form

$$\delta_d = \psi_d(X) \cdot \phi_{\delta_d}(\eta_*, Z), \tag{2.26}$$

which is consistent with the expected general form (2.18).

2.3.3 Geometric characteristics of ripples

Ripple length

Since ripples forming on the bed of wide channels are (by definition) independent of h and B, we have for the ripple length Λ_r (on the basis of (2.12))

$$\frac{\Lambda_r}{D} = \phi''_{\Lambda_r}(X, Y) = \phi'_{\Lambda_r}(X, \Xi) = \phi_{\Lambda_r}(\eta_*, \Xi). \tag{2.27}$$

The family of experimental curves implying $\phi'_{\Lambda_r}(X, \Xi)$ is shown in Figure 2.26. Observe that the most frequently encountered value of Λ_r/D is around 1000; hence the reason for the approximate relation

$$\Lambda_r \approx 1000D \tag{2.28}$$

proposed in the earlier works of the second author (Yalin 1964, 1977).

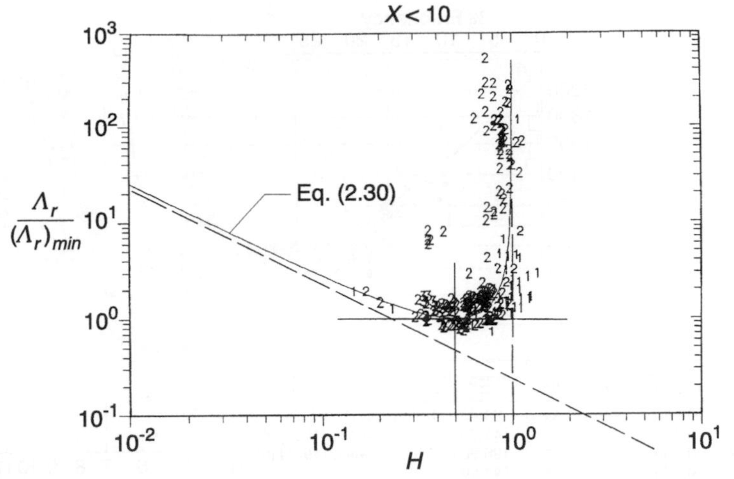

Figure 2.27 Unified plot of relative ripple length (adapted from Yalin 1992).

From the analysis of the data in Yalin (1992), it appears that the ordinate $(\Lambda_r)_{min}/D$ of the lowest points of the curves in Figure 2.26 is a function of only Ξ, and that this function can be expressed as

$$\frac{(\Lambda_r)_{min}}{D} = \frac{2650}{\Xi^{0.88}} . \tag{2.29}$$

Moreover, the data indicates that ripples disappear (their length Λ_r increases indefinitely) when η_* reaches the value $\eta_* \approx 21$ – for all materials.

It has been found (Yalin 1992) that the curves in Figure 2.26 can be unified into a single curve with the aid of the relation

$$\frac{\Lambda_r}{(\Lambda_r)_{min}} = [4H(1-H)]^{-1} \tag{2.30}$$

where

$$H = \sqrt{\frac{\eta_*}{\eta_{*max}}} = \sqrt{\frac{\eta_*}{21}} . \tag{2.31}$$

The curve implied by Eq. (2.30) is shown in Figure 2.27, together with the data.[13] Substituting Eqs. (2.29) and (2.31) in Eq. (2.30), one determines

$$\frac{\Lambda_r}{D} \approx \frac{3000}{\Xi^{0.88}\sqrt{\eta_*}(1-0.22\sqrt{\eta_*})} \qquad (=\phi_{\Lambda_r}(\eta_*, \Xi)) , \tag{2.32}$$

which can be used for the computation of Λ_r.

[13]The sources of the data-points in Figures 2.27 and 2.28 are given, among others, in Appendix A:
Point-symbol 1 ($0.02mm \le D_{50} \le 0.04mm$): [14a], [23a], [24a]
Point-symbol 2 ($0.1mm \le D_{50} \le 0.3mm$): [2a], [4a], [5a], [10a], [14a], [17a], [18a], [21a], [22a], [23a], [29a], [30a], [32a], [36a]
Point-symbol 3 ($0.32mm \le D_{50} \le 0.47mm$): [12a], [14a], [26a], [29a], [30a].

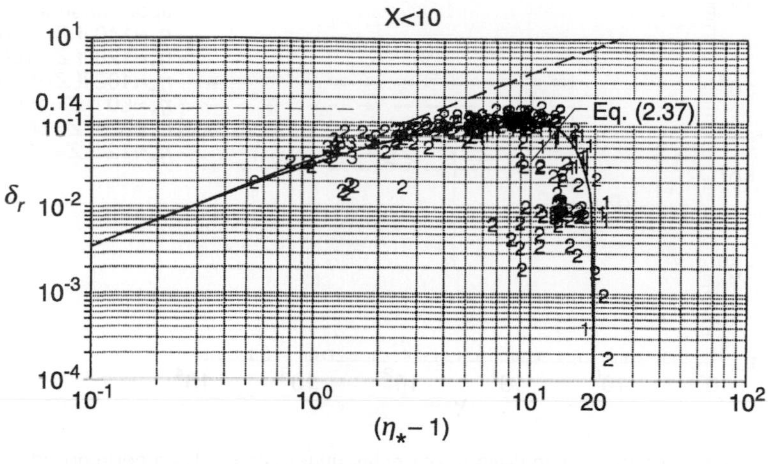

Figure 2.28 Plot of ripple steepness versus relative flow intensity (adapted from Yalin 1992).

Ripple steepness

For the ripple steepness $\delta_r = \Delta_r/\Lambda_r$, which does not depend on Z, we have

$$\delta_r = \phi'''_{\delta_r}(X, Y) = \phi''_{\delta_r}(X, \eta_*) = \phi'_{\delta_r}(\Xi, \eta_*). \tag{2.33}$$

1- It will be assumed first that the (initial) flow contacting the bed is viscous ($X <\approx 2.5$), and thus that the development of δ_r to the full extent is not impeded.

Consider the pattern of the data-points obtained for δ_r in flows having small values of X (Figure 2.28). Note that at the advanced stages ripples disappear (i.e. we have $\delta_r = 0$) when $\eta_* = \eta_{*max} \approx 21$ (as has already been mentioned), and that the maximum ripple steepness, viz $(\delta_r)_{max} \approx 0.14$, occurs when $\eta_* = \hat{\eta}_{*r} \approx 11$. These values do not appear to be affected by Ξ or X. Hence the "unimpeded" ripple steepness δ_r can be treated as a function of η_* alone, and one can adopt for this function, $\phi_{\delta_r}(\eta_*)$ say, the similar to (2.22) form

$$\delta_r = (\delta_r)_{max}\, r\, \zeta_r\, e^{1-\zeta_r} \qquad (= \phi_{\delta_r}(\eta_*)). \tag{2.34}$$

In this expression (which has already been introduced in Yalin 1992),

$$\zeta_r = \frac{\eta_* - 1}{\hat{\eta}_{*r} - 1} \tag{2.35}$$

and

$$r = 1 \quad \text{if } \zeta_r \leq 1; \quad r = \zeta_r(2 - \zeta_r) \quad \text{if } 1 < \zeta_r \leq 2. \tag{2.36}$$

Since $(\delta_r)_{max} \approx 0.14$ and $\hat{\eta}_{*r} \approx 11$ (which yields $\zeta_r = 0.1(\eta_* - 1)$), the relation (2.34) can be expressed as

$$\delta_r = 0.014\, r\, (\eta_* - 1) e^{(1.1 - 0.1\eta_*)} \qquad (= \phi_{\delta_r}(\eta_*)), \tag{2.37}$$

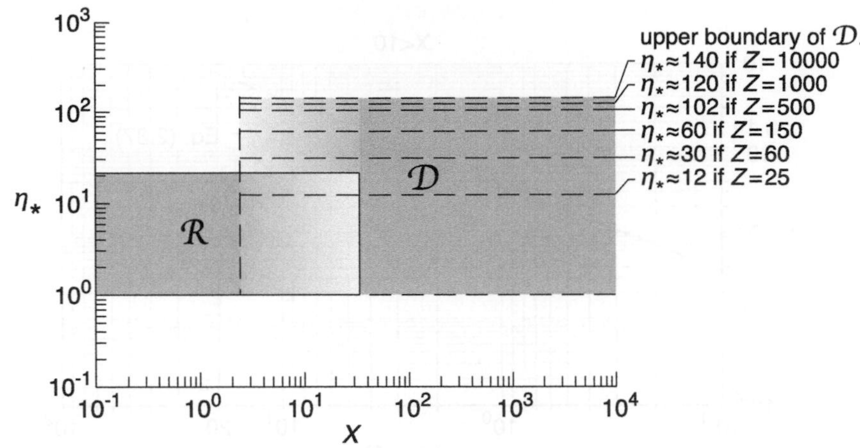

Figure 2.29 Existence region diagram of ripples, dunes and ripples superimposed on dunes.

which is valid for $\eta_* \in [1; 21]$. (Note that $\delta_r = 0$ when $\eta_* < 1$ and $\eta_* > 21$). The solid curve in Figure 2.28 is the graph of Eq. (2.37).

2- With the increment of X from ≈ 2.5 onwards, the viscous influence at the bed progressively decreases and the value of δ_r progressively decreases as well (as to vanish completely when $X \approx 35$). This fact can be taken into account (as it has been done for the dune steepness) by multiplying $(\delta_r)_{max} \approx 0.14$ with a (smaller than unity) function

$$\psi_r(X) = \begin{cases} e^{-[(X-2.5)/14]^2} & \text{if } X > 2.5 \\ 1 & \text{if } X \leq 2.5 \end{cases} \tag{2.38}$$

The graph of $\psi_r(X)$ is shown (alongside its dune counterpart $\psi_d(X)$) in Figure 2.25.

Hence, in total, we have for the ripple steepness

$$\delta_r = \psi_r(X) \cdot \phi_{\delta_r}(\eta_*), \tag{2.39}$$

which is in line with $\phi''_{\delta_r}(X, \eta_*)$ in Eq. (2.33).

2.3.4 Existence regions of dunes and ripples

A bed form "is present" only if it has a detectable non-zero steepness. Thus the *existence region* ($\mathcal{D}$) of dunes, having $\delta_d = \phi'_{\delta_d}(X, \eta_*, Z)$ (Eq. (2.18)), is the set of those X; η_*; Z which yields $\delta_d > 0$. Similarly, the existence region ($\mathcal{R}$) of ripples, having $\delta_r = \phi''_{\delta_r}(X, \eta_*)$ (Eq. (2.33)), is the set of those X; η_* for which $\delta_r > 0$. Since X and η_* are common for both dunes and ripples, it would be appropriate to consider $\mathcal{R}$ and $\mathcal{D}$ in the same $(X; \eta_*)$-plane, and use Z as parameter. Figure 2.29 shows $\mathcal{R}$ and $\mathcal{D}$ as they emerge from the information contained in Sub-sections 2.2.2 and 2.2.3.

The lower boundary of $\mathcal{R}$ is $\eta_* = 1$, its upper boundary being $\eta_* = 21$. The region $\mathcal{D}$, confined by broken lines, varies depending on Z. The lower boundary of $\mathcal{D}$ is the same as that of $\mathcal{R}$, viz $\eta_* = 1$ for all Z. The upper boundary of $\mathcal{D}$ varies as an increasing function of Z and η_*. The η_*-values of these upper boundaries, viz ≈ 12, 30, 60, ..., etc., were identified with those η_* for which the dune steepness is as small as $\delta_d = 10^{-3}$ (η_*-values on the abcissa of Figures 2.22a,b). The ripple steepness δ_r decreases from $X \approx 2.5$ onwards, as to vanish completely at $X \approx 35$ – which is the right-hand side boundary of $\mathcal{R}$. The dune steepness δ_d decreases from $X \approx 35$ downwards, as to vanish at $X \approx 2.5$ – which is the left-hand side boundary of $\mathcal{D}$. In the interval $\approx 2.5 < X < \approx 35$ ripples and dunes can be present simultaneously, as ripples superimposed on dunes. Clearly, such a superimposition can be realized only if Λ_d is by multiple times larger than Λ_r. Identifying "multiple times" with e.g. three or more, we can write, roughly,

$$\frac{\Lambda_d}{\Lambda_r} \approx \frac{6h}{1000D} = 0.006Z \geq 3 \quad \text{i.e.} \quad Z \geq 500, \tag{2.40}$$

which can be regarded as an additional (to $\approx 2.5 < X < \approx 35$) condition for the simultaneous occurrence of dunes and ripples.

2.4 GEOMETRY AND EXISTENCE REGIONS OF BARS

2.4.1 Existence regions of alternate and multiple bars

Consider the functional relationship (2.10), according to which the number of n-rows of the configuration (in flow plan) of the LSHCS's of a fully rough flow past an initially flat bed, must be expected to depend on B/h and h/D. From this relation it follows that

$$\left(\frac{B}{h}\right)_{n,n+1} = \psi_n\left(n, \frac{h}{D}\right). \tag{2.41}$$

where $(B/h)_{n,n+1}$ is the value of B/h at the boundary from the n to the $n+1$ plan configuration of LSHCS's.

This means that, in a diagram having h/D as abscissa and B/h as ordinate, the different configurations of the LSHCS's imply different existence regions, or zones. Following da Silva (1991), such a diagram will henceforth be termed the $(B/h, h/D)$-plane. [Note that, as demonstrated in da Silva (1991), the form (2.41) is also valid for the boundary $(B/h)_{0,1}$ in the sense of Section 2.2 (see Figure 2.10)].

If the configuration of bars in flow plan is determined by the configuration of the LSHCS's of the initial flow (Section 2.2), then the geometric nature of bars (i.e. their plan configuration) must also be determined by B/h and h/D. But if so, then it is appropriate to locate the existence regions of alternate and multiple bars on the $(B/h; h/D)$-plane – Eq. (2.41) thus implying the equation of the boundary-lines separating the different regions.[14] It should be noted that the variables B/h and h/D

[14]Clearly, in the $(B/h; h/D)$-plane the boundary-lines separating the n from $(n+1)$-row bars are, at the same time, the boundaries separating the n from $(n+1)$-row configuration of LSHCS's.

were independently proposed as classifying parameters of the different types of bars by Fujita and Muramoto (1980) and the Public Works Research Institute, Ministry of Construction of Japan (1982). They emerge also, albeit with an additional variable (namely Y), from stability analysis (see e.g. Colombini et al. 1987).

In accordance with the considerations above, the available field and laboratory bar-data were plotted on the $(B/h; h/D)$-plane as shown in Figure 2.30.[15] Despite some "diffusion" of the data from one region to another, the zones of alternate and multiple bars are clearly detectable. On the basis of Figure 2.30, and following Ahmari and da Silva (2011), the existence regions of the different types of bars are thus defined as follows:

1- The alternate bar region is that bounded at the bottom by line $\mathcal{L}_{0,1}$, at the top by line $\mathcal{L}_{1,2}$ and on the left by line $\mathcal{L}_{P,A}$. The line $\mathcal{L}_{1,2}$ is described by the following relations:

$$\text{If } \left(\tfrac{h}{D}\right) <\approx 200, \text{ then } \quad \left(\tfrac{B}{h}\right)_{1,2} = 25\left(\tfrac{h}{D}\right)^{1/3} \tag{2.42}$$

$$\text{If } \left(\tfrac{h}{D}\right) >\approx 200, \text{ then } \quad \left(\tfrac{B}{h}\right)_{1,2} = 146 \ (\approx 150), \tag{2.43}$$

where $(B/h)_{1,2}$ are the ordinates of $\mathcal{L}_{1,2}$. The lines $\mathcal{L}_{P,A}$ and $\mathcal{L}_{0,1}$ can be viewed as a single lower boundary-line of the A-region, given by

$$\text{If } \tfrac{h}{D} <\approx 25, \text{ then } \quad \left(\tfrac{B}{h}\right)_{0,1} = 25\left(\tfrac{h}{D}\right)^{-0.55} \tag{2.44}$$

$$\text{If } \approx 25 < \tfrac{h}{D} <\approx 130, \text{ then } \quad \left(\tfrac{B}{h}\right)_{0,1} = \tfrac{2}{13}\left(\tfrac{h}{D}\right) \tag{2.45}$$

$$\text{If } \tfrac{h}{D} >\approx 130, \text{ then } \quad \left(\tfrac{B}{h}\right)_{0,1} = 20 \tag{2.46}$$

where $(B/h)_{0,1}$ are the ordinates of $\mathcal{L}_{P,A}$ for $h/D <\approx 25$ and $\mathcal{L}_{0,1}$ for $h/D >\approx 25$.[16]

2- The existence region of multiple bars is the part of the graph above line $\mathcal{L}_{1,2}$. The existing data suggest that in this region, the zone between the lines $\mathcal{L}_{1,2}$ and $\mathcal{L}_{2,3}$ is the existence region of 2-row bars. Since not enough information on the plan configuration of the reported multiple bars is available, it is not possible at present to define the boundaries separating 3- from 4-row bars ($\mathcal{L}_{3,4}$), etc. For this reason, the region above the line $\mathcal{L}_{2,3}$ is identified with the existence region of multiple bars having three or more rows (C3+ region).

[15]The sources of the data-points in Figure 2.30 are given in Appendix B. These sources include: [1b], [2b], [4b], [5b], [6b], [7b], [8b], [10b], [12b], [14b], [15b], [16b], [17b], [18b], [19b], [20b], [21b], [22b], [23b], [24b], [26b], [27b], [29b].
The plot contains also the data from various Japanese rivers – kindly provided by Dr. S. Ikeda (Tokyo Institute of Technology).

[16]The lines $\mathcal{L}_{0,1}$, $\mathcal{L}_{1,2}$, ..., formed by straight line segments in Figure 2.30, are but the schematic representations of their curvilinear counterparts. No explanation can be offered by the authors for why these boundary-curves should become progressively flatter with the increment of h/D.

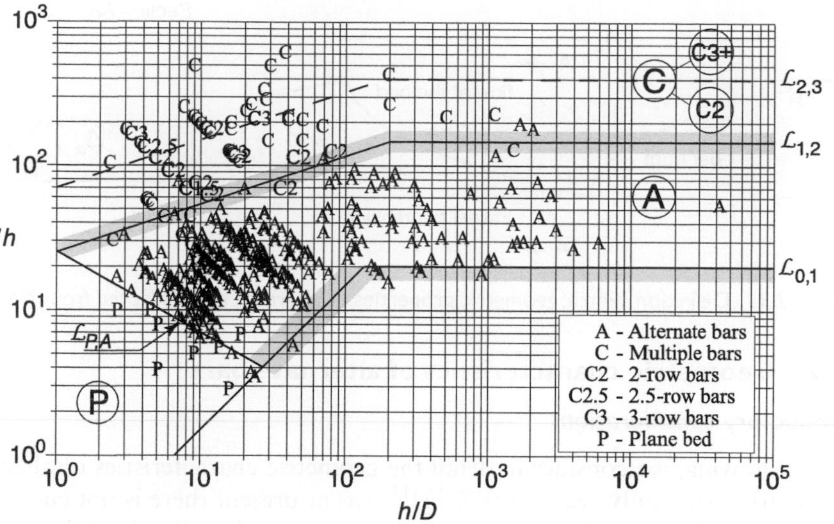

Figure 2.30 The (*B/h*; *h/D*)-plane and the related existence regions diagram of alternate and multiple bars (from Ahmari and da Silva 2011, reprinted with permission from IAHR).

The regions mentioned are labeled in Figure 2.30 by the symbols A, C, C2, C3+ and P, implying alternate bars, multiple bars, 2-row bars, 3 or more row bars, and plane bed. This latest version of the $(B/h; h/D)$-plane supersedes its earlier versions in da Silva (1991), Yalin (1992) and Yalin and da Silva (2001).

From the content of this chapter so far, it should be clear that the adoption of B/h and h/D only as classifying parameters results from considerations strictly valid for fully rough turbulent flows and an initial flat bed whose roughness can be identified with the skin granular roughness $k_s \sim D$. However, if the flow is not fully rough, then the conditions must be expected to depend also on Re_*; furthermore, in the presence of other bed forms such as dunes or ripples, the bed roughness can no longer be characterized by k_s alone. This means that each of the boundary-lines of the different regions in the $(B/h; h/D)$-plane is not really a single line, but rather a family of lines, each line of such family corresponding to a different value of the additional parameter or parameters. No research has been carried out to date to reveal the family of lines forming each boundary. However, as pointed out by Ahmari and da Silva (2011), the extent to which the data corresponding to a certain bar configuration "diffuses" into a different region, nonetheless provides an indication of the width of the "ribbons" spanned by such, as yet unknown, families of curves. Such "ribbons" are tentatively indicated in Figure 2.30 by the shaded regions around the lines $\mathcal{L}_{0,1}$ and $\mathcal{L}_{1,2}$. For practical purposes, such shaded regions are to be viewed as regions of uncertainty regarding the type of bar that in reality will be observed. That is, if a flow plots in the shaded region around the line $\mathcal{L}_{0,1}$ it is possible that its outcome is either no bars or alternate bars; if it plots in the shaded region around the line $\mathcal{L}_{1,2}$, then the outcome may be either alternate bars or 2-row bars.

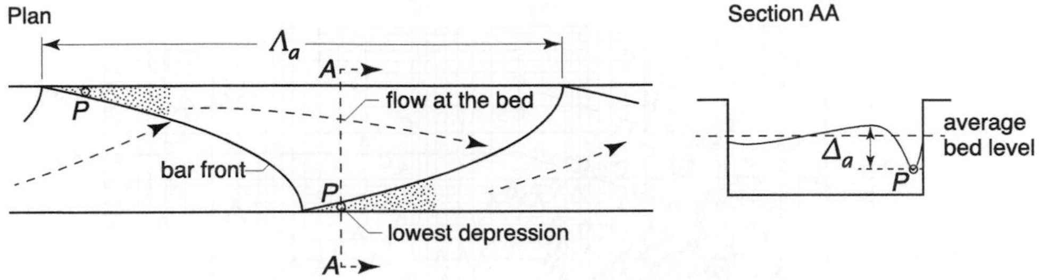

Figure 2.31 Definition sketch: geometric properties of alternate bars (adapted from Ikeda 1984).

2.4.2 Geometric characteristics of alternate bars

Introductory considerations

In the following, we consider in detail the geometric characteristics of alternate bars (or one-row bars) only (see Figure 2.31)[17] – as at present there is not enough data to conduct a similar analysis of the geometric characteristics of multiple bars.

On the basis of Eq. (2.13), it follows that an appropriate dimensionless counterpart of Λ_a, as well as δ_a can be expected to be determined by, at most, X, Y, Z and B/h (or X, η_*, Z and B/h). Since Λ_a is directly proportional to the flow width B, we can thus write

$$\frac{\Lambda_a}{B} = \phi''_{\Lambda_a}(X, \eta_*, Z, B/h) \qquad \text{and} \qquad \delta_a = \phi'_{\delta_a}(X, \eta_*, Z, B/h). \tag{2.47}$$

Alternate bar length

An analysis of the available bar length data in view of the first relation in Eq. (2.47) has recently been presented in Boraey and da Silva (2014) (see also Boraey 2014). While a dependency on η_* was not noticeable from the data, differences in the length of bars produced under fully and transitionally rough flows pointed to a clear dependency on X.

The data used in the just mentioned study are plotted in the $(B/h; h/D)$-plane in Figure 2.32 (the meaning of the lines marked $(h/D)_{min}$ and $(h/D)_{max}$ in this figure will be clarified later on).[18] With the exception of one point corresponding to the Naka River in Japan, the data result from laboratory experiments. These were carried out by numerous authors, who in their experiments kept B and D constant, while varying h. Hence the data are naturally grouped into $B/D = const$ values and plot along 1/1-declining straight lines in Figure 2.32 (see Problem 3.9). As a result of such grouping, one can more precisely infer from the data how Λ_a/B varies along $B/D = const$ lines, than how it varies along $h/D = const$ lines. Taking this into account, as well as the

[17]With regard to Figure 2.31, and for practical purposes, it should be noted that the near bank scour depth due to alternate bars is equal to $\approx 75\%$ of the bar height Δ_a (Jaeggi 1984). Here the scour depth is measured with regard to the average bed level.

[18]The data were derived from the following data-sources in Appendix B: [2b], [3b], [4b], [9b], [11b], [13b], [14b], [15b], [16b], [17b], [21b], [22b], [23b], [24b], [25b], [28b], [29b].

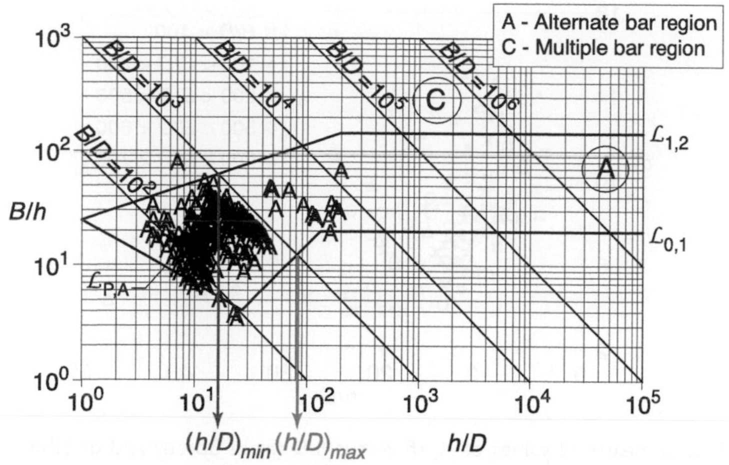

Figure 2.32 Plot on the $(B/h; h/D)$-plane of the dataset used for the development of bar length equation (from Boraey and da Silva 2014, reprinted with permission from CRC Press).

findings described in the previous paragraph, Boraey and da Silva (2014) sought for Λ_a/B an expression of the form

$$\frac{\Lambda_a}{B} = \phi'_{\Lambda_a}(X, Z, B/D). \tag{2.48}$$

Note that the inclusion of B/D instead of B/h in Eq. (2.48) is justified, as $B/D = (B/h) \cdot (h/D)$.[19]

In the following, the subsequent analysis by Boraey and da Silva (2014) is briefly reviewed.

1- As previously noted, if the turbulent flow is fully rough ($X > \approx 35$), then the Reynolds number X is no longer a variable and Eq. (2.48) reduces to

$$\frac{\Lambda_a}{B} = \phi_{\Lambda_a}(Z, B/D), \tag{2.49}$$

which can be represented by a family of curves: Z-abcissa, Λ_a/B-ordinate, B/D-curve-classifying parameter. Accordingly, the measured Λ_a/B-values were plotted versus Z $(= h/D)$ as shown in Figure 2.33. A sorting of the data with B/D is clear from this figure. Observe also that for any given value of B/D, Λ_a/B continuously decreases with increasing values of h/D. The data-pattern in Figure 2.33 is adequately reflected by the comp-eq.

$$\frac{\Lambda_a}{B} = 6 \cdot \left[\phi_1(B/D) \cdot e^{[-\phi_2(B/D)\cdot(h/D)]} + \phi_3(B/D)\right], \tag{2.50}$$

[19]Just because the replacement of B/h by B/D is justified from a dimensional standpoint and is of no consequences where the results of the final equation are concerned, this (convenient) transformation *should not* be interpreted as implying that bar length depends on B/D instead of B/h: for more on the topic, see Yalin (1977), pp. 66-68.

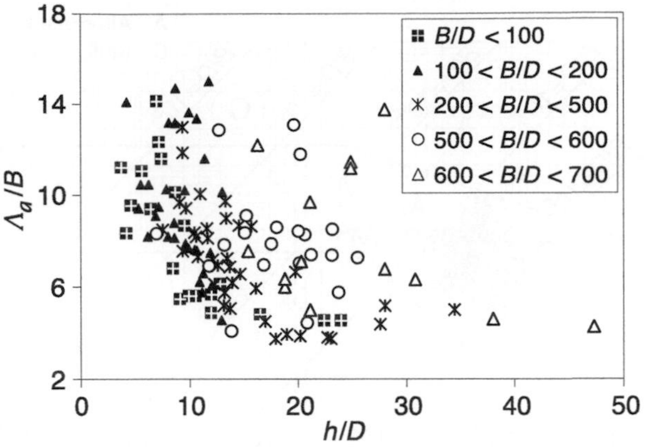

Figure 2.33 Plot of measured values of Λ_a/B versus h/D (from Boraey and da Silva 2014, reprinted with permission from CRC Press).

in which

$$\phi_1(B/D) = 1.028 \ln(B/D) - 1.393, \tag{2.51}$$

$$\phi_2(B/D) = 0.5765(B/D)^{-0.266}, \tag{2.52}$$

and[20]

$$\phi_3(B/D) = 1 - \frac{1}{2.05 + e^{[0.001(B/D) - 3.75]}}. \tag{2.53}$$

The curve-family implied by Eq. (2.50) is shown, together with the data, in Figure 2.34.

2- Let us now consider the case of transitionally rough flows ($\approx 2.5 < X < \approx 35$). Following the approach adopted for dunes, an expression for Λ_a/B was sought by modifying Eq. (2.49) through the introduction of a multiplier function $\psi_a(c_f)$. As follows from Eq. (2.48), such a function must be expected to depend solely on X or, possibly, also on either one or both of Z and B/D. The nature of such a function was investigated by plotting the ratio between the values of Λ_a/B measured in transitionally rough flows and their counterparts as given by Eq. (2.50) for fully rough flows, versus X, Z and B/D. The resulting plots suggest that the multiplier function under consideration depends on both X and Z (see Figures 5 and 6 in Boraey and da Silva 2014). However, a considerably better correlation was found with the dimensionless Chézy friction factor c_f (see Figure 2.35). Since the latter is a function of X and Z (see Chapter 1), it is perfectly valid to replace the consideration of X and Z by that of the single variable c_f. The following form was adopted (as comp-eq.) for $\psi_a(c_f)$:

$$\psi_a(c_f) = 0.25 + 3.0 \cdot \frac{1}{1 + e^{c_f - 12.75}}. \tag{2.54}$$

[20] Note that the coefficient 2.05 in Eq. (2.53) is incorrectly typed as 0.5 in Eq. (11) in Boraey and da Silva (2014).

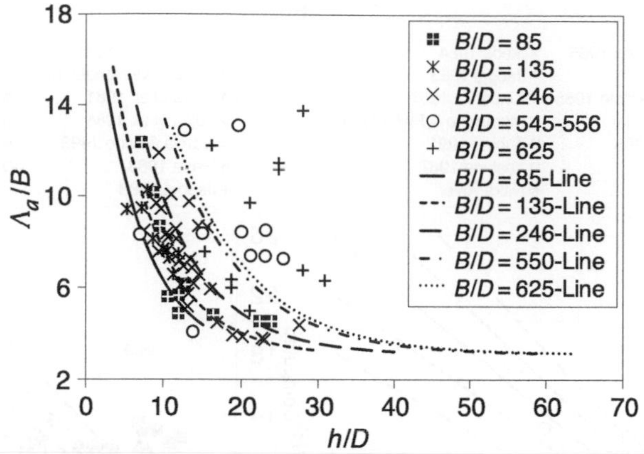

Figure 2.34 Plot of selected curves of the curve-family implied by Eq. (2.50), together with the data (from Boraey and da Silva 2014, reprinted with permission from CRC Press).

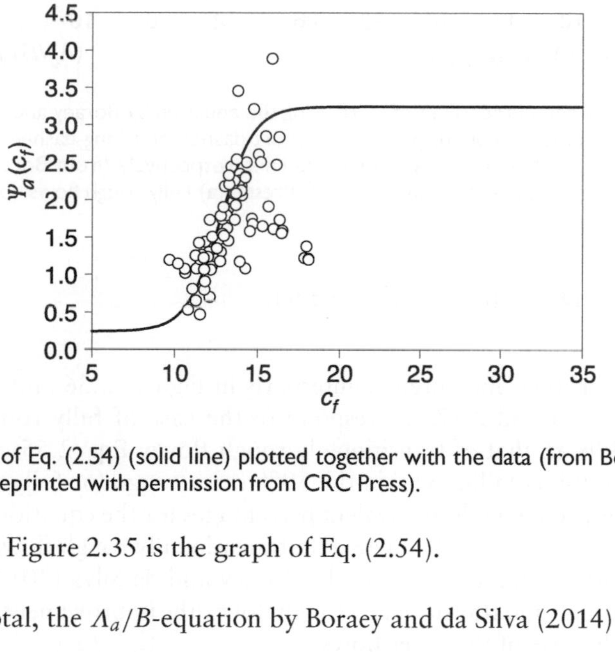

Figure 2.35 Graph of Eq. (2.54) (solid line) plotted together with the data (from Boraey and da Silva 2014, reprinted with permission from CRC Press).

The solid line in Figure 2.35 is the graph of Eq. (2.54).

Hence, in total, the Λ_a/B-equation by Boraey and da Silva (2014) is of the form

$$\frac{\Lambda_a}{B} = \psi_a(c_f) \cdot \phi_{\Lambda_a}(Z, B/D), \qquad (2.55)$$

in which ϕ_{Λ_a} is given by Eqs. (2.50)-(2.53), and $\psi_a(c_f)$ is given by Eq. (2.54) if $\approx 2.5 < X <\approx 35$, and $\psi_a(c_f) = 1$ if $X >\approx 35$.

Values of Λ_a/B obtained with the aid of the equations above and those by Ikeda (1984), namely

$$\frac{\Lambda_a}{B} = 5c(B/h)^{-0.5} \quad \text{if } Fr < 0.64 \qquad (2.56)$$

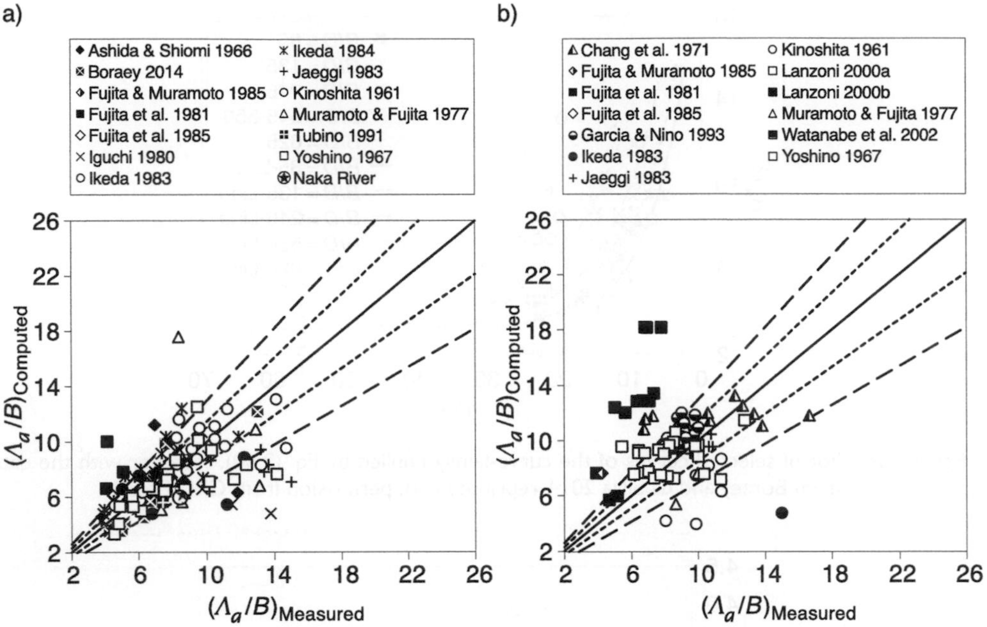

Figure 2.36 Plots of computed values of Λ_a/B using the equation by Boraey and da Silva (2014) versus their measured counterparts; solid, short-dashed and long-dashed lines are the perfect agreement, 15% and 30% error range lines, respectively (from Boraey and da Silva 2014, reprinted with permission from CRC Press). (a) Fully rough flows; (b) transitionally rough flows.

$$\frac{\Lambda_a}{B} = 5.3(B/D)^{-0.45}(B/h) \quad \text{if } Fr \geq 0.64, \tag{2.57}$$

are plotted versus their measured counterparts in Figures 2.36 and 2.37, respectively. Here Figures 2.36a and 2.37a correspond to the case of fully rough flows; Figures 2.36b and 2.37b, to that of transitionally rough flows. Eq. (2.55) yields 73.1% and 69.3% of data-points falling within the 30% error range for fully and transitionally rough flows, respectively; the equivalent percentages for the equations by Ikeda (1984) are 61.8% (fully rough flows) and 46.2% (transitionally rough flows). That is, for the existing laboratory data, the equation by Boraey and da Silva (2014) performs better for both fully rough and transitionally rough flows, the improvement being particularly noticeable in the case of the latter flows.

A significant difference between Eq. (2.55) and Eqs. (2.56)-(2.57) relates to their performance when extrapolated to large values of B/D (and, consequently, large values of h/D) as encountered in real rivers, and in particular large sand rivers. This is explained with the aid of Figure 2.38, showing the graph of the Λ_a/B-equation by Ikeda (1984) ($Fr \geq 0.64$; Eq. (2.57)) for $B/D = 10^4$. Here $(h/D)_{min}$ and $(h/D)_{max}$ are the values of h/D where the $B/D = const$ lines intersect the boundary-lines $\mathcal{L}_{1,2}$ and $\mathcal{L}_{0,1}$, respectively (see Figure 2.32). As follows from Figure 2.38, for $B/D = 10^4$, the equation by Ikeda (1984) already yields unrealistically small values of Λ_a/B ($\Lambda_a/B \leq 3$)

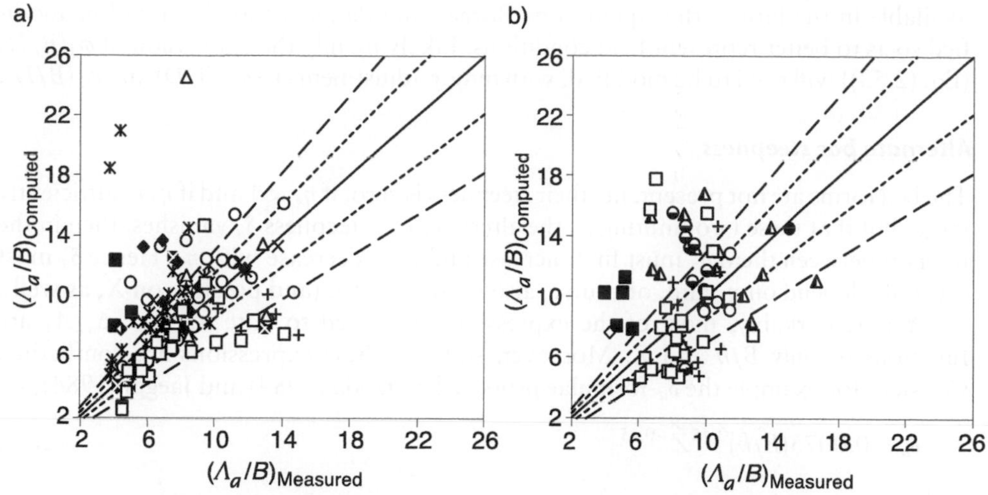

Figure 2.37 Plots of computed values of Λ_a/B using the equations by Ikeda (1984) versus their measured counterparts; solid, short-dashed and long-dashed lines are the perfect agreement, 15% and 30% error range lines, respectively (from Boraey and da Silva 2014, reprinted with permission from CRC Press). (a) Fully rough flows; (b) transitionally rough flows.

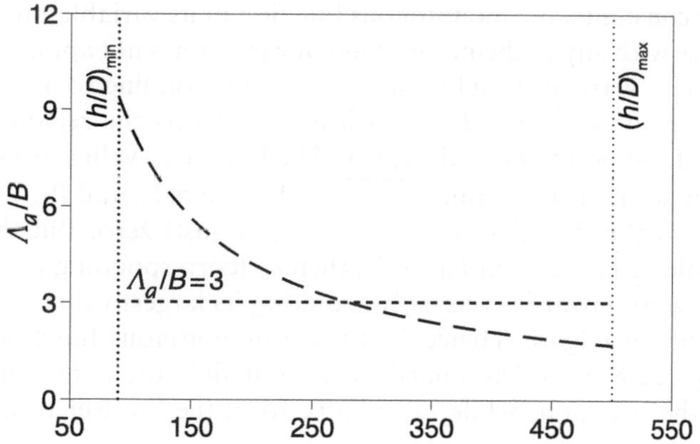

Figure 2.38 Graph of the bar length equation (2.57) by Ikeda (1984) for $B/D = 10^4$ (long-dashed line); short-dashed line marks the value $\Lambda_a/B = 3$ (from Boraey and da Silva 2014, reprinted with permission from CRC Press).

for a substantial part of the alternate bar region (in the $(B/h; h/D)$-plane) – and does the more so, the larger B/D.

On the other hand, for large values of B/D, $\phi_1(B/D) \cdot \exp\left[-\phi_2(B/D) \cdot (h/D)\right] \to 0$, while $\phi_3(B/D) \to 1$ (see Eq. (2.50)). Hence, for large B/D, Eqs. (2.50)-(2.53) invariably yield $\Lambda_a/B = 6$, irrespective of the value of h/D. While this result is more realistic than that obtained from Eqs. (2.56)-(2.57), it should be noted that the limiting value of

$6B$ for large B/D was adopted due to the lack of field data. Should such data become available in the future, the equation by Boraey and da Silva (2014) should be modified so as to better represent field conditions. Likely mainly the expression of $\phi_3(B/D)$ (Eq. (2.53)) will need to be modified, with minor adjustments to $\phi_1(B/D)$ and $\phi_2(B/D)$.

Alternate bar steepness

The bed forms are not present, i.e. their steepness is zero, if $\eta_* < 1$ and if η_* is sufficiently large. But if at these two limiting η_* the alternate bar steepness δ_a vanishes, then in the interval between them δ_a must first increase and then decrease with η_*. Hence δ_a must certainly depend on η_* and, of course, also on B/h and Z (and possibly on X, as well).

Yet, remarkably, most of the expressions produced to date for $\delta_a = \Delta_a/\Lambda_a$ are functions of only B/h and Z. Moreover, some of these expressions are conflicting. Consider for example the δ_a-formulae proposed by Ikeda (1984) and Jaeggi (1984), viz

$$\delta_a = 0.0073(B/h)^{0.45} Z^{-0.45} \tag{2.58}$$

and

$$\delta_a = 0.0365(B/h)^{-0.15} Z^{-0.15}, \tag{2.59}$$

respectively. According to the first of these relations, δ_a increases with B/h; according to the second, δ_a decreases with it. The reason for this is due to the intention to express δ_a as a power-product of its variables B/h and Z. A power product $(B/h)^k \cdot Z^l$, where k and l are some constants, is a monotonous function of its variables: it can only increase or only decrease with any of them. This kind of variation is not applicable to δ_a. Indeed, take an imaginary vertical straight line, λ say, corresponding to a given h/D value in the $(B/h; h/D)$-plane in Figure 2.30. Such line intersects the region's boundary-lines $\mathcal{L}_{0,1}$ and $\mathcal{L}_{1,2}$ at the points P_0 and P_1, say. The largest δ_a-values occur somewhere in the midst of the straight-line segment $\overline{P_0P_1}$; at the points P_0 and P_1, which are situated at the "edges" of the A-region, the value of δ_a is (just) zero. But this means that if one "moves" along $\overline{P_0P_1}$, from P_0 to P_1, then δ_a (corresponding to a $Z = const$) first increases from zero onwards eventually acquiring its largest value, and then decreases so as to become zero again. Hence δ_a is *not* a monotonous function of B/h. And it seems that Eq. (2.58) was determined by using mainly the data from the "nearer to $\mathcal{L}_{0,1}$" part of the A-region, while Eq. (2.59), from the "nearer to $\mathcal{L}_{1,2}$" part of that region.

The aforementioned conveys that the determination of bar steepness is still a topic for future research. Fortunately, the magnitude of δ_a does not exceed ≈ 0.015, say (see e.g. Figures 3.30, 3.31, 3.32 in Yalin 1992), and therefore the contribution of bars to the value of the flow resistance factor is negligible.

2.5 FINAL REMARKS

For the sake of simplicity, in the following we consider the case of fully rough flows only.

As should be clear from the content of this chapter, for such flows the (average) length of dunes reduces to $\Lambda_d \approx 6h$. Yet, in the case of alternate bars developed under fully rough flows, Λ_a/B is more adequately described by an equation of the type (see Eq. (2.50), again noting that $B/D = (B/h) \cdot (h/D)$)

$$\frac{\Lambda_a}{B} = 6 \cdot \psi_{\Lambda_a}(Z, B/h), \tag{2.60}$$

instead of simply $\Lambda_a/B \approx 6$. Here the function $\psi_{\Lambda_a}(Z, B/h)$ varies so that it yields Λ_a/B monotonously increasing from $\Lambda_a/B \approx 4$ at the lower boundary of the alternate bar region $\mathcal{L}_{0,1}$, to ≈ 14 or so at the upper boundary $\mathcal{L}_{1,2}$ (see Figure 2.34). As can be inferred from Figure 2.34, $\Lambda_a/B \approx 6$ (and thus $\psi_{\Lambda_a}(Z, B/h) \approx 1$) is characteristic of the length of bars occurring in the midst of the alternate bar region (i.e. approximately half way between the lines $\mathcal{L}_{0,1}$ and $\mathcal{L}_{1,2}$ in Figure 2.30).

With regard to the origin of bars (see Sub-section 2.2.1), the aforementioned raises the question of whether a similar situation is present where the length scale L_H of the LSHCS's is concerned, including for (fully rough) flows past the initial flat beds. That is, does L_H/B also increase from ≈ 4 to ≈ 14 as one moves from the lower boundary of the alternate bar region to the upper boundary? At present, it is not possible to answer this question, as the existing measurements of L_H (as described in Sub-section 2.2.1) were invariably carried out for cases falling in the midst of the alternate bar region. On the other hand, it is conceivable, and in the authors' view more likely, that L_H of the (fully rough) flow past the initial flat bed is always close to $6B$, i.e. irrespective of the values of B/h and h/D, and that changes to the length scale of the LSHCS's occur as a result of their interaction with the (deforming) bed. Indeed, the emergence of the alternate bars, with the largest depositions near the banks, may create such conditions that allow the LSHCS's to travel longer distances in the x-direction before the structures come into contact with both banks and start to disintegrate. Such explanation would be consistent with the fact that alternate bars also exhibit the largest values of Δ_a/h towards the upper boundary of the alternate bar region $\mathcal{L}_{1,2}$ (see e.g. Boraey 2014). Clearly, further research is needed to properly evaluate the length scale L_H of the LSHCS's under varying flow conditions (varying B/h and Z), as well as understand the effect of the bars themselves on L_H.

REFERENCES

Adrian, R. J. and Marusic, I. (2012). Coherent structures in flow over hydraulic engineering surfaces. *Journal of Hydraulic Research*, 50(5), 451-464.

Ahmari, H. and da Silva, A.M.F. (2011). Regions of bars, meandering and braiding in da Silva and Yalin's plan. *Journal of Hydraulic Research*, 49(6), 718-727.

Allen, J. (1985). *Principles of physical sedimentology*. Chapman & Hall, London.

Aubry, N., Holmes, P., Lumley, J.L. and Stone, E. (1988). The dynamics of coherent structures in the wall region of a turbulent boundary layer. *Journal of Fluid Mechanics*, Vol. 192, 115-173.

Baas, J.H. (1994). A flume study on the development and equilibrium morphology of current ripples in very fine sand. *Sedimentology*, 41, 185-209.

Balakumar, B.J. and Adrian, R.J. (2007). Large- and very-large-scale motions in channel and boundary-layer flows. *Philosophical Transactions of the Royal Society A: Mathematical, Physical and Engineering Sciences*, 365, 665-681.

Blackwelder, R.F. (1983). Analogies between transitional and turbulent boundary layers. *Physics of Fluids*, 26(10), 2807-2815.
Blackwelder, R.F. and Eckelmann, H. (1979). Streamwise vortices associated with the bursting phenomenon. *Journal of Fluid Mechanics*, Vol. 94, 577-594.
Blackwelder, R.F. (1978). The bursting process in turbulent boundary layers. In *Coherent Structures in Turbulent Boundary Layers*, edited by C.R. Smith and D.E. Abbott, AFOSR/Lehigh University Workshop, Bethlehem, Penn., 211.
Boraey, A.A. and da Silva, A.M.F. (2014). A new equation for alternate bar length. *Proceedings of River Flow 2014, 7th International Conference on Fluvial Hydraulics*, Lausanne, Switzerland, edited by Anton J. Schleiss, Giovanni de Cesare, Mário J. Franca and Michael Pfister, CRC Press, Taylor & Francis Group, London, 1195-1202.
Boraey, A.A. (2014). *Alternate bars under steady state flows: time of development and geometric characteristics*. Ph.D. Thesis, Department of Civil Engineering, Queen's University, Kingston, Canada.
Cantwell, B.J. (1981). Organised motion in turbulent flow. *Annual Review of Fluid Mechanics*, Vol. 13, 457-515.
Coleman, S.E. and Nikora, V.I. (2011). Fluvial dunes: initiation, characterization, flow structure. *Earth Surface Processes and Landforms*, 36, 39-57.
Coleman, S.E. and Eling, B. (2000). Sand wavelets in laminar open-channel flows. *Journal of Hydraulic Research*, 38(5), 331-338.
Colombini, M., Seminara, G. and Tubino, M. (1987). Finite-amplitude alternate bars. *Journal of Fluid Mechanics*, Vol. 181, 213-232.
Constantinescu, G., Miyawaki, S., Rhoads, B. and Sukhodolov, A. (2012). Numerical analysis of the effect of momentum ratio on the dynamics and sediment entrainment capacity of coherent flow structures at a stream confluence. *Journal of Geophysical Research*, F04028, doi:10.1029/2012JF002452.
da Silva, A.M.F., Ahmari, H. and Kanani, A. (2012). Characteristic scales and consequences of large-scale horizontal coherent structures in shallow open-channel flows. Chapter 5, in *Environmental Fluid Mechanics - Memorial volume in honour of Prof. Gerhard H. Jirka*, IAHR Monograph, edited by W. Rodi and M. Uhlmann, CRC Press, Taylor & Francis Group, 85-105.
da Silva, A.M.F. and Ahmari, H. (2009). Size and effect on the mean flow of large-scale horizontal coherent structures in open-channel flows: an experimental study. *Canadian Journal of Civil Engineering*, 36(10), 1643-1655.
da Silva, A.M.F. (2006). On why and how do rivers meander. *Journal of Hydraulic Research*, 44(5), 579-590.
da Silva, A.M.F. and Zhang, Y. (1999). On the steepness of dunes and determination of alluvial stream friction factor. *Proceeding XXVIII IAHR Congress*, Graz, Austria, Aug. 22-27.
da Silva, A.M.F. (1991). *Alternate bars and related alluvial processes*. M.Sc. Thesis, Department of Civil Engineering, Queen's University, Kingston, Canada.
del Álamo, J.C. and Jiménez, J. (2003). Spectra of the very large anisotropic scales in turbulent channels. *Physics of Fluids*, 15(6), L41-L44.
Dementiev, M.A. (1962). *Investigation of flow velocity fluctuations and their influences on the flow rate of mountainous rivers*. (In Russian) Technical Report of the State Hydro-Geological Institute (GGI), Vol. 98.

Devauchelle, O., Malverti, L., Lajeunesse, É, Josserand, C., Lagrée, P.-Y. and Métivier, F. (2010). Rhomboid beach pattern: A laboratory investigation. *Journal of Geophysical Research*, 115, F02017, doi:10.1029/2009JF001471.

Fok, A.T.K. (1975). *On the development of ripples by an open channel flow*. M.Sc. Thesis, Department of Civil Engineering, Queen's University, Kingston, Canada.

Franca, M.J. and Lemmin, U. (2015). Detection and reconstruction of large-scale coherent flow structures in gravel-bed rivers. *Earth Surface Processes and Landforms*, 40, 93-104.

Franca, M.J. and Lemmin, U. (2008). Using empirical mode decomposition to detect large-scale coherent structures in river flows. *Proceedings of River Flow 2008, 4th International Conference on Fluvial Hydraulics*, Cesme-Izmir, Turkey, edited by M.S. Altinakar, M.A. Kokpinar, I. Aydin, S. Cokgor and S. Kirkgoz, Ankara: Kubaba Congress Department and Travel Services, 67-74.

Franca, M.J. (2005). *A field study of turbulent flows in shallow gravel-bed rivers*. Ph.D. Dissertation, École Polytechnique Fédérale de Lausanne (EPFL), doi: 10.5075/epfl-thesis-3393.

Fujita, Y. and Muramoto, Y. (1988). Multiple bars and stream braiding. *Proceedings of the International Conference on River Regime*, edited by W.R. White, published on behalf of Hydraulics Research Ltd., Wallingford, John Wiley and Sons, 289-300.

Fujita, Y. and Muramoto, Y. (1985). Study on the process of development of alternate bars. *Disaster Prevention Research Institute Bulletin*, Kyoto University, Japan, 35(3), 55-86.

Fujita, Y. and Muramoto, Y. (1980). On the formation of stream channel patterns and its processes. *Disaster Prevention Research Institute Annuals*, Kyoto University, Japan, No. 23, Part B-2, 475-492.

Gad-el-Hak, M. and Hussain, A.K.M.F. (1986). Coherent structures in a turbulent boundary layer. Part 1: Generation of "artificial" bursts. *Physics of Fluids*, 29(7), 2124-2139.

Grishanin, K.V. (1979). *Dynamics of alluvial streams*. Gidrometeoizdat, Leningrad.

Hayashi, T. (1971). Study of the cause of meandering rivers. *Transactions of the Japan Society of Civil Engineers*, Vol. 2, Part 2.

Hino, M. (1968). Equilibrium-range spectra of sand waves formed by running water. *Journal of Fluid Mechanics*, Vol. 34, Part 3, 565-573.

Hussain, A.F. (1983). Coherent structures – reality and myth. *Physics of Fluids*, 26(10), 2816-2850.

Ikeda, H. (1983). *Experiments on bed load transport, bed forms and sedimentary structures using fine gravel in the 4-meter-wide flume*. Environmental Research Center Papers, No. 2, The University of Tsukuba, Japan.

Ikeda, S. (1984). Prediction of alternate bar wavelength and height. *Journal of Hydraulic Engineering*, 110(4), 371-386.

Jackson, R.G. (1976). Sedimentological and fluid-dynamics implications of the turbulent bursting phenomenon in geophysical flows. *Journal of Fluid Mechanics*, Vol. 77, 531-560.

Jaeggi, M. (1984). Formation and effects of alternate bars. *Journal of Hydraulic Engineering*, 110(2), 142-156.

Jiménez, J. (1998). The largest structures in turbulent wall flows. *Center for Turbulence Research Annual Research Briefs*, Stanford University.

Jirka, G.H. and Uijttewall, W.S.J., *Editors*, (2004). *Shallow flows: selected papers of the International Symposium on Shallow Flows*, 16-18 June 2003, Delft, The Netherlands, A.A. Balkema, Rotterdam, The Netherlands.

Johansson, A.V., Alfredsson, P.H. and Kim, J. (1989). Velocity and pressure fields associated with near-wall turbulence structures. In *Near Wall Turbulence, Proceedings of the Zaric Meml. Conference 1988*, edited by S.J. Kline and N.H. Afgan, New York: Hemisphere, 368-380.

JSCE Task Committee on the Bed Configuration and Hydraulic Resistance of Alluvial Streams (1973). The bed configuration and roughness of alluvial streams. (In Japanese) *Proceedings of the Japan Society of Civil Engineers*, No. 210, 65-91.

Kadota, A. and Nezu, I. (1999). Three-dimensional structure of space-time correlation on coherent vortices generated behind dune crest. *Journal of Hydraulic Research*, 37(1), 59-80.

Kanani, A. and da Silva, A.M.F. (2015). Application of continuous wavelet transform to the study of large-scale coherent structures. *Journal of Environmental Fluid Mechanics*, 15(6), 1293-1319.

Kim, K. and Adrian, R. (1999). Very large-scale motion in the outer layer. *Physics of Fluids*, 11(2), 417-422.

Klaven, A.B. and Kopaliani, Z.D. (1973). *Laboratory investigations of the kinematic structure of turbulent flow past a very rough bed.* Technical Report of the State Hydro-Geological Institute (GGI), USSR, Vol. 209.

Klaven, A.B. (1966). *Investigation of structure of turbulent streams.* Technical Report of the State Hydro-Geological Institute (GGI), USSR, Vol. 136.

Kondratiev, N., Popov, I. and Snishchenko, B. (1982). *Foundations of hydromorphological theory of fluvial processes.* (In Russian) Gidrometeoizdat, Leningrad.

Kondratiev, N., Lyapin, A.N., Popov, I.V., Pinikovskii, S.I., Fedorov, N.N. and Yakunin, I.I. (1959). *Channel processes.* Gidrometeoizdat, Leningrad.

Konsoer, K.M. and Rhoads, B.L. (2014). Spatial-temporal structure of mixing interface turbulence at two large river confluences. *Journal of Environmental Fluid Mechanics*, 14(5), 1043-1070.

Lajeunesse, E., Malverti, L., Lancien, P., Armstrong, L., Métivier, F., Coleman, S., Smith, C.E., Davies, T., Cantelli, A. and Parker, G. (2010). Fluvial and submarine morphodynamics of laminar and near-laminar flows: a synthesis. *Sedimentology*, 57, 1-26.

Leliavsky, S. (1959). *An introduction to fluvial hydraulics.* 2nd edition, Constable and Company Ltd., London.

Lu, L.J. and Smith, C.R. (1991). Use of flow visualization data to examine spatial-temporal velocity and burst-type characteristics in a boundary layer. *Journal of Fluid Mechanics*, Vol. 232, 303-340.

Matthes, G.H. (1947). Macroturbulence in natural stream flow. *Transactions of the American Geophysical Union*, 28(2), 255-262.

McLelland, S.J., Ashworth, P. and Best, J.L. (1996). The origin and downstream development of coherent flow structures at channel junctions. In *Coherent flow structures in open channels*, John Wiley and Sons, Chichester, UK, 459-490.

Nezu, I. and Nakagawa, H. (1993). *Turbulence in open channel flow.* IAHR Monograph, A.A. Balkema, Rotterdam, The Netherlands.

Pope, S.B. (2000). *Turbulent flows*. Cambridge University Press, UK.
Public Works Research Institute, Ministry of Construction of Japan (1982). *Meandering phenomenon and design of river channels*. (In Japanese) PWRI Technical Research Centre, Tsukuba.
Quadrio, M. and and Luchini, P. (2003). Integral space-time scales in turbulent wall flows. *Physics of Fluids*, 15(8), 2219-2227.
Rao, K.N., Narasimha, R. and Narayanan, M.A.B. (1971). The 'bursting' phenomenon in a turbulent boundary layer. *Journal of Fluid Mechanics*, Vol. 4, Part 2, 339-352.
Rashidi, M. and Banerjee, S. (1990). Streak characteristics and behaviour near wall and interface in open channel flows. *Journal of Fluids Engineering*, 112(2), 164-170.
Rashidi, M. and Banerjee, S. (1988). Turbulence structures in free-surface channel flows. *Physics of Fluids*, 31(9), 2491-2501.
Robinson, S.K. (1991). Coherent motions in the turbulent boundary layer. *Annual Review of Fluid Mechanics*, Vol. 23, 601-639.
Rodi, W., Constantinescu, G. and Stoesser, T. (2013). *Large-Eddy Simulation in hydraulics*. IAHR Monograph, CRC Press, Taylor & Francis Group.
Roy, A.G., Buffin-Bélanger, T., Lamarre, H. and Kirkbride, A.D. (2004). Size, shape and dynamics of large-scale turbulent flow structures in a gravel-bed river. *Journal of Fluid Mechanics*, Vol. 500, 1-27.
Schmid, A. (1985). *Wandnahe turbulente Bewegungsabläufe und ihre Bedeutung für die Riffelbildung*. R 22-85, Institut für Hydromechanik und Wasserwirtschaft, ETH Zurich.
Sedov, L.I. (1960). *Similarity and dimensional methods in mechanics*. Academic Press Inc., New York.
Seminara, G. (1998). Stability and Morphodynamics. *Meccanica*, 33, Kluwer Academic Publishers, The Netherlands, 59-99.
Smith, C.R., Walker, J.D.A., Haidari, A.H. and Sobrun, U. (1991). On the dynamics of near-wall turbulence. *Philosophical Transactions of the Royal Society A: Mathematical, Physical and Engineering Sciences*, 336, 131-175.
Sukhodolov, A.N., Nikora, V.I. and Katolikov, V.M. (2011). Flow dynamics in alluvial channels: the legacy of Kirill V. Grishanin. *Journal of Hydraulic Research*, 49(3), 285-292.
Sukhodolov, A.N. and Uijttewaal, W.S. (2010). Assessment of a river reach for environmental fluid dynamics studies. *Journal of Hydraulic Engineering*, 136(11), 880-888.
Tison, L.J. (1949). Origine des ondes de sable et des bancs de sable sous l'action des courants. *3rd IAHR Congress*, Grenoble, France, Report II-13.
Utami, T. and Ueno, T. (1977). Lagrangian and Eulerian measurement of large-scale turbulence. *Proceedings of the International Symposium on Flow Visualization, Flow Visualization I*, edited by T. Asanuma, Hemisphere, Wash.
Valance, A. and Langlois, V. (2005). Ripple formation over a sand bed submitted to a laminar shear flow. *The European Physical Journal B*, 43, 283-294.
van Rijn, L.C. (1984). Sediment transport. Part III: bed forms and alluvial roughness. *Journal of Hydraulic Engineering*, 110(12), 1733-1753.
Velikanov, M.A. (1955). *Dynamics of alluvial streams. Vol. II: sediment and bed flow*. State Publishing House for Theoretical and Technical Literature, Moscow.

Yalin, M.S. (2006). Large-scale turbulence and river morphology. *Proceedings of River Flow 2006, 3rd International Conference on Fluvial Hydraulics*, edited by Rui M.L. Ferreira, Elsa C.T.L. Alves, João G.A.B. Leal and António H. Cardoso, CRC Press, Taylor & Francis Group, 1243-1249.

Yalin, M.S. and da Silva, A.M.F. (2001). *Fluvial processes*. IAHR Monograph, IAHR, The Netherlands.

Yalin, M.S. (1992). *River mechanics*. Pergamon Press, Oxford.

Yalin, M.S. and da Silva, A.M.F. (1992). Horizontal turbulence and alternate bars. *Journal of Hydroscience and Hydraulic Engineering*, JSCE, 9(2), 47-58.

Yalin, M.S. (1987). On the formation mechanism of dunes and ripples. *Proceedings of Euromech 261*, Genoa, Sept.

Yalin, M.S. (1985). On the determination of ripple geometry. *Journal of Hydraulic Engineering*, 111(8), 1148-1155.

Yalin, M.S. and Karahan, E. (1979). Steepness of sedimentary dunes. *Journal of the Hydraulics Division*, ASCE, 105(4), 381-392.

Yalin, M.S. (1977). *Mechanics of sediment transport*. 2nd edition, Pergamon Press, Oxford.

Yalin, M.S. (1971). *Theory of hydraulic models*. MacMillan and Co., Ltd., London.

Yalin, M.S. (1964). Geometrical properties of sand waves. *Journal of the Hydraulics Division*, ASCE, 90(5), 105-119.

Yokosi, S. (1967). The structure of river turbulence. *Disaster Prevention Research Institute Bulletin*, Kyoto University, Japan, Vol. 17, Part 2, 1-29.

Zhang, Y. (1999). *Bed form geometry and friction factor over a bed covered by dunes*. M.A.Sc. Thesis, Department of Civil and Environmental Engineering, University of Windsor, Windsor, Canada.

Zhou, D. and Mendoza, C. (2009). On bedform growth and nonequilibrium sediment transport. *Canadian Journal of Civil Engineering*, 36(10), 1634-1642.

Chapter 3

Flow past undulated beds

3.1 FLOW RESISTANCE FACTOR

3.1.1 General

Consider a uniform two-phase motion (in a wide open-channel) which can be treated as two-dimensional: it is assumed that the bed is covered by bed forms. The presence of bed forms means the elevation of the effective bed roughness. Therefore the resistance factor c corresponding to this (general) case must be smaller than c_f corresponding to the (special) case of a flat bed. Clearly, the difference between c and c_f is solely due to the "geometry" of bed forms. We will assume first that only one mode of bed forms (i.e. only dunes, or only ripples, etc.) is present (Figure 3.1) and that they are two-dimensional – their longitudinal section ABA' does not vary along the "third dimension" y.

3.1.2 Resistance factor formula

Derivation

Since forces acting in the same direction are algebraically additive, the total shear stress $\tau_0 = \gamma hS$ acting on the undulated bed surface usually is considered to be given by the sum

$$\tau_0 = (\tau_0)_f + (\tau_0)_\Delta. \tag{3.1}$$

Here, $(\tau_0)_f$ is due to the factual friction ("skin friction") between the flow and the bed surface (which is assumed to possess a granular surface roughness ("skin roughness") $k_s \approx 2D$), and $(\tau_0)_\Delta$ is due to the drag-force acting on the (supposedly frictionless) bed forms.

Dividing both sides of Eq. (3.1) by $\rho\bar{u}^2$, we obtain

$$\frac{\tau_0}{\rho\bar{u}^2} = \frac{(\tau_0)_f}{\rho\bar{u}^2} + \frac{(\tau_0)_\Delta}{\rho\bar{u}^2}, \tag{3.2}$$

which can be denoted as

$$\frac{1}{c^2} = \frac{1}{c_f^2} + \frac{1}{c_\Delta^2}. \tag{3.3}$$

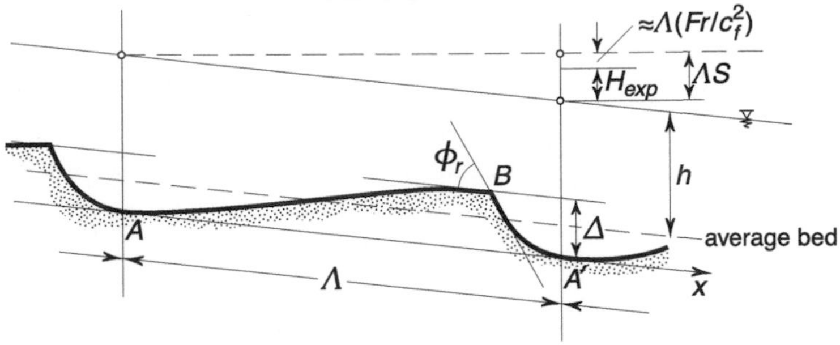

Figure 3.1 Definition sketch: geometric and flow characteristics pertinent for the expression of the flow resistance factor.

In this expression, c is the *total resistance factor* of an undulated bed, c_f is the *friction factor* of the undulated bed surface, and c_Δ is the *bed-form resistance* factor.

Since the steepness and curvature of bed forms are never excessive, it became a convention to evaluate c_f (of an undulated bed) with the aid of the (flat bed) relation (1.15), i.e. as

$$c_f = \frac{1}{\kappa}\ln\left(0.368\frac{h}{k_s}\right) + B_s. \tag{3.4}$$

Consider c_Δ. Given that the angle of repose ϕ_r (at B in Figure 3.1) does not vary significantly from one granular material to another, while the upstream face of bed forms (curve AB) has at A and B tangents parallel to the x-axis, the geometry presented by the bed form ABA' can be regarded as determined solely by the two length ratios Δ/Λ $(=\delta)$ and Λ/h. Hence c_Δ, which is due to this geometry, should be given by

$$\frac{1}{c_\Delta^2} = \phi_\Delta\left(\delta, \frac{\Lambda}{h}\right) \qquad \text{(where } \phi_\Delta(0, \Lambda/h) = 0\text{)}. \tag{3.5}$$

In order to reveal the form of the function ϕ_Δ, we consider Figure 3.1, bearing in mind that bed forms are rather flat features – the largest possible Δ/Λ is only ≈ 0.14 (see Section 2.3). The hydraulic energy loss $\Lambda S = \Lambda(Fr/c^2)$ of the flow along one bed form length Λ (between sections A and A') is the sum of the friction loss $\approx \Lambda(Fr/c_f^2)$, and the energy loss H_{exp} due to the sudden expansion of flow at B (Figure 3.1) – the energy loss due to the "gradual contraction" of flow along AB is negligible. i.e.

$$\Lambda\frac{Fr}{c^2} = \Lambda\frac{Fr}{c_f^2} + H_{exp} \tag{3.6}$$

or

$$\frac{1}{c^2} = \frac{1}{c_f^2} + \frac{H_{exp}}{\Lambda Fr}. \tag{3.7}$$

From Eqs. (3.3), (3.5) and (3.7) it follows that

$$\phi_\Delta\left(\delta, \frac{\Lambda}{h}\right) = \frac{H_{exp}}{\Lambda\, Fr}. \tag{3.8}$$

As is well known, the sudden-expansion loss H_{exp} is determined by the Borda-theorem

$$\begin{aligned} H_{exp} = \frac{(\overline{u}_{A'} - \overline{u}_B)^2}{2g} &= \frac{\overline{u}^2}{2g}\left(\frac{1}{1-\frac{\Delta}{2h}} - \frac{1}{1+\frac{\Delta}{2h}}\right) \\ &= \frac{\overline{u}^2}{2g} \cdot \frac{(\Delta/h)^2}{[1-\frac{1}{4}(\Delta/h)^2]^2} \approx \frac{\overline{u}^2}{2g}\left(\frac{\Delta}{h}\right)^2. \end{aligned} \tag{3.9}$$

Even if Δ/h is as large as $\approx 1/3$, the denominator of $(\Delta/h)^2$ in Eq. (3.9) is as near to unity as 0.95; hence the reason for the last step in Eq. (3.9). Eliminating H_{exp} between Eqs. (3.8) and (3.9) and replacing ($\approx$) by ($=$), we determine

$$\phi_\Delta\left(\delta, \frac{\Lambda}{h}\right) = \frac{1}{2}\,\delta^2\,\frac{\Lambda}{h} \tag{3.10}$$

which yields, in conjunction with Eq. (3.3),

$$\frac{1}{c^2} = \frac{1}{c_f^2} + \frac{1}{2}\,\delta^2\,\frac{\Lambda}{h}. \tag{3.11}$$

(Note that $\delta = 0$ implies $c = c_f$, while $\delta \neq 0$ means $c < c_f$). Eq. (3.11), which was first introduced, independently, by Yalin (1964) and Engelund (1966), can be generalized (by virtue of the additivity of losses and thus of the $1/c^2$-values) to two modes of bed forms, i.e. to the case of ripples (r) superimposed on dunes (d) as

$$\frac{1}{c^2} = \frac{1}{c_f^2} + \frac{1}{2h}\left(\delta_d^2\,\Lambda_d + \delta_r^2\,\Lambda_r\right). \tag{3.12}$$

Although a third mode of bed forms, viz bars (e.g. alternate bars), can also be present, their contribution to the c-value can be ignored (for usually $\delta_a <\approx 1.5 \cdot 10^{-2}$, which makes their $\delta_a^2\,\Lambda_a$-values approximately by one order of magnitude smaller than $\delta_d^2\,\Lambda_d$ or $\delta_r^2\,\Lambda_r$).

Comparison with the data

The resistance factor c, which involves in its expression c_f, δ_i and Λ_i, where $i = d, r$ (Eq. (3.12)), is a function of three dimensionless variables, which can be expressed in terms of e.g. X, Y, Z, or Ξ, η_*, Z, etc. In this paragraph, the version

$$c = \phi_c(\Xi, \eta_*, Z) \tag{3.13}$$

is used.

In the following, the family of the family of curves (3.13) is computed by specifying each curve-family by a constant value of Ξ, and each curve of a family, by a constant value of Z. As examples, the resulting curve-families corresponding to $\Xi = 7.6$, 12.6 and 35.4 ($D = 0.3$, 0.5 and 1.40mm) are shown in Figures 3.2a-c, respectively. The data plotted on these graphs stem from the well-known and extensive database of W.R. Brownlie (Ref. [2c] in Appendix C). The *Z-values* of the computed curves, as well as the *Z-ranges* of the points plotted, are also given on these graphs. The curves having the Z-values which fall into the Z-ranges of the points plotted are shown by solid lines. In spite of the gross scatter (which is only typical of the sediment related data including natural rivers), the point-patterns appear to agree satisfactorily with the corresponding computed c-curves.

For sand or gravel and water, the knowledge of h, S and D is sufficient to determine Ξ, η_*, and Z needed for the computation of c according to the present method. i.e. $c = \phi_c(\Xi, \eta_*, Z)$ can be shown as $c = \psi_c(h, S, D)$. Substituting this symbolic expression of c into the resistance equation $Q = Bhc\sqrt{gSh}$, one determines $Q = Bh\psi_c(h, S, D)\sqrt{gSh}$, which indicates that $h = f_h(Q, B, S, D)$. Figure 3.3 shows how the values of h computed from $Q = Bh\psi_c(h, S, D)\sqrt{gSh}$ (with the aid of the present method of determination of c) compare with their measured counterparts.[1]

We go over to illustrate the application of the present method of determination of c with a numerical example.

Numerical example

Consider a quasi-uniform two-dimensional two-phase motion determined by the following values of its characteristic parameters

Flow:	Fluid:	Bed material:
$h = 1.25m$	$\rho = 1000kg/m^3$	$\gamma_s = 16186.5N/m^3$
$S = 0.0002$	$\nu = 10^{-6}m^2/s$	$D = D_{50} = 0.33mm$

1- Knowing these values, we determine

$$\upsilon_* = \sqrt{gSh} = 0.0495m/s, \tag{3.14}$$

and compute

$$X = \frac{\upsilon_* D}{\nu} = 16.3, \quad Y = \frac{\rho \upsilon_*^2}{\gamma_s D} = 0.459, \quad Z = \frac{h}{D} = 3788. \tag{3.15}$$

Consequently

$$\Xi = \left(\frac{X^2}{Y}\right)^{1/3} = 8.3. \tag{3.16}$$

[1]The sources of the data-points in Figure 3.3 (as well as in Figures 3.4a,b) are given, among others, in Appendices A and C. These sources include
Laboratory data: [3a], [6a], [7a], [9a], [12a], [13a], [15a], [28a], [35a], [1c], [3c], [5c], [6c], [7c], [9c], [11c], [12c], [14c], [15c], and
Field data: [8a], [27a], [31a], [37a], [39a], [4c], [8c], [10c], [13c].
All of the points plotted in Figures 3.3 and 3.4 correspond to $X > 25$, i.e. to dunes.

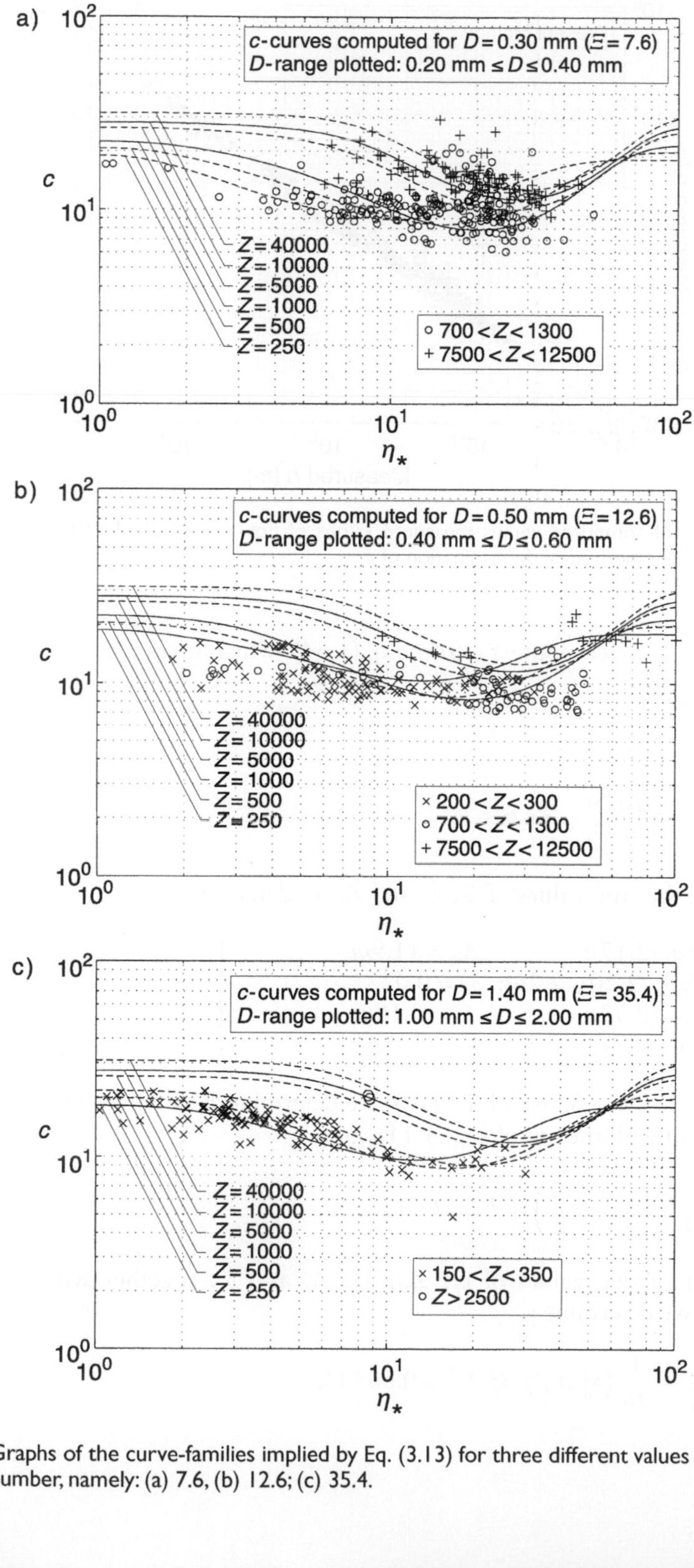

Figure 3.2 Graphs of the curve-families implied by Eq. (3.13) for three different values of the material number, namely: (a) 7.6, (b) 12.6; (c) 35.4.

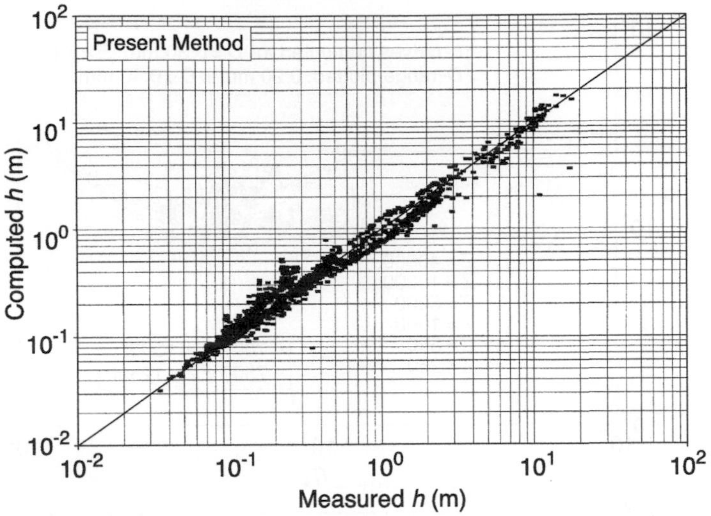

Figure 3.3 Plot of computed versus measured values of flow depth, produced on the basis of the present method of determination of c.

2- Knowing Ξ, we compute Y_{cr} from Eq. (1.34), which gives

$$Y_{cr} = 0.039. \tag{3.17}$$

Hence

$$\eta_* = \frac{Y}{Y_{cr}} = \frac{0.459}{0.039} = 11.7. \tag{3.18}$$

3- Knowing thus the values of Ξ, η_* and Z, we determine

$$\left.\begin{array}{ll} \text{from Eq. (2.17):} & \Lambda_d = 11.9m \\ \text{from Eqs. (2.20)-(2.26):} & \delta_d = 0.027 \\ \text{from Eq. (2.32):} & \Lambda_r = 0.18m \\ \text{from Eqs. (2.37)-(2.39):} & \delta_r = 0.052 \end{array}\right\} \tag{3.19}$$

4- Using $Re_* = 2X = 32.6$ in Eq. (1.10) we determine $B_s = 9.06$ which, together with $Z = 3788$ and with the aid of Eq. (1.11), gives

$$c_f = \frac{1}{0.4} \ln\left(0.368 \frac{Z}{2}\right) + B_s = 25.4. \tag{3.20}$$

5- Using the above values of c_f, Λ_d, δ_d, Λ_r and δ_r, together with $h = 1.25m$, in Eq. (3.12), we determine

$$\frac{1}{c^2} = \frac{1}{c_f^2} + \frac{1}{2h} \left(\delta_d^2 \Lambda_d + \delta_r^2 \Lambda_r\right) = 0.00533, \tag{3.21}$$

and thus

$$c = 13.7. \tag{3.22}$$

The computer program RFACTOR (both Fortran and MATLAB versions) for the computation of c by the method above can be downloaded from the CRC Press website.

3.1.3 Other methods for determination of c

Numerous methods have been produced for the determination of c (see Garde and Ranga Raju 1966, Raudkivi 1967, Alam and Kennedy 1969, Ranga Raju 1970, White et al. 1979, Paris 1980, Karim and Kennedy 1981, Brownlie 1983, van Rijn 1984b, etc.): out of them, only two will be explained here.

Method by White, Paris and Bettess (1979)

In this work, a procedure for the determination of c has been developed on the basis of extensive experimental field and laboratory data. This procedure, which gives the value of c as a function of the dimensionless variables Ξ, Y, Z, can be summarized as follows.

Consider the following two relations,

$$F_{gr} = \left(\frac{\sqrt{32}\log(10Z)}{c} \right)^{n-1} \sqrt{Y} \tag{3.23}$$

and

$$\frac{F_{gr} - A}{\sqrt{Y} - A} = \phi(\Xi). \tag{3.24}$$

The following relation is given in White et al. (1979) for the function $\phi(\Xi)$ in Eq. (3.24):

$$\phi(\Xi) = \begin{cases} 1 - 0.76(1 - e^{-(\log \Xi)^{1.7}}) & \text{if } D = D_{35} \text{ (bed material)} \\ 1 - 0.70(1 - e^{-1.4(\log \Xi)^{2.65}}) & \text{if } D = D_{65} \text{ (surface material),} \end{cases} \tag{3.25}$$

while for n and A the following relations are suggested:

$$\text{If } 1 \le \Xi < 60, \text{ then: } \quad n = 1 - 0.56 \log \Xi \quad \text{and} \quad A = \frac{0.23}{\sqrt{\Xi}} + 0.14, \tag{3.26}$$

$$\text{If } \Xi \ge 60, \text{ then: } \quad n = 0 \quad \text{and} \quad A = 0.17. \tag{3.27}$$

Hence:

1 Knowing the values of the characteristic parameters, determine Ξ, Y and Z.
2 Determine $\phi(\Xi)$, n and A from Eqs. (3.25), (3.26) and (3.27).
3 Determine F_{gr} from Eq. (3.24).
4 Using this F_{gr} in Eq. (3.23), determine c.

Method by van Rijn (1984b)

In this work, the total resistance of the mobile bed covered by bed forms is reflected by the "equivalent bed-roughness" K_s, which is related to the resistance factor c by the logarithmic form

$$c = 2.5 \ln\left(11 \frac{\mathcal{R}}{K_s}\right). \tag{3.28}$$

From regression analysis of a large number of flume and field data corresponding to $B/h > 5$ and $h/K_s > 10$, van Rijn (1984b) arrived at the following expression for K_s

$$K_s = 3D + 1.1\Delta(1 - e^{-25\delta}) \qquad \text{(with } 0.01 \leq \delta \leq 0.2\text{)}. \tag{3.29}$$

In the just mentioned work, this expression is evaluated for the case of dunes only ($\Delta = \Delta_d$, $\delta = \delta_d$), and the following relations are suggested:

$$\frac{\Delta_d}{h} = 0.11 Z^{-0.3}(1 - e^{-0.5T})(25 - T) \tag{3.30}$$

and

$$\delta_d = 0.015 Z^{-0.3}(1 - e^{-0.5T})(25 - T), \tag{3.31}$$

where

$$T = \frac{(\overline{u}/c_f)^2 - v_{*cr}^2}{v_{*cr}^2}. \tag{3.32}$$

Comparison with the data

Figures 3.4a,b show how the values of h computed on the basis of the methods by White et al. (1979) and van Rijn (1984b), respectively, compare with experimental data. The data plotted in Figures 3.4a,b are the same as in Figure 3.3. Since the data correspond to dunes, its utilization to test van Rijn's method, which corresponds to dunes only, is in order.

3.1.4 Flow rate as one of the parameters determining c

In the numerical example introduced at the end of Sub-section 3.1.2, the flow depth h was selected as one of the flow-defining parameters, and c was determined (in accordance with Eq. (1.23)) as a function of three variables

$$c = \phi_c(X, Y, Z), \tag{3.33}$$

where the influence of h is reflected by Z. However, in some cases it is the flow rate Q, rather than h, which is used as one of the parameters defining the flow (regime channels, meandering streams, etc.), and in such cases c must be expressed by an appropriately modified counterpart of Eq. (3.33).

Consider the resistance equation (1.17), which can be expressed as follows

$$Q = Bhc\sqrt{gSh} \quad (= Bhcv_*). \tag{3.34}$$

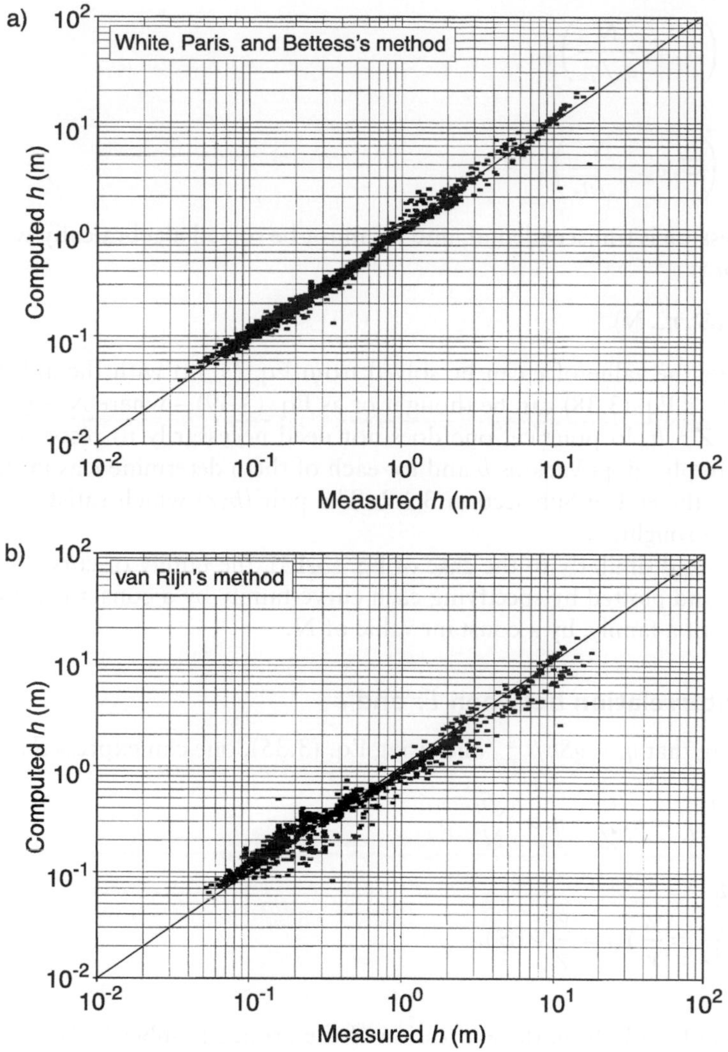

Figure 3.4 Plots of computed versus measured values of flow depth produced on the basis of the following methods of determination of the resistance factor c: (a) White et al. (1979); (b) van Rijn (1984b).

Dividing both sides by $BD\upsilon_{*cr}$, we obtain

$$N = Zc\sqrt{\eta_*} \quad \text{i.e.} \quad Z = \frac{N}{c\sqrt{\eta_*}}, \tag{3.35}$$

where

$$N = \frac{Q}{BD\upsilon_{*cr}}. \tag{3.36}$$

Substituting the value of Z given by Eq. (3.35) into Eq. (3.33), one determines

$$c = \phi_c\left(X, Y, \frac{N}{c\sqrt{\eta_*}}\right), \tag{3.37}$$

and thus

$$c = \phi'_c\left(\Xi, \eta_*, \frac{N}{c\sqrt{\eta_*}}\right). \tag{3.38}$$

This expression is transcendental, and it cannot be solved algebraically with respect to c in the form

$$c = \phi_c(\Xi, \eta_*, N). \tag{3.39}$$

Nonetheless the value of c can be solved from Eq. (3.38) with the aid of numerical methods, and Eq. (3.38) can be thought of as Eq. (3.39) – where $N \sim Q/B$ is present instead of $Z \sim h$. In practice, one does not need necessarily to resort to Eq. (3.38). One can simply adopt various h and for each of them determine c as in the numerical example at the end of Sub-section 3.1.2. The pair $(h; c)$ which satisfies Eq. (3.34) is the solution sought.

Clearly, and similarly to the case of Eq. (3.13), the family of curves (3.39) can be computed and plotted by specifying each curve-family by a constant value of Ξ, and each curve of a family, by a constant value of N.

3.1.5 Interrelation between Fr and c

Considering that $\eta_* = gSh/v_{*cr}^2$ and using Eq. (3.35), one can express S as

$$S = \eta_* \frac{v_{*cr}^2}{gh} = \eta_* \frac{\alpha}{Z} = \eta_* \alpha \frac{c\sqrt{\eta_*}}{N} \tag{3.40}$$

where

$$\alpha = \frac{v_{*cr}^2}{gD} = \frac{\gamma_s}{\gamma} Y_{cr} = \frac{\gamma_s}{\gamma} \Psi(\Xi). \tag{3.41}$$

Substituting Eq. (3.40) in the expression of the Froude number (1.18), viz

$$Fr = c^2 S, \tag{3.42}$$

one determines

$$Fr = \frac{\alpha}{N}(c^2 \eta_*)^{3/2}. \tag{3.43}$$

Eliminating (in principle) c between Eqs. (3.43) and (3.39), one realizes that (for a specified γ_s/γ)

$$Fr = \phi_{Fr}(\Xi, \eta_*, N). \tag{3.44}$$

The variables of the function (3.44) are exactly the same as those of the function (3.39). Hence Fr can be plotted in exactly the same manner as c (i.e. as described at the end of Sub-section 3.1.4); see Sub-section 4.4.4 (ii) later on.

3.2 SEDIMENT TRANSPORT RATE OF FLOW PAST UNDULATED BED

We assume, following the convention (van Rijn 1984a, Raudkivi 1990, etc.), that bed forms do not alter the mathematical form of a transport equation as such; their influence on the transport rate can be taken into account by an appropriate modification of the flow velocities and shear stresses (entering that equation).

i) *Modification of* τ_0

The sediment transport rate expressions introduced in Section 1.4, viz Eqs. (1.39) and (1.45), correspond to the case of a flat bed, where the detachment of grains is due to the skin friction only ($(\tau_0)_\Delta = 0$ and $\tau_0 = (\tau_0)_f$). If, however, the bed is undulated, then it is only the part $(\tau_0)_f$ of the total bed shear stress τ_0 which is responsible for the grain motion. Hence in the case of an undulated bed, τ_0, which appears in Eqs. (1.39) and (1.45), must be replaced by (the smaller)

$$(\tau_0)_f = \tau_0 \left(\frac{c}{c_f}\right)^2 = \tau_0 \lambda_c^2, \tag{3.45}$$

where c and c_f, forming $\lambda_c = c/c_f$, are given by Eqs. (3.12) and (3.4), respectively.

ii) *Modification of* $\bar{u}$

Imagine two sediment transporting flows which differ from each other *only* because one of them takes place past a flat bed, while the other, past an undulated bed. Marking the characteristics of the former flow by the subscript "f", and not marking those of the latter at all, we obtain

$$\bar{u}_f = c_f \sqrt{\tau_0/\rho}; \qquad \bar{u} = c\sqrt{\tau_0/\rho} \tag{3.46}$$

and thus

$$\frac{\bar{u}}{\bar{u}_f} = \frac{c}{c_f} = \lambda_c \qquad (<1), \tag{3.47}$$

which indicates how $\bar{u}$ (in Eq. (1.45)), which according to the notation just mentioned is to be viewed as $\bar{u}_f$, is to be modified (reduced).

iii) *Modification of* q_{sb}

According to the notation above, the original *flat bed* Bagnold's formula (1.45) is to be expressed as

$$(q_{sb})_f = \beta' \bar{u}_f (\tau_0 - (\tau_0)_{cr})/\gamma_s, \tag{3.48}$$

The relations (3.45) and (3.47) indicate that for the case of an *undulated bed*, the original Bagnold's formula must be modified into

$$q_{sb} = \beta' \lambda_c \bar{u}_f (\lambda_c^2 \tau_0 - (\tau_0)_{cr})/\gamma_s = \beta' \bar{u} (\lambda_c^2 \tau_0 - (\tau_0)_{cr})/\gamma_s. \tag{3.49}$$

This relation indicates, in turn, that the flat bed expression (1.78) of $\psi_q = q_s/h\overline{U}$ (see Eq. (1.77)) must be augmented into

$$\psi_q = \beta' \left(\frac{\lambda_c^2 \tau_0 - (\tau_0)_{cr}}{\gamma_s h} \right) + \alpha_c \overline{C}. \tag{3.50}$$

Dividing Eq. (3.49) by Eq. (3.48), one determines

$$\frac{q_{sb}}{(q_{sb})_f} = \lambda_c \frac{\eta_* \lambda_c^2 - 1}{\eta_* - 1}. \tag{3.51}$$

As follows from Eq. (1.46), the ratio β/β' is not exactly a constant, and therefore, strictly speaking, $\overline{u}$ and u_b are not always related to each other by the same constant proportion – even in the case of a flat bed. Yet, it is often assumed that $\overline{u}$ and u_b decrease (due to the presence of bed forms) by nearly the same factor (van Rijn 1984a, Raudkivi 1990, etc.). Owing to this reason, the relations above, which emerge from the Bagnold's form (1.45), are used also for the case of the Bagnold's form (1.39) – simply by replacing $\beta', \overline{u}, \overline{u}_f$ with $\beta, u_b, (u_b)_f$.

3.3 MECHANICAL STRUCTURE OF FLOW OVER DUNES

The mechanical structure of flow over dunes has been the focus of numerous laboratory and numerical studies for many years (see e.g. Nelson et al. 1993, Nezu and Nakagawa 1993, Bennett and Best 1995, Best 2005, Balachandar and Patel 2007, 2008) – with a number of recent works involving Large Eddy Simulation and contributing substantially to the understanding of the coherent structures in such flows (e.g. Stoesser et al. 2008, Chang and Constantinescu 2013, Omidyeganeh and Piomelli 2013; see also Rodi et al. 2013).

The flow over a dune can be briefly described as follows (see Figure 3.5). As the flow suddenly expands downstream of the dune crest, it separates from the boundary, and reattaches at a distance of ≈ 4 to 6 times the dune height. The fluid is then conveyed downstream over the stoss side of the dune, inducing the growth of an internal boundary layer. The flow continuously accelerates as it moves downstream in this (contracting) region. The maximum flow velocities occur towards the free surface at a section at or near the dune crest.

Spanwise vortices are generated at the shear layer separating from the dune crest due to a Kelvin-Helmholtz instability. The spanwise vortices are either conveyed downstream along the shear layer eventually dissipating, or travel towards the free surface. Some of the vortices (eddies) generated along the shear layer separating from the dune, or perhaps around the reattachment point, undergo a process as described in Section 2.2, eventually becoming large-scale vertical coherent structures (LSVCS's) extending throughout the flow depth (Figure 3.6). As these impinge on the water surface, they lead to the appearance of boils on it (for more on the topic, see Nezu and Nakagawa 1993). As a result of eddy generation, the region just above the shear layer separating from the dune crest is characterized by large values of turbulence intensity

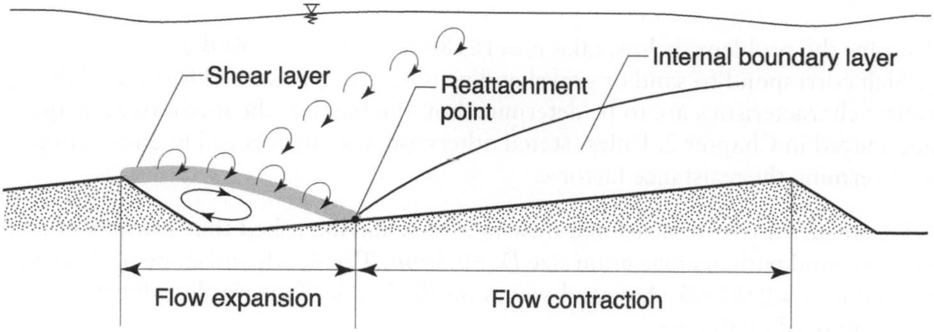

Figure 3.5 Schematic representation of different flow regions over a dune.

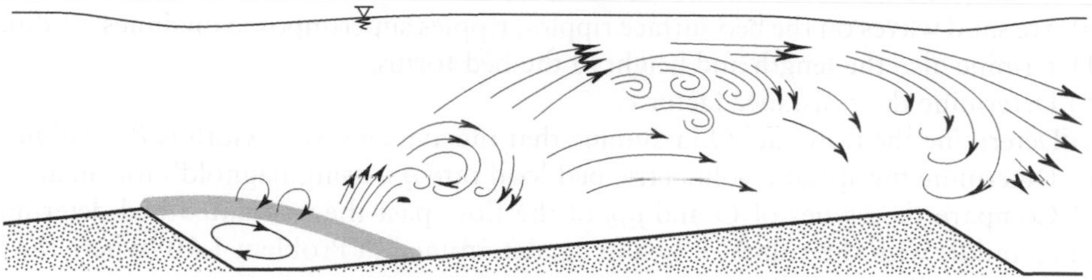

Figure 3.6 Large-scale vertical coherent structure (LSVCS) evolving over a dune.

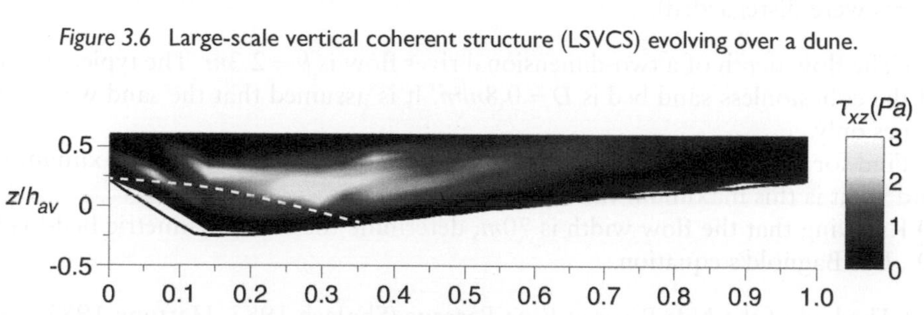

Figure 3.7 Contour-plot of measured turbulence stresses in the vertical (x;z)-planes (courtesy of Mr. Kenneth Lockwood, Ph.D. candidate, Queen's University).

and turbulence stresses. Such a region is evident in Figure 3.7, showing the contour-plots of measured turbulence stresses τ_{xz}.[2]

Even though the presence of dunes is not necessary for the generation and growth of LSVCS's, the dunes regularize and stabilize them.

[2]Figure 3.7 results from measurements recently carried out at Queen's University: $\Lambda_d = 1.3m$, $\Delta_d = 0.074m$, lee side slope $= 30°$, $h_{av} = 0.20m$, $u_{av} = 0.48m/s$. The dunes were formed by sand with $D_{50} = 1.1mm$; the grains forming the upper layer of the bed surface were immobilized.

PROBLEMS

To solve the problems below, take $\gamma_s = 16186.5N/m^3$, $\rho = 1000kg/m^3$, $\nu = 10^{-6}m^2/s$ (which correspond to sand or gravel and water). The type of bed forms and their geometric characteristics are to be determined on the basis of the methods and equations introduced in Chapter 2. Unless stated otherwise, use the method in Sub-section 3.1.2 to determine the resistance factor c.

3.1 Consider a two-dimensional flow in a river having a bed that consists of a cohesionless sand with average grain size $D = 0.3mm$. The depth and slope of this river are $h = 3.0m$, $S = 0.00005$. Are sand waves on the bed surface ripples, dunes, or ripples superimposed on dunes?

3.2 In a certain region of the Pembina River, we have $D = 0.4mm$, $h = 5m$ and $S = 0.00025$ (see Problem 1.10).
a) Are sand waves on the bed surface ripples, ripples superimposed on dunes, or dunes? Determine also the length and height of the bed forms.
b) Determine the resistance factor c.
c) Determine the flow rate Q (assuming that the average river width is $B = 100m$).
d) Determine the specific volumetric bed-load rate q_{sb} using Bagnold's formula.
e) Compare the values of Q and q_{sb} of the flow past the undulated bed determined in c) and d) with the values of Q and q_{sb} determined in Problem 1.10 (where the bed forms were disregarded).

3.3 The flow depth of a two-dimensional river flow is $h = 2.3m$. The typical grain size of the cohesionless sand bed is $D = 0.8mm$. It is assumed that the sand waves can be dunes only.
a) Find for which value of the slope S the steepness of dunes has its maximum value, and what is this maximum value.
b) Knowing that the flow width is $70m$, determine the total volumetric bed-load rate Q_s. Use Bagnold's equation.

3.4 The bed of the Nile River at Esna Barrage (Shalash 1983, Hartung 1987; see also Yalin and Scheuerlein 1988) consists of a cohesionless sand having the average grain size $D = 0.28mm$. The slope is $S = 0.000077$.
a) For the flow depths $h_1 = 2.70m$, $h_2 = 4.00m$ and $h_3 = 5.70m$, determine the type of bed forms (ripples, ripples superimposed on dunes, or dunes), the length and steepness of the bed forms, and the resistance factor.
b) Use the information on the type and geometric characteristics of the bed forms to explain the drop of the resistance factor with increasing flow stage.

3.5 The Yangtze River at Da-tong, halfway between Wuhan and Nanjing, has the following characteristics at average flood peak (Luo et al. 1980, Lin and Li 1986; see also Yalin and Scheuerlein 1988): $D = 0.2mm$, $S = 0.0000277$, $h = 16.1m$.
a) Determine whether the bed forms are ripples, ripples superimposed on dunes, or dunes.
b) Determine the resistance factor c.
c) Use the resistance factor c determined in part b) to estimate the value of Manning's n.

3.6 Consider the Mississippi River (at Natchez) at its bankfull stage (Winkley 1977, Schumm 1977; see also Yalin and Scheuerlein 1988): $D = 0.4mm$, $h = 14.63m$, $S = 0.00007$, $B = 1771m$.

a) Determine the resistance factor c and the flow rate Q using:
   1 the method of Sub-section 3.1.2;
   2 the method of White, Paris and Bettess (1979);
   3 the method of van Rijn (1984b) (knowing that $\overline{u} = 1.39m/s$).

b) Compare your estimates of the flow rate Q with the measured value $Q = 35963m^3/s$.

3.7 Consider the following conditions observed on the Zaire River at Ntua Nkulu (Peters 1978; see also Yalin and Scheuerlein 1988), where $D = 0.75mm$: $B = 500m$, $S = 0.000058$, $Q = 12300m^3/s$. Estimate the flow depth h.

3.8 In the $0.76m$-wide, $21m$-long sediment transport flume of Queen's University Coastal Engineering Research Laboratory, several experimental runs were conducted in order to investigate the formation of bars: the bed material was a cohesionless sand, $D = 1.1mm$. All runs were carried out starting from a flat initial bed. The initial flows (at $t = 0$) were uniform; the values of Q, h and S were as indicated in the table below.

Run	Q (l/s)	h (cm)	$S \times 10^3$
CS-1	2.11	1.0	8.1
CS-2	2.86	1.2	8.1
CS-3	4.15	1.5	8.1
CS-4	6.12	2.2	8.1
CS-5	9.73	2.5	8.1
CS-6	11.00	4.1	3.0

a) Plot these initial flows as data-points on the $(B/h; h/D)$-plane in Figure 2.30.

b) What was the configuration in plan view of the bars (one-row or multiple-row) observed in each run?

c) Determine the bar length for the conditions of Run CS-5 using both the equations by Boraey and da Silva (2014) and Ikeda (1984).

3.9 Explain why the flume-data (corresponding to given $B = const$ and $D = const$) determined for various h and/or S group on the $(B/h; h/D)$-plane around the 1/1-declining straight lines (see the data in Figure 2.30; see also Problem 3.8, part (a)).

3.10 In a wide tilting flume, whose bed is covered by a layer of sand having $D = 1.5mm$, a flow having $Q = 0.44m^3/s$, $S = 1/600$, and $\eta_* = 3.5$ is produced. Will this flow cause the appearence of dunes, ripples, alternate bars or multiple bars?

REFERENCES

Alam, A.M.Z. and Kennedy, J.F. (1969). Friction factors for flow in sand bed channels. *Journal of the Hydraulics Division*, ASCE, 95(6), 1973-1992.

Balachandar, R. and Patel, V.C. (2008). Flow over a fixed rough dune. *Canadian Journal of Civil Engineering*, 35(5), 511-520.

Balachandar, R. and Patel, V.C. (2007). Effect of depth on flow over a fixed dune. *Canadian Journal of Civil Engineering*, 34(12), 1587-1599.

Bennett, S.J. and Best, J.L. (1995). Mean flow and turbulence structure over fixed, two- dimensional dunes: implications for sediment transport and bedform stability. *Sedimentology*, 42, 491-513.

Best, J. (2005). The fluid dynamics of river dunes: a review and some future research directions. *Journal of Geophysical Research*, 110, F04S02, doi:10.1029/2004J F000218.

Brownlie, W.R. (1983). Flow depth in sand bed channels. *Journal of Hydraulic Engineering*, 109(7), 959-990.

Chang, K. and Constantinescu, G. (2013). Coherent structures in flow over two-dimensional dunes. *Water Resources Research*, 49(5), 2446-2460.

Engelund, F. (1966). Hydraulic resistance of alluvial streams. *Journal of the Hydraulics Division*, ASCE, 92(2), 315-326.

Garde, R.J. and Ranga Raju, K.G. (1966). Resistance relationships for alluvial channel flow. *Journal of the Hydraulics Division*, ASCE, 92(4), 77-99.

Hartung, F. (1987). *Der Assuan-Hochdamm – Fehlplanung oder unvollendet?*. Wasser und Boden, Heft 9.

Karim, M.F. and Kennedy, J.F. (1981). *Computer-based predictors for sediment discharge and friction factor of alluvial streams*. Report No. 242, Iowa Institute of Hydraulic Research, Iowa.

Lin, B. and Li, G. (1986). *The Changjiang and the Huanghe – two leading rivers of China*. IRTCES Series of Publications, Circular No. 1.

Luo, H., Zhou, X., You, L. and Jin, D. (1980). On the cause of formation of braided river in the middle and lower reaches of the Yangtze River. *Proceedings of the First International Symposium on River Sedimentation*, Beijing, Vol. 1, Paper C5.

Nelson, J.M., McLean, S.R. and Wolfe, S.R. (1993). Mean flow and turbulence fields over two-dimensional bedforms. *Water Resources Research*, 29(12), 3935-3953.

Nezu, I. and Nakagawa, H. (1993). *Turbulence in open-channel flows*. IAHR Monograph, A.A. Balkema, Rotterdam, The Netherlands.

Omidyeganeh, M. and Piomelli, U. (2013). Large-eddy simulation of three-dimensional dunes in a steady, unidirectional flow. Part 2. Flow structures. *Journal of Fluid Mechanics*, Vol. 734, 509-534.

Paris, E. (1980). New criteria for predicting the frictional characteristics in alluvial streams: a comparison. *Proceedings of the IAHR Symposium on River Engineering and its Interaction with Hydrological and Hydraulic Research*, Belgrade, May.

Peters, J.J. (1978). Discharge and sand transport in the braided zone of the Zaire Estuary. *Netherlands Journal of Sea Research*, 12 (3/4), 273-292.

Ranga Raju, K.G. (1970). Resistance relation for alluvial streams. *La Houille Blanche*, No. 1, 51-54.

Raudkivi, A.J. (1990). *Loose boundary hydraulics*. 3rd edition, Pergamon Press, Oxford.

Raudkivi, A.J. (1967). Analysis of resistance in fluvial channels. *Journal of the Hydraulics Division*, ASCE, 93(5), 73-84.

Rodi, W., Constantinescu, G. and Stoesser, T. (2013). *Large Eddy Simulation in Hydraulics*. IAHR Monograph, CRC Press, Taylor & Francis Group.

Schumm, S.A. (1977). *The fluvial system*. John Wiley and Sons, New York.

Shalash, S. (1983). *Degradation of the River Nile*. Water Power and Dam Construction, July/August.

Stoesser, T., Braun, C., Garcia-Villalba, M. and Rodi, W. (2008). Turbulence structures in flow over two-dimensional dunes. *Journal of Hydraulic Engineering*, 134(1), 42-55.

van Rijn, L.C. (1984a). Sediment transport. Part I: bed-load transport. *Journal of Hydraulic Engineering*, 110(10), 1431-1456.

van Rijn, L.C. (1984b). Sediment transport. Part III: bed forms and alluvial roughness. *Journal of Hydraulic Engineering*, 110(12), 1733-1753.

White, W.R., Paris, E. and Bettess, R. (1979). *A new general method for predicting the frictional characteristics of alluvial streams*. Hydraulics Research Station, Report No. IT 187, Wallingford, England, July.

Winkley, B.R. (1977). *Man-made cutoffs on the Lower Mississippi River, conception, construction, and river management*. Report 300-2, U.S. Army Corps of Engineers, Vicksburg, Mississippi.

Yalin, M.S. and Scheuerlein, H. (1988). *Friction factors in alluvial rivers*. Oskar v. Miller Institut in Obernach, Technische Universitat Munchen, Bericht Nr. 59.

Yalin, M.S. (1964). On the average velocity of flow over a mobile bed. *La Houille Blanche*, No. 1, 45-51.

Chapter 4

Regime channels and their computation

4.1 INTRODUCTION

It has been realized since a long time that, as a rule, an alluvial stream does not "accept" any channel provided for it by nature or men; the stream endeavours to modify the provided channel so as to create a certain channel of "its own" – which is referred to as the *regime channel.* Under ideal conditions, the regime channel created by the stream does not vary with time any longer.

Being thus the channel a river "would like to have", the regime channel requires minimum protection, and thus minimum expenses for its maintenance. This explains why regime channels are of such an importance in river engineering, and why their prediction became such a popular research topic.

It will be assumed in this chapter (in agreement with the statements in the "Scope of the Monograph") that the alluvial channel is wide, the alluvium is cohesionless, and the flow is turbulent. Moreover, complying with convention, it will be assumed that $Q = const$, and that the (wide) alluvial channel is specified by its three characteristics B, h, S (which, in fact, imply B_{av}, h_{av}, S_{av}).

In this chapter we will be dealing only with the non-braiding ("single-channel") streams. The regime formation by braiding, i.e. by splitting of the original channel into a multitude of channels, will be considered in Chapter 5.

4.1.1 Empirical regime formulae

The study of regime channels was initiated in the first half of the 20th century by a number of British engineers working in the design and maintenance of major sand and silt irrigation canals in India and what is now Pakistan (R.G. Kennedy, E.S. Lindley, G. Lacey, C.C. Inglis, T. Blench). On the basis of the data resulting from observations and measurements in such canals, these authors developed purely empirical equations for the regime characteristics B_R, h_R, S_R and/or some other related characteristics (such as wetted perimeter, velocity, etc.).

It appears that for some decades after its introduction, the concept of regime channel was used exclusively in irrigation channel research and practice. However, by the mid-1960's, and following both analysis of river data in view of the earlier regime equations and careful observations of nature, such as those carried out by Leopold and Maddock (1953), the idea had already settled that the "principle of self-adjustment" (Blench 1966) is a fundamental principle of river behaviour, and that the related regime concept is fully applicable to real rivers. This prompted the emergence of a large number of (empirical) equations developed on the basis also of river data.

The original equations produced for B_R, h_R and S_R (in short for A_R) by the aforementioned British engineers and their numerous followers are almost invariably of the form

$$A_R = \alpha_A Q^{n_A}. \tag{4.1}$$

As will be seen in Section 4.2, a regime channel is specified by at least six characteristic parameters, and any quantity (A_R) related to this channel must thus be expected to vary as a function of these parameters. Yet, in most of the expressions of the type (4.1) proposed to date α_A and n_A are treated as some empirical constants (Chitale 1973, ASCE Task Committee 1982, Garde and Raju 1985, Yalin 1992, etc.). In spite of this, the empirical regime formulae led to the emergence of the following important relations (which must not be contradicted by the expressions of any theory):

1 B_R is practically proportional to the square root of Q in all cases:

$$B_R \sim Q^{1/2}. \tag{4.2}$$

2 The value of the Q-exponent in $h_R \sim Q^{n_h}$ for the case of sand streams does not differ considerably from that for the case of gravel streams, but nonetheless n_h appears to be affected by the grain size D ($\sim \Xi$):

$$n_h \approx 1/3 \text{ for fine sand;} \quad n_h \approx 0.43 \text{ for gravel.} \tag{4.3}$$

3 The Q-exponent in $S_R \sim Q^{n_S}$ is certainly affected by D ($\sim \Xi$):

$$n_S \approx -0.1 \text{ for fine sand;} \quad n_S \approx -0.43 \text{ for gravel.} \tag{4.4}$$

Thus, in all cases, B_R and h_R increase with the increment of Q, while S_R decreases when Q increases. Note also that the absolute values of n_h and n_S are comparable for gravel bed channels.

The validity of the statements and relations above can be inferred from the existing regime formulae (many of which are displayed in e.g. Table 4.1 in Yalin 1992 and Table 1 in Shu-you and Knight 2002). It can also be inferred from Figures 4.1a-c, showing plots of measured values of B_R, h_R and S_R, respectively, versus (bankfull) flow rate Q. These figures were produced on the basis of all the regime data available to the authors (see Refs. [1d] to [20d] in Appendix D for the data sources). The data cover measurements in 497 self-formed streams in materials ranging from fine sand to gravel (255 sand streams with $0.08mm \leq D \leq 2.0mm$; 190 gravel streams with $2.0mm < D \leq 63mm$; and 52 streams with $63mm < D \leq 175mm$). Laboratory streams, irrigation canals, and real streams of all sizes, including rivers as large as the Mississippi River, are included. Both straight (or nearly straight) as well as meandering streams are represented. [Since all points correspond to sand or gravel ($\gamma_s = 16186.5N/m^3$) and water ($\rho = 1000kg/m^3$ and $\nu = 10^{-6}m^2/s$), the values of Ξ in Figures 4.1a-c are related to D as $\Xi = 25296D$.]

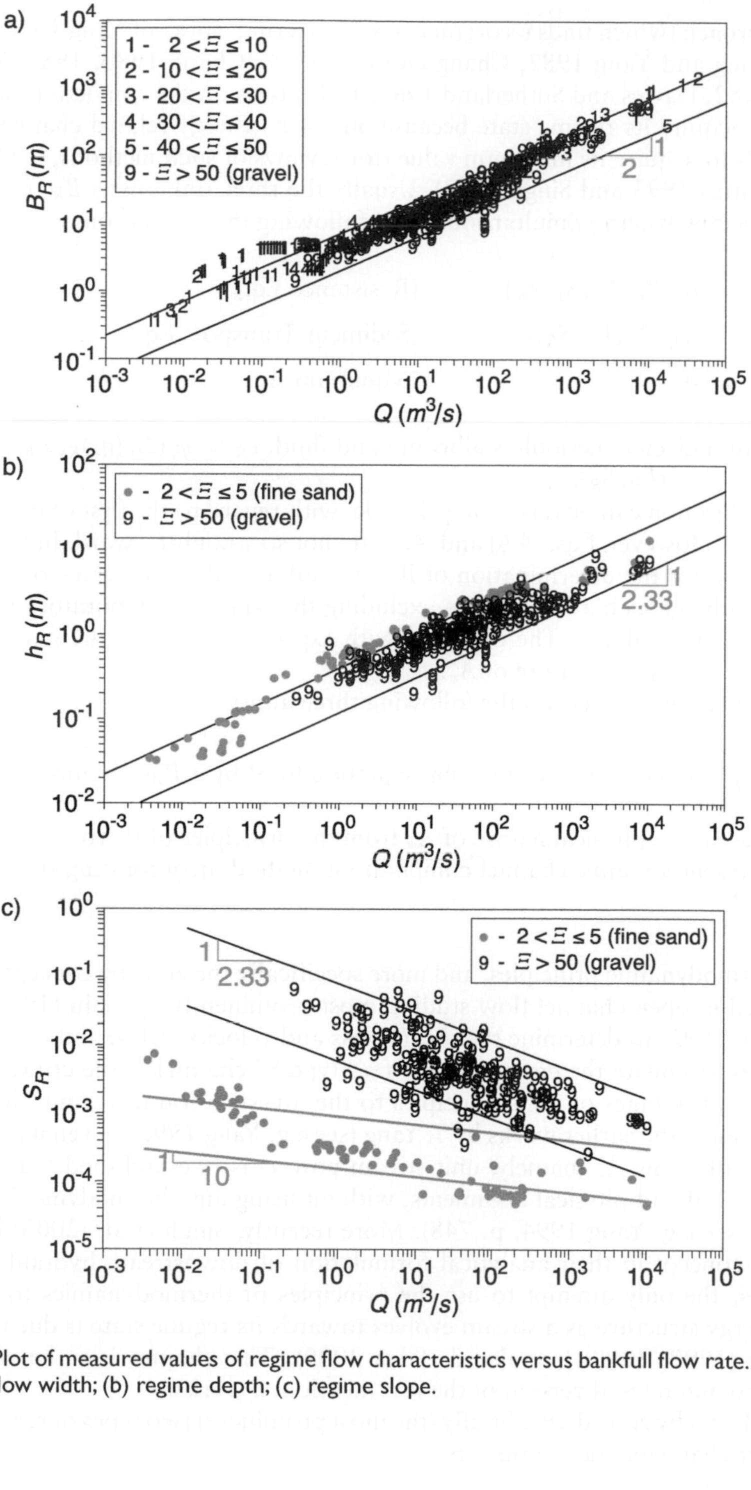

Figure 4.1 Plot of measured values of regime flow characteristics versus bankfull flow rate. (a) Regime flow width; (b) regime depth; (c) regime slope.

4.1.2 Rational methods based on extremal hypotheses

This approach (which finds its origin in the pioneering works of Yang 1976, Yang et al. 1981, Song and Yang 1982, Chang and Hill 1977, Chang 1980, 1988, White et al. 1981, 1982, Davies and Sutherland 1980, 1983) rests on the conviction that a stream tends to acquire its regime state because one of its energy-related characteristics, A_* say, tends to acquire its minimum value (for reviews of such methods, see White et al. 1986, Farias 1993 and Singh 2003). Usually the three unknowns B_R, h_R and S_R are determined by solving simultaneously the following three equations

$$Q = f_Q(B_R, h_R, S_R, c_R) \qquad \text{(Resistance Eq.)} \qquad (4.5)$$

$$Q_s = f_{Q_s}(B_R, h_R, S_R, c_R) \qquad \text{(Sediment Transport Eq.)} \qquad (4.6)$$

$$dA_* = 0 \qquad \text{(Minimum } A_*\text{)} \qquad (4.7)$$

where, for a given cohesionless alluvium and fluid, c_R $(=\phi_c(\Xi, (\eta_*)_R, Z_R))$ is a known function $c_R = f_c(h_R, S_R)$.

No objection can be raised, in principle, with regard to the first of these equations (Eq. (4.5)). However, Eqs. (4.6) and (4.7) are not so straightforward. Indeed, Eq. (4.6) can be used for the determination of B_R, h_R and S_R only if the value of Q_s is known beforehand: yet such a knowledge, excluding the controlled laboratory experiments, is not readily available. The difficulty with Eq. (4.7) is due to the uncertainty with regard to the physical nature of A_*.

The present chapter has the following three aims:

1. to replace the sediment transport equation (4.6) by a B_R-relation which does not involve Q_s;
2. to reveal the physical nature of A_* from the principles of thermodynamics;
3. to present a regime channel computation method incorporating the findings in 1 and 2.

Thermodynamic principles, and more specifically the entropy concept has already been used in open-channel flow studies, most prominently by Chiu (1991) and Chiu and Said (1995) to determine the shear stress and velocity fields; and Cao and Knight (1997) to formulate the cross-section of a "type-S" channel at the critical stage. The potential advantages of such principles to the investigation of regime channels were pointed out in the earlier works by T. Yang (see e.g. Yang 1992) – even if in later works by this author his A_* (namely, unit stream power) is *re*-established "on the basis of mathematical and physical arguments, without using any thermodynamic or entropy theory" (see e.g. Yang 1994, p. 748). More recently, Singh et al. (2003) invoked the entropy concept in their analytical formulation of downstream hydraulic geometry. However, the only attempt to use the principles of thermodynamics to analyse the flow energy structure as a stream evolves towards its regime state is due to Yalin and da Silva (1999, 2000) (see also da Silva 2009). This chapter is to be viewed as an expanded and revised version of the just mentioned works.

We begin by considering briefly (the most prominent) two types of regime channels and their characteristic parameters.

4.2 REGIME-CHANNEL DEFINING PARAMETERS

A particular regime channel is specified by some *constant* values of the characteristic parameters defining it. It is important to bear in mind that these parameters must maintain their constant values throughout the duration of the regime channel development. For the sake of simplicity, we will assume in this section that the channel remains (reasonably) straight in the course of its regime development, and that the variation of its slope S is due to the bed degradation alone.

4.2.1 Regime channel R_1

Consider the longitudinal section of the experimental set-up shown schematically in Figure 4.2a, where L_1 is the "effective length" of the channel. At the channel entrance (section I), in addition to Q, sediment is also continually fed into the channel: the feeding-rate Q_s is (just like Q) constant. If the selected $Q_s = const$ is realistic, i.e. if it can indeed be transported by the given Q, then after the passage of a certain time T_{R_1}, say, the initial channel $[B_0, h_0, S_0]$ will deform into a regime channel $[B_{R_1}, h_{R_1}, S_{R_1}]$ – which will be referred to (following Yalin 1992) as the Regime Channel R_1 (region L_1 in Figure 4.2a). If, for the same remaining conditions, the value of $Q_s = const$ is altered and the experiment is repeated, then the resulting regime channel will be different. But this means that Q_s is certainly one of the characteristic parameters of the regime channel R_1, the total set of *seven* parameters determining R_1 being

$$Q,\ Q_s,\ \rho,\ \nu,\ \gamma_s \text{ or } \upsilon_{*cr},\ D,\ g. \tag{4.8}$$

It should thus be clear that the simultaneous solution of the three equations (4.5)-(4.7) means, in fact, the determination of the regime channel R_1 (and B_R, h_R, S_R mentioned in Sub-section 4.1.2 are but the characteristics B_{R_1}, h_{R_1}, S_{R_1} of R_1).

Since the regime channel R_1 is supposed to convey both Q and Q_s, these two parameters are of equal importance, and their numerical values must be known with the same accuracy.

4.2.2 Regime channel R

Consider now the scheme in Figure 4.2b: no sediment is fed at the entrance of the channel, which is assumed to be very long. Since the fluid entering the channel (at I) is clear, while the fluid leaving the channel (at II) carries the sediment (for $S_0 > S_{cr}$), the volume of material forming the channel must continually decrease with the passage of time. This decrement manifests itself by the continual displacement of the "top of the hill" (point a_0) in the direction downwards-to the right. "Downwards", because of the regime formation ($S_0 \to S_R$) along L; "to the right", because of the erosion of material at the channel entrance (due to $dq_s/dx > 0$). Hence the effective length L *continually decreases* during $0 < t < T_R$ – and this is why the channel is assumed to be "very long" (in the sense that L/h is sufficiently large even at $t = T_R$).

Clearly, the sediment transport rate $(Q_s)_R$ conveyed by the regime-flow along L is solely due to this flow itself. Therefore $(Q_s)_R$ cannot be anything else but a certain function of S_R, h_R, and B_R – as well as of the related characteristic parameters:

$$(Q_s)_R = f_{Q_s}(B_R, h_R, S_R, \ldots). \tag{4.9}$$

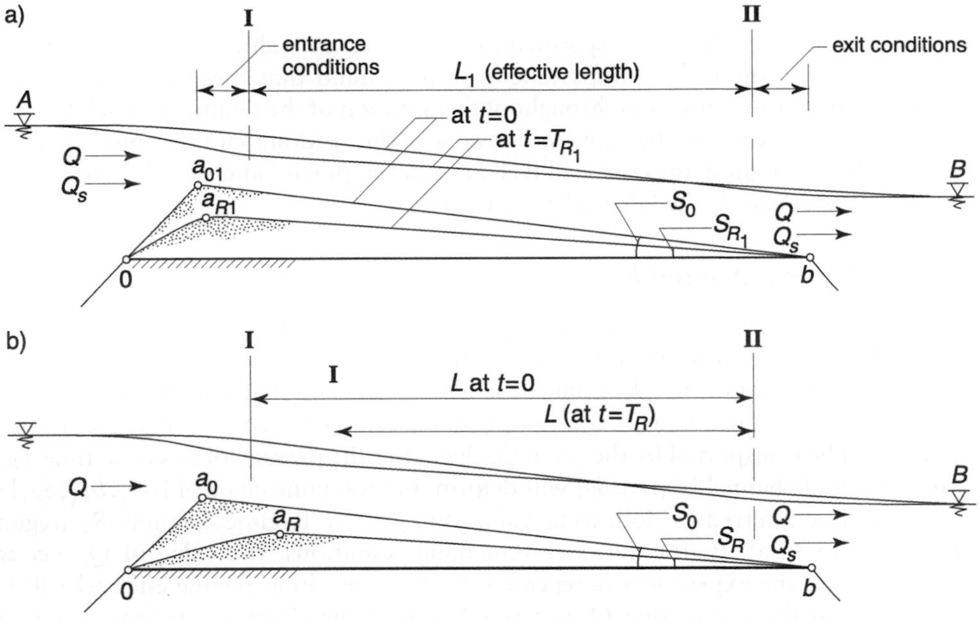

Figure 4.2 Longitudinal sections of imaginary experimental setups illustrating two different types of regime channel development. (a) Regime channel R_1; (b) regime channel R.

The regime channel described above will be referred to as the Regime Channel R. This channel is determined by the *six* characteristic parameters

$$Q,\ \rho,\ \nu,\ \gamma_s \text{ or } \upsilon_{*cr},\ D,\ g. \tag{4.10}$$

Hence, in the case of the regime channel R the sediment transport rate $(Q_s)_R$ is but an "outcome" of the channel-formation process (just like B_R, h_R, S_R, c_R, ..., etc.), and its value is thus determined as a certain function of the parameters (4.10). (The symbolic relation (4.9) is consistent with this statement, for B_R, h_R, S_R are functions of the parameters (4.10)).

At $t = T_R$ the formation of R terminates and the slope $S = S_R$ of the region L remains invariant for all $t > T_R$. Yet, the "top of the hill" (a_R) will still continue to move downstream, and the length L will still continue to decrease (for at the channel entrance I, $Q_s \equiv 0$, at any t). The further displacement of a_R, and the consequent decrement of L, can be prevented if, at a time t ($\geq T_R$), one would start to feed the sediment, at the channel entrance (I), with the rate $(Q_s)_R$; i.e. with the same (constant) rate that is passing through the channel exit II at any $t > T_R$.[1]

[1] It should be noted that the required length of a laboratory channel depends also on the difference between the slopes S_0 and S_R. The nearer is (the selected) S_0 to S_R, the shorter will be the development duration T_R, and consequently, the smaller will be the displacement of the point a_R downstream. Hence the decrement of L will also be smaller. Moreover, if the transition $S_0 \rightarrow S_R$ is by meandering rather than degradation, then the elevation of the "hill-tops" a_{01} or

4.2.3 Relation between R_1 and R

Consider the set of all regime channels R_1 which differ from each other *only* because each of them transports a different Q_s. Out of them, the channel which transports $Q_s = (Q_s)_R$ is the same as the regime channel R. In other words, R is that particular R_1 whose Q_s is $(Q_s)_R$. [It is assumed, of course, that both R and R_1 have identical values of the parameters (4.10), and that they are determined by the same resistance equation and the same A_*].

In this text we will be dealing exclusively with the regime channels R.

4.3 REGIME FLOW-WIDTH AND DEPTH

We go over now to reveal the relation which will be used in the regime channel computations *instead* of the Q_s-equation (4.6).

4.3.1 Basic dimensionless formulations

(i) Let A_R be any quantity related to the regime channel R. On the basis of (4.10), we have

$$A_R = f_{A_R}(Q, \rho, \nu, \upsilon_{*cr}, D, g). \tag{4.11}$$

Using Q, υ_{*cr} and ρ as basic quantities, one determines the set of three dimensionless variables

$$X_D = \frac{D}{\lambda}, \quad X_g = \frac{\upsilon_{*cr}^2}{g\lambda}, \quad X_\nu = \frac{\upsilon_{*cr}\lambda}{\nu}, \tag{4.12}$$

where

$$\lambda = \sqrt{\frac{Q}{\upsilon_{*cr}}} \tag{4.13}$$

is the "typical length" of the phenomenon (Yalin 1992).

Hence for the dimensionless counterpart Π_{A_R} of A_R we have

$$\Pi_{A_R} = \phi''_{A_R}(X_D, X_g, X_\nu). \tag{4.14}$$

The material number Ξ^3 implies X_{cr}^2/Y_{cr} (see Eq. (1.32)) and it can be expressed, with the aid of the relations (4.12), as

$$\Xi^3 = \frac{X_\nu^2 X_D^2}{Y_{cr}} \tag{4.15}$$

which, considering that $Y_{cr} = \Psi(\Xi)$ (see Eq. (1.33)), yields

$$X_\nu = \sqrt{\Xi^3\, \Psi(\Xi)}/X_D. \tag{4.16}$$

---

a_0 (in Figures 4.2a,b) will not change significantly, for the transition $S_0 \to S_R$ will be caused by the increment of the channel length, and not by the lowering of a_{01} or a_0.

Substituting Eq. (4.16) in Eq. (4.14), we determine the following equivalent of Eq. (4.14)

$$\Pi_{A_R} = \phi'_{A_R}(\Xi, X_D, X_g). \tag{4.17}$$

Using Eq. (4.17), one can express the regime-channel determining characteristics as

$$\Pi_{B_R} = \frac{B_R}{\lambda} = \phi'_{B_R}(\Xi, X_D, X_g) \tag{4.18}$$

$$\Pi_{h_R} = \frac{h_R}{\lambda} = \phi'_{h_R}(\Xi, X_D, X_g) \tag{4.19}$$

$$\Pi_{S_R} = S_R = \phi'_{S_R}(\Xi, X_D, X_g). \tag{4.20}$$

The following should be mentioned here. Since A_R can be *any* characteristic of the regime channel, its dimensionless counterpart Π_{A_R} can be taken as c_R or as $(Fr)_R$. One may find it strange that in the preceding chapter c and Fr were treated as functions of Ξ, η_*, N (Eqs. (3.39) and (3.44)), whereas here they (or, more precisely, their regime values) are considered as functions of Ξ, X_D, X_g. It should be noted, however, that

$$(\eta_*)_R = \frac{gS_R h_R}{\upsilon_{*cr}^2} = \frac{g\lambda h_R}{\upsilon_{*cr}^2 \lambda} S_R = \frac{\Pi_{h_R} S_R}{X_g} = \phi'_{(\eta_*)_R}(\Xi, X_D, X_g), \tag{4.21}$$

where the last step is by virtue of Eqs. (4.19) and (4.20). Similarly, considering Eq. (4.13) and the first relation of (4.12), we determine

$$N_R = \frac{Q}{B_R D \upsilon_{*cr}} = \frac{\lambda}{B_R} \cdot \frac{\lambda}{D} = \frac{1}{\Pi_{B_R}} \cdot \frac{1}{X_D} = \phi'_{N_R}(\Xi, X_D, X_g). \tag{4.22}$$

Therefore, any function of Ξ, $(\eta_*)_R$, N_R (such as c_R and $(Fr)_R$) can be considered indeed as a function of Ξ, X_D, X_g. The important aspect of any dimensionless formulation is the number and independence of the variables adopted, their composition can always be adjusted depending on the convenience of handling the topic under study (Yalin 1977, 1992).

(ii) Consider yet another version of formulation of the regime characteristics. The resistance factor c and the friction factor c_f are independent (neither of them is expressible as a function of the other alone). Hence their regime values can be expressed, on the basis of Eq. (4.17), as

$$c_R = \phi'_{c_R}(\Xi, X_D, X_g) \quad \text{and} \quad (c_f)_R = \phi'_{(c_f)_R}(\Xi, X_D, X_g), \tag{4.23}$$

which yield

$$X_D = \phi_D(\Xi, c_R, (c_f)_R) \quad \text{and} \quad X_g = \phi_g(\Xi, c_R, (c_f)_R). \tag{4.24}$$

Substituting Eq. (4.24) in Eq. (4.17), one determines

$$\Pi_{A_R} = \phi_{A_R}(\Xi, c_R, (c_f)_R), \tag{4.25}$$

which indicates that a dimensionless characteristic of a regime flow can be considered also as a function of Ξ, c_R and $(c_f)_R$.

4.3.2 The expressions of B_R and h_R

(i) Using Eq. (4.25), we can express Π_{B_R} as follows

$$\Pi_{B_R} = B_R\sqrt{\frac{v_{*cr}}{Q}} = \phi_{B_R}(\Xi, c_R, (c_f)_R) \quad (=\alpha_B). \tag{4.26}$$

As pointed out in Sub-section 4.1.1, the width B_R of all regime channels (sand and gravel) must be proportional to $\sqrt{Q}$. This proportionality will be ensured if α_B is evaluated as implied by Eq. (4.26), for none of Ξ, c_R and $(c_f)_R$ is dependent on Q explicitly. This would not be the case if α_B were expressed as implied by the relation (4.17), for X_D and X_g are explicit functions of $\lambda \sim \sqrt{Q}$.

Experimenting with the field and laboratory data of various sources, it has been found that the function (4.26) can best be expressed as

$$\alpha_B = \phi_1(\Xi) \cdot \phi_2(c_R, (c_f)_R) \tag{4.27}$$

where

$$\phi_1(\Xi) = 0.639\Xi^{0.3}, \text{ if } \Xi \leq 15; \quad \phi_1(\Xi) = 1.42, \text{ if } \Xi > 15, \tag{4.28}$$

and

$$\phi_2(c_R, (c_f)_R) = \left[n \cdot 0.2\left(1 - e^{-0.35|(c_f)_R - 27.5|^{1.2}}\right) + 1.2\right] \cdot \frac{c_R}{(c_f)_R}. \tag{4.29}$$

Here $n = +1$ if $(c_f)_R \geq 27.5$, $n = -1$ if $(c_f)_R < 27.5$, and $|(c_f)_R - 27.5|$ implies the (positive) absolute value of the difference between $(c_f)_R$ and 27.5.

(ii) Substituting Eq. (4.26) into the resistance equation (3.34), which for the regime state is to be expressed as

$$Q = B_R h_R c_R \sqrt{g S_R h_R}, \tag{4.30}$$

we determine for h_R

$$h_R = \left[\alpha_B^2 (c_R^2 S_R)\right]^{-1/3} \left(\frac{Q v_{*cr}}{g}\right)^{1/3}, \tag{4.31}$$

where $c_R^2 S_R$ is the regime value of the Froude number $Fr = c^2 S$.

Using Eq. (4.31) in conjunction with Eq. (4.26), we determine the aspect ratio of the regime channel

$$\frac{B_R}{h_R} = \left[\alpha_B^5 (c_R^2 S_R)\right]^{1/3} \left(\frac{Q g^2}{v_{*cr}^5}\right)^{1/6}. \tag{4.32}$$

No usual distinction between sand and gravel channels has been made in the derivation of the expressions above (*unified approach*). Any difference due to the "small" and "large" values of $D \sim \Xi$ is reflected by different numerical values of the variables involved – not by different forms of the functions of these variables.

Eqs. (4.26) and (4.31) were tested by using the available regime-data, as shown in Figures 4.3 and 4.4 (where, like in Figures 4.1a-c, the larger is the number of the

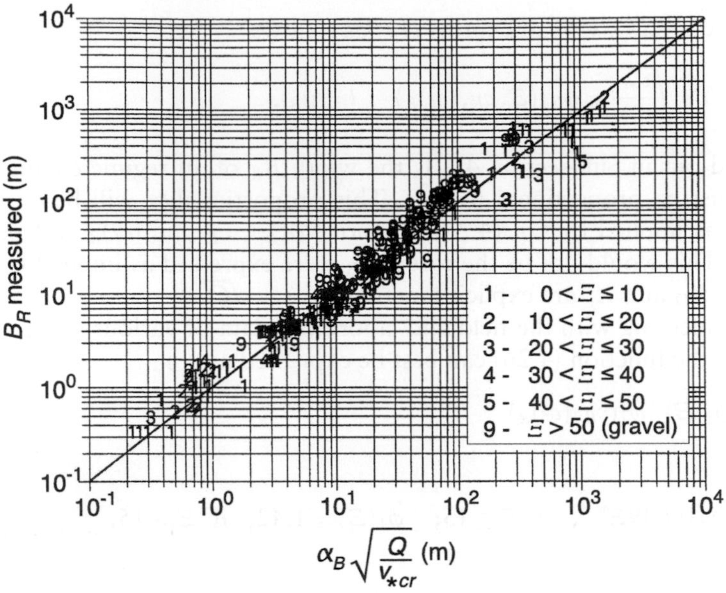

Figure 4.3 Plot of measured values of regime width versus their computed counterparts obtained from Eq. (4.26).

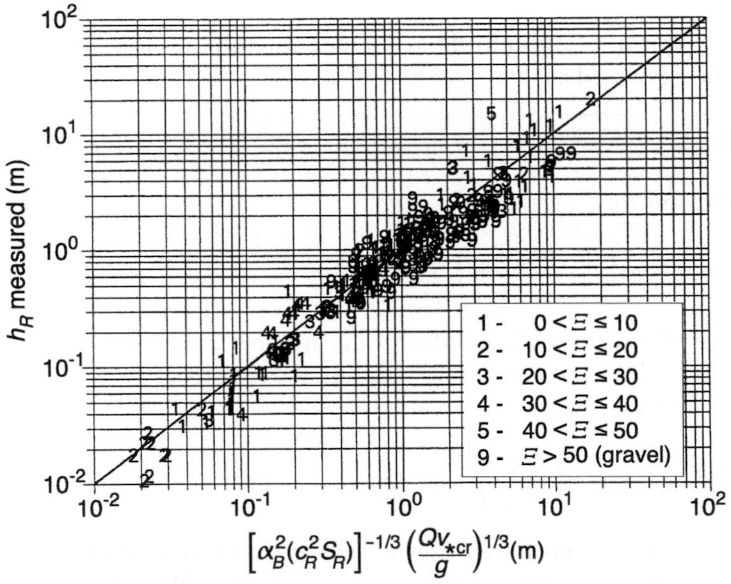

Figure 4.4 Plot of measured values of regime depth versus their computed counterparts obtained from Eq. (4.31).

point-symbol, the larger is the grain size $D \sim \Xi$). The scatter of the data-points in Figures 4.3 and 4.4 is purely experimental: the points do not exhibit any tendency to form separate lines for sand and gravel (which justifies the unified approach adopted).[2]

4.4 DETERMINATION OF THE REGIME MINIMIZATION CRITERION

As follows from Sub-section 4.1.2, no agreement has been reached yet as to what the energy-related characteristic A_* must be, and various authors propose and use various A_*. For example, $A_* = S$ according to Chang (1979, 1980), Chang and Hill (1977), Song and Yang (1982); $A_* = Su_{av}$ according to Yang (1976, 1984), Yang and Molinas (1982); $A_* = Q_s^{-1}$ according to Bettess and White (1987), White et al. (1981, 1982); $A_* = c^{-1}$ according to Davies and Sutherland (1983); $A_* = SL$ according to Yang (1987), Yang et al. (1981); etc.

4.4.1 Basic relations and assumptions

(i) A natural alluvial stream will be idealized in the following by the regime channel R – determined by the set of six characteristic parameters (4.10). The flow commences at the time $t = 0$ in a straight (unstable) *initial channel* $[B_0, h_0, S_0]$ which, with the passage of time, deforms so as to become at $t = T_R$ a (stable) *regime channel* $[B_R, h_R, S_R]$.[3] The channel is assumed to be wide at all stages of its formation. Since $Q = const$, the non-steadiness of the flow during $0 < t < T_R$ is entirely due to the time-variation of its channel. A particular stage of the regime channel development will be reflected by the (long-range) normalized dimensionless time $\Theta = t/T_R$ $(\in [0; 1])$.

(ii) A wide alluvial channel *itself* is determined by its width B and slope S. The flow depth h is a characteristic of flow – not of the "hardware" channel *per se*. Indeed, h establishes itself, depending on the existing B and S, by the fulfillment of the resistance equation.

From observations and measurements one infers that if a tranquil alluvial stream is "left to itself" (no artificial interference of any kind, no input of sediment from "external" sources), then its slope S either decreases with the passage of time, or it remains as it is – it never increases spontaneously. Similarly, the stream width B either increases or remains the same. Hence, it will be assumed throughout the present text that S *decreases*, while B *increases* during T_R.

The laboratory research (Leopold and Wolman 1957, Ackers 1964) indicates that the variation of B, h and S during T_R takes place as illustrated schematically

---

[2] In order not to overload Figures 4.3 and 4.4 with data and avoid making it difficult to distinguish the points falling under the different ranges of Ξ, only a subset of the total available regime-data is plotted in these figures. This covers a total of 397 streams (laboratory streams and real rivers). The data are derived from the following data sources in Appendix D: [1d], [2d], [3d], [4d], [5d], [6d], [7d], [9d], [10d], [11d], [12d], [13d], [14d], [16d], [17d], [19d].

[3] Here, and in this text in general, the meaning of the word "unstable" is not quite the same as its conventional meaning. This word is used more like a synonym of "unsettled", i.e. "*destined* to change by virtue of its own physical nature" (and not "*prone* to change because of a perturbation").

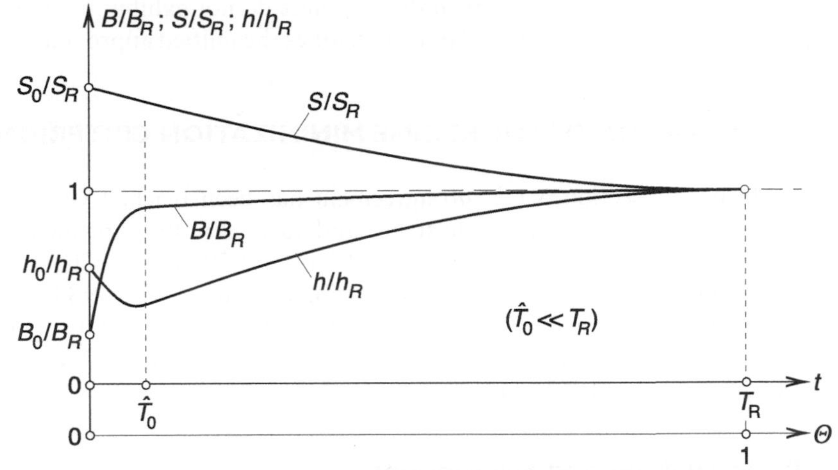

Figure 4.5 Schematic plot showing the variation with time of flow width B, flow depth h and slope S as a stream evolves towards its regime state.

in Figure 4.5. In the (very short) initial part $\hat{T}_0$ of T_R, where B and h vary substantially, while S remains nearly constant ($S \approx S_0$), no regime development as such takes place. The part $\hat{T}_0$ of T_R is merely the duration needed to alter (the arbitrary) B_0 and h_0 into such $\hat{B}_0$ and $\hat{h}_0$, say, that are (nearly) in equilibrium with the existing $S \approx S_0$. (The meaning of "equilibrium" in this context will be clarified in Sub-section 5.3.1 (ii)). The regime development in the proper sense of the word takes place only after the *adjustment period* $\hat{T}_0$. Indeed, only for $t > \hat{T}_0$ do the channel characteristics B, h, and S, and their time-derivatives, vary with t monotonously (as indicated in Figure 4.5 by the respective monotonous curves in the interval $\hat{T}_0 < t < T_R$).

In Leopold and Wolman (1957), it is reported that in more than 20 runs (with $B_0 < B_R$) "the increment of the channel width by bank erosion and the consequent change of mean depth took place, on average, in less than 5 minutes" (i.e. $\hat{T}_0 \leq\approx 5min$); although "the runs lasted not less than 5 hours and often as long as 30 hours. After the adjustment in channel width within the first 5 minutes, no subsequent change in width took place during the rest of the run" (Leopold and Wolman 1957, p. 64). These experiments indicate that $\hat{T}_0 <\approx 0.01 T_R$, i.e. that $\hat{T}_0$ is "very short" indeed. Moreover, the experiments of Leopold and Wolman (1957), where $\hat{B}_0$ is even identified with B_R, also confirm the fact, known from observations, that $\hat{B}_0$ does not differ much from B_R, and thus that the variation of B in the interval $T_R - \hat{T}_0$ (which is nearly as large as T_R itself) is rather feeble (as indicated in Figure 4.5).

Figure 4.6 shows schematically the "adjustment" of the initial channel during $\hat{T}_0$. The material eroded from the banks is deposited on the bed. Since $S \approx S_0$, the flow velocity u_{av} does not vary significantly during $\hat{T}_0$, and therefore the increment $B_0 \rightarrow \hat{B}_0$ is accompanied with the decrement $h_0 \rightarrow \hat{h}_0$. The duration $\hat{T}_0$, however short, is assumed to be sufficient for the initial bed to become covered by bed forms.

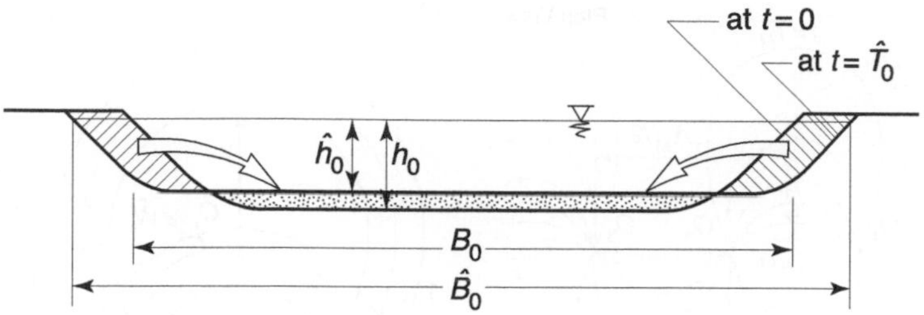

Figure 4.6 Cross-sectional stream adjustment during $\hat{T}_0$.

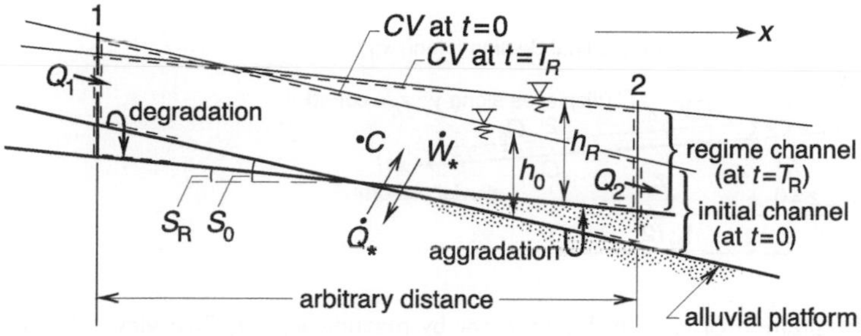

Figure 4.7 Idealization of regime development by degradation-aggradation, showing pertinent thermodynamic system (control volume CV).

The decrement of the slope S during T_R is either by degradation-aggradation (short channels; large D/h – Yalin 1992) or by meandering (long channels; small D/h – Chang 1980, Bettess and White 1983, Yalin 1992); or by both. The case of degradation-aggradation is idealized in Figure 4.7, that of meandering in Figure 4.8. (The data in Figure 4.9 indicate, though only vaguely, that the stream sinuosity σ tends to decrease, and thus that the degradation-aggradation tends to prevail, with the increment of D/h).[4] The present analysis is interpreted in terms of meandering only (although it can equally well be interpreted by (the simpler) degradation-aggradation).

(iii) No systematic displacement of the inflection points O_1, O_2, ..., O_i (see Figures 4.8a,b) is observed in plan view during the growth of meander loops. And if the meandering channel as a whole is migrating (along x with a migration velocity $\mathcal{W}_x$), then the distance between the points O_1, O_2, ..., O_i does not vary in a systematic manner. Considering this, it will be assumed that the idealized meandering channel under study consists of a series of anti-symmetrically identical loops which, with the passage of long-range dimensionless time Θ, expand, in concert, around the *fixed* inflection points O_1, O_2, ..., O_i.[5] Thus the length $L/2$ of the channel loops increases

[4]The data in Figure 4.9 are from Refs. [3e] and [32e] in Appendix E.

[5]If the channel as a whole is migrating, then the points O_i are still *fixed* with regard to an observer travelling with the velocity $\mathcal{W}_x$, and all the statements of this section are to be interpreted with reference to that observer.

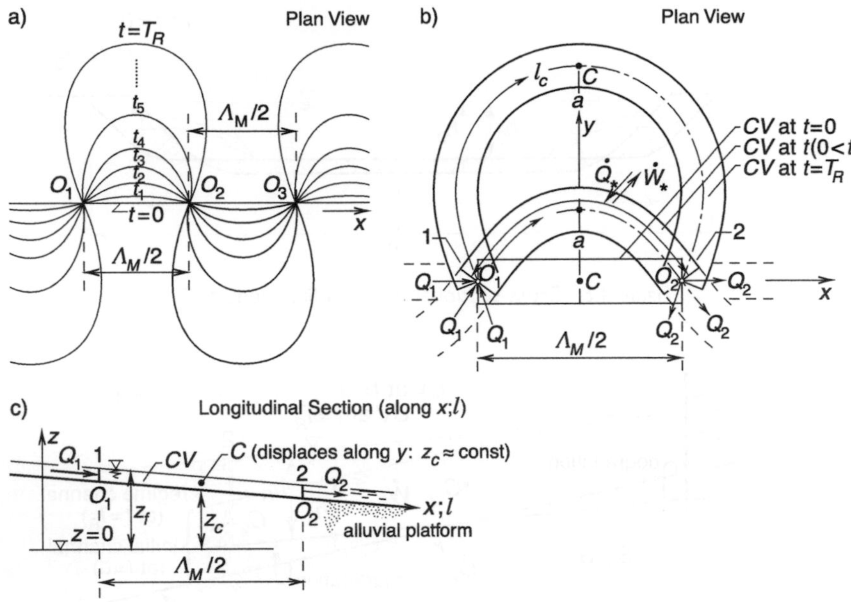

Figure 4.8 Idealization of regime development by meandering. (a) Plan view of growth of meander loops; (b) plan view of pertinent thermodynamic system (control volume CV); (c) longitudinal view of control volume CV.

and $S \sim 1/(L/2)$ decreases. The loops under study are supposed to be "far" from the upstream- and downstream-ends of the stream; hence, not only the channel loops themselves but also the flows in them are (at any Θ) anti-symmetrically identical. The analysis that follows is based on the study of only one of the (identical) loops in Figure 4.8a: we select the loop O_1O_2. The flow cross-sections passing, at any Θ, through the (fixed) points O_1 and O_2 will be referred to as the "in" and "out" sections (of the loop O_1O_2), or as sections 1 and 2, respectively.

It has already been mentioned that the flow width B does not increase significantly with time after $\hat{T}_0$. And at this stage it may also be added that at any given time t (at any development stage Θ) the value of B does not appear to vary in any systematic manner in the flow direction l_c either. This can be inferred, for example, from the aerial photo of a meandering stream near Kreole, Mississippi, in Figure 4.10, as well as from Figure 6.4; hence the reason for B to be the same for any l_c and t in Figure 4.8b.

(iv) Let u_{av} be the average flow velocity, and V the fluid volume in a meander loop: both at the same development stage Θ. The typical "flow-time" can be identified e.g. with the duration $(L/2)/u_{av}$ – needed for an average fluid particle to traverse the loop length $L/2$. As is well known, the duration of development T_R of a regime channel usually is by orders of magnitude larger than $(L/2)/u_{av}$. Considering this, it will be assumed, in accordance with reality, that the fluid travels from section 1 to section 2 in a *virtually rigid* channel ($dV/dt = 0$ during $(L/2)/u_{av}$).

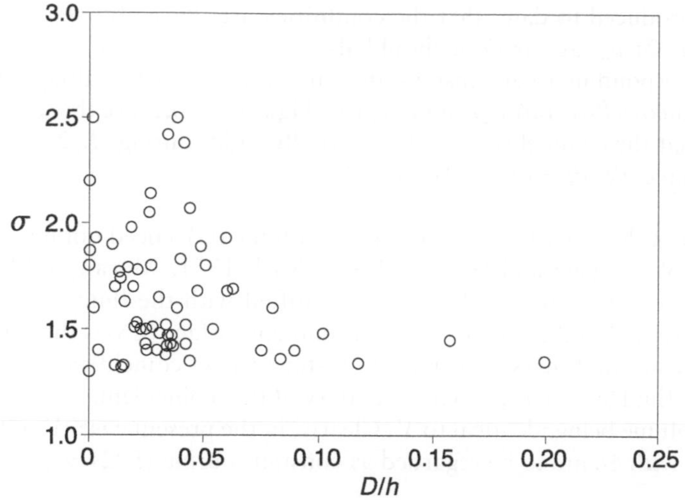

Figure 4.9 Plot of measured values of sinuosity versus the inverse of relative flow depth.

Figure 4.10 Aerial photo of a meandering stream near Kreole, Mississippi.

Owing to this reason, it will further be assumed (as in all the related research contributions produced to date) that the conditions are (virtually) *stationary* during the flow-time $(L/2)/u_{av}$ at any Θ. It should also be noted here that the usual assumption that Q, corresponding to any instant, does not vary (decrease) along x (and thus that the instantaneous flow rates Q_1 and Q_2 (see Figures 4.8b,c) are the same) means automatically that the channel is treated as virtually rigid (during $(L/2)/u_{av}$) – otherwise, we would have $dV/dt = Q_1 - Q_2 \neq 0$.

(v) The part of the space used as the control volume (CV) need not necessarily be non-deforming (Van Wylen and Sonntag 1965, Wark 1971, Moran and Shapiro 1992). And in the present analysis CV will be identified with the fluid volume V confined between sections 1 and 2 (Figure 4.8). The control surface (CS) is the surface enclosing V. The moving fluid mass, which at an instant t is in coincidence with CV, will be taken as the fluid System (*Sys*). The boundary of the (coincidental) *Sys* is thus the same as *CS*, its volume being identical to V. Clearly, in the present case, V, CV, *CS* and *Sys*, which vary with Θ, are to be regarded as constant during $(L/2)/u_{av}$ (at any Θ).

(vi) Let A_1 and A_2 be the (identical) areas of the flow sections 1 and 2, and $(CS)_a$ be that remaining part of *CS* which separates CV from alluvium and air: $(CS)_a = CS - (A_1 + A_2)$. The net heat and work-time rates "crossing" $(CS)_a$ will be denoted by $\dot{Q}_*$ and $\dot{W}_*$, respectively. The heat exchange between the stream and its surroundings does not exhibit any standard (applicable to all flows) pattern; nor there is any work exchange between the stream and its environment (in analogy to the "shaft work"). Owing to these reasons, the thermodynamic studies of flows in open or closed conduits are invariably carried out for

$$\dot{Q}_* \equiv 0; \qquad \dot{W}_* \equiv 0, \tag{4.33}$$

(see Van Wylen and Sonntag 1965, Pefley and Murray 1966, Wark 1971, Kirillin et al. 1976, etc.) and so it will be done in the following. Accordingly, it will be assumed that the channel under study is embedded in a thermodynamically neutral "alluvial platform" which is inclined along x by the "valley slope" S_0; the inclination of this platform along y is zero (Figures 4.7 and 4.8b,c).

(vii) Here we will be dealing only with those cases where the suspended-load (if present) is acquired by the flow "naturally", i.e. by its dynamic action on the mobile bed (no external sources). But this means that the suspended-load concentration C cannot exceed ≈ 0.02, say – "even in the neighbourhood of the bed" (Soo 1967, Yalin 1977). Considering this, the density ρ, the velocity u and the pressure p of the fluid-sediment mixture, at a space point, will be identified with those of a clear fluid. Accordingly, we will refer to the fluid-sediment mixture simply as "fluid".

4.4.2 Variation of flow energy structure in the flow direction l_c at an instant

(i) Before entering the determination of A_* it might be worthwhile to clarify the role of the "internal energy" with regard to the flow under study. For this purpose, we consider the *First Law* of Thermodynamics. This universal energy-conservation law can be expressed for CV and the coincidental Sys as follows

$$\frac{DE_{Sys}}{Dt} = \frac{dE_{CV}}{dt} + (\mathcal{F}_2 - \mathcal{F}_1) = \dot{Q}_* - \dot{W}_*. \tag{4.34}$$

Here E_{CV} and E_{Sys} are the total energy contents of CV and (of the coincidental) Sys. They can be expressed as

$$E_{Sys} = E_{CV} = \rho \int_V e dV, \tag{4.35}$$

where e is the total energy content per unit fluid mass (*specific total energy*) at a space-point $P(x;y;z)$. $\mathcal{F}_1$ and $\mathcal{F}_2$ signify the sum

$$\mathcal{F}_j = \int_{A_j} e_j \rho u dA_j + \int_{A_j} \frac{p}{\rho} u dA_j \quad (= \dot{E}_j + W_j) \tag{4.36}$$

which corresponds to the "in" and "out" sections A_j ($j = 1$ and 2). The first integral in Eq. (4.36), viz $\dot{E}_j$, is the *energy flux* through the cross-sectional area A_j; the second, viz W_j, is the *fluid displacement work* at A_j.

The relation (4.34) reflects the variation of the quantities involved per unit time. Since the flow is subcritical ($u_{av}^2 < gh \ll gL/2$), the average "distance" u_{av} travelled by the fluid per unit time is much smaller than the loop length $L/2$. But this means that the conditions in CV are virtually steady (Sub-section 4.4.1 (iv)), and thus that $dE_{CV}/dt = 0$. Moreover, Eq. (4.33) is valid, while the flow characteristics in A_1 and A_2 are anti-symmetrically identical (Sub-section 4.4.1 (iii)). Hence the first law (4.34) yields, for the present case,

$$(\dot{E}_2 - \dot{E}_1) = 0 \qquad \text{i.e.} \qquad e_2 - e_1 = 0 \tag{4.37}$$

and

$$\frac{DE_{Sys}}{Dt} = \frac{dE_{CV}}{dt} \qquad \text{i.e.} \qquad E_{Sys} = E_{CV} \quad (= E = const), \tag{4.38}$$

which are valid for any instant t of the flow-time $(L/2)/u_{av}$ (corresponding to any development stage Θ).

The total energy E is the sum

$$E = E_k + E_p + E_i = \rho \int_V (e_k + e_p + e_i) dV, \tag{4.39}$$

where E_k, E_p and E_i are kinetic, potential and internal energy components of E; and

$$e = e_k + e_p + e_i = \frac{u^2}{2} + gz + e_i\,. \tag{4.40}$$

Consider Figure 4.11, where s and s_f are time-average streamlines (at a stage Θ). The point-pairs such as P_1 and P_2, P_{f1} and P_{f2}, lying on the same streamlines (s and s_f) will be referred to as the "corresponding points of the cross-sectional areas A_1 and A_2". Using Eq. (4.40) for the corresponding points of the (anti-symmetrically identical)

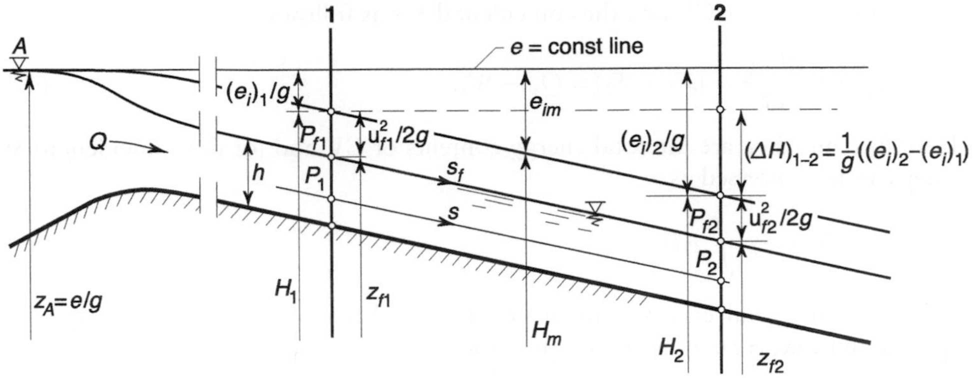

Figure 4.11 Longitudinal view of a stream originating at a lake with indication of energy components at sections 1 and 2.

areas A_1 and A_2, and taking into account that $u_1 = u_2$ while $(z_2 - z_1) = (z_{f2} - z_{f1})$, one determines from the second equation of (4.37)

$$\frac{1}{g}[(e_i)_2 - (e_i)_1] = (z_{f1} - z_{f2}) \quad [= (\Delta H)_{1-2}], \tag{4.41}$$

where $(\Delta H)_{1-2}$ is the hydraulic energy loss between sections 1 and 2. In general, e_i is not the same at all points of a flow cross-section. Yet, as is clear from Eq. (4.41), its increment $(e_i)_2 - (e_i)_1$ is the same (viz $(\Delta H)_{1-2}$) for all the corresponding point-pairs. It may also be noted here that, in analogy to z, the value of the internal energy e_i or E_i (and of the entropy s_* or S_* which will be dealt with in Sub-section 4.4.3) can be measured from any arbitrary datum. For it is only the space or time-*increments* (and not the absolute values) of z, E_i and S_* which are physically meaningful.

(ii) From the aforementioned it should thus be clear that the "hydraulic energy loss $(\Delta H)_{1-2}$" – a term used by hydraulic engineers – is, in fact, no "loss" of any kind at all. From the standpoint of physics in general and thermodynamics in particular, $(\Delta H)_{1-2}$ signifies merely an "energy-conversion". The total energy of the moving *Sys* is *conserved*, and Eq. (4.41) (which can be found also e.g. in Pefley and Murray 1966, Kirillin et al. 1976, Munson et al. 1994) merely indicates that the specific potential energy gz_f of the *down*-stream moving fluid becomes continually converted into its (specific) internal energy e_i (i.e. z_f decreases while e_i increases along l_c). This energy conversion is by means of the cumulative Joulean work done by *all* forces opposing the fluid motion within CV.[6]

The time-rate of the *increment* of the internal energy of CV, viz dE_i/dt (>0), caused by the Joulean work, compensates exactly the time-rate of its *decrement* $(\dot{E}_i)_1 - (\dot{E}_i)_2$

[6]This work includes the work of the viscous and turbulent stresses τ_{ij}, the hydrodynamic forces interacting between the transported grains and transporting fluid, the form-drag and skin friction forces interacting between the fluid and bed forms, etc.

(<0); see Pefley and Murray 1966, Kirillin et al. 1976, Munson et al. 1994, etc. Thus the "level" of the internal energy of CV is maintained (during a time interval comparable with $(L/2)/u_{av}$). (Here, $(\dot{E}_i)_1$ and $(\dot{E}_i)_2$ are the amounts of E_i entering and leaving CV per unit time, respectively).

4.4.3 Variation of flow energy structure with the passage of time Θ

In Figure 4.11 the constant value of e/g is identified with the elevation z_A of the free surface of the lake forming the source of the stream ($e/g = z_A$). No systematic variation (applicable to all cases) can be ascribed to the yearly average elevation of a natural lake. Considering this, we will treat $e = gz_A$ as a constant (during $0 < t < T_R$, i.e. $0 < \Theta < 1$). This, however, does not mean that each of the components $[e_k, e_p, e_i]$ of e can also be treated as such, and the following is an attempt to reveal how these components vary with Θ ($\in [0;1]$).

(i) Averaging Eq. (4.39) over CV, differentiating with respect to Θ, and taking into account that $e_{av} = e$ is a constant, we obtain

$$\frac{dE}{d\Theta} = \frac{d}{d\Theta}[\rho V e_{av}] = \frac{d}{d\Theta}[\rho V(e_{k,av} + e_{p,av} + e_{i,av})] =$$

$$= \frac{d}{d\Theta}\left[\rho V\left(\frac{\alpha u_{av}^2}{2} + gz_c\right)\right] + \frac{dE_i}{d\Theta} = 0 \quad (E_i = \rho V e_{i,av}). \tag{4.42}$$

Here z_c ($= z_{av}$) is the elevation of the centroid of CV, and α (≈ 1.1) is the Coriolis coefficient (which will be ignored henceforward). When a meander loop expands, its centroid C displaces in a direction close to the (horizontal) y-axis, and the more so the more symmetrical is the loop with regard to the axis of bend (see Figure 4.8). Hence it would be reasonable to consider z_b as $z_c = const$, or even as $z_c = 0$. Indeed, the z-datum is arbitrary, and it may just as well be selected so as to coincide with the (practically constant) elevation of C. This will reduce Eq. (4.42) into

$$\left(\frac{dE}{d\Theta} = \right) \quad \frac{dE_i}{d\Theta} + \frac{1}{2}\frac{d}{d\Theta}(\rho V u_{av}^2) = 0\,. \tag{4.43}$$

As is well known (see e.g. Wark 1971, Spalding and Cole 1973, Kirillin et al. 1976), the increment of the internal energy E_i of a System or CV is accompanied by the corresponding increment of its entropy S_* of that System or CV. These increments are interrelated by the Gibbs' equation, which, for the present case of an incompressible fluid, can be expressed as

$$T^{\circ}\frac{dS_*}{d\Theta} = \frac{dE_i}{d\Theta}. \tag{4.44}$$

Here, T° is the (always positive) absolute temperature.[7] Substituting Eq. (4.44) in Eq. (4.43), one determines

$$2T^{\circ}\frac{dS_*}{d\Theta} + \frac{d}{d\Theta}(\rho V u_{av}^2) = 0. \tag{4.45}$$

[7]The Gibbs' relation is valid irrespective of whether the process is "reversible or irreversible. Gravity and motion may also be present" (Spalding and Cole 1973). The fact that the fluid containing the sediment is not exactly a "pure substance" (as required by the Gibbs' equation)

(ii) Consider now the Second Law of Thermodynamics. It can be stated in a variety of (equivalent) ways, and among them we choose the "principle of the increase of entropy" (R. Clausius), which can be summarized as follows:

The Entropy of an irreversible isolated system monotonously increases with the passage of time.

In Van Wylen and Sonntag (1965) it is rigorously demonstrated that "the same general conclusion is reached (also) in the case of a control volume" (pp. 219-220).

The frictional real-fluid flow under study is obviously irreversible, and the fact that the present CV can be treated as an isolated system can be explained as follows.

(iii) The "earth-surface" can be identified with our planet's outer shell whose thickness is "small" in comparison to the earth's radius. The earth surface, together with the atmosphere around it, can be viewed as a thermodynamic system (Υ) having a double-boundary $\mathcal{B}$, which consists of two (inner and outer) spherical surfaces $\mathcal{B}_1$ and $\mathcal{B}_2$ (Figure 4.12a). No mass crosses $\mathcal{B}_1$ and $\mathcal{B}_2$, and therefore Υ is a *closed system*; no work-rate $\dot{\mathcal{W}}$ crosses the system's boundary $\mathcal{B}$ either. Consider the net heat-rate $\dot{\mathcal{Q}}$ which may cross $\mathcal{B}$. "The surface of the earth is warmed partly by the sun and partly by the heat of its interior. Surface temperatures on the earth vary daily, seasonally, and with latitude, but the overall average, about 300°K, hardly changes over hundreds and even thousands of years. Since the earth receives large quantities of energy from the sun, its average temperature would not remain so nearly constant were it not also losing energy by radiation, approximately the same amount as it receives" (Goldstein and Goldstein 1993, p. 251). Let $\dot{\mathcal{Q}}_{sol}$, $\dot{\mathcal{Q}}_{int}$ and $\dot{\mathcal{Q}}_{rad}$ be the solar, interior, and radiation heat-rates, respectively. From the aforementioned it follows that $\dot{\mathcal{Q}} = \dot{\mathcal{Q}}_{sol} + \dot{\mathcal{Q}}_{int} + \dot{\mathcal{Q}}_{rad} = 0$ can be considered to be valid (for any duration "less than thousands of years"). Substituting $\dot{\mathcal{W}} = 0$ and $\dot{\mathcal{Q}} = 0$ in the expression of the first law $\dot{\mathcal{E}} = \dot{\mathcal{Q}} - \dot{\mathcal{W}}$ (where $\dot{\mathcal{E}}$ is the time-rate of change of the energy $\mathcal{E}$ of Υ), we obtain $\dot{\mathcal{E}} = 0$ and thus $\mathcal{E} = const$. Hence the closed system Υ can be treated as an *isolated system*. And since this system is also obviously frictional, and thus irreversible, its entropy $S_{*\Upsilon}$ must continually increase with the passage of time ($\dot{S}_{*\Upsilon} > 0$).

Suppose now that Υ is divided into a large number n of the constituent sub-systems Υ_i ($i = 1, 2, ..., n$), each of which being as shown in Figure 4.12a. The fact that $\mathcal{E}$ $(= \sum_{i=1}^{n} \mathcal{E}_i)$ is time-invariant is, of course, no indication that each $\mathcal{E}_i$ must also be such (for the time increment of $\mathcal{E}_j$ can always be neutralized by the time decrement of $\mathcal{E}_k$). Yet, it would be only reasonable to treat each $\mathcal{E}_i$ as time-invariant. For, although each Υ_i is interacting with its "neighbours" and its total side boundary σ_i is crossed by the heat- and work-rates $\dot{\mathcal{Q}}_i$ and $\dot{\mathcal{W}}_i$, the long-range time average values of these rates (averaged over several years) is zero and so must thus be the long-range energy-rate $\dot{\mathcal{E}}_i$ (the climate of France does not become progressively warmer at the cost of cooling

and that $p(dV/d\Theta)$ is not exactly zero, may slightly affect the proportionality between $dS_*/d\Theta$ and $dE_i/d\Theta$ implied by Eq. (4.44). This, however, has no bearing on the deductions that will be made later on in paragraph (iv), which rest solely on the fact that $dS_*/d\Theta$ *increases* (and *not* decreases) with $dE_i/d\Theta$ – the exact amount of the increment being immaterial.

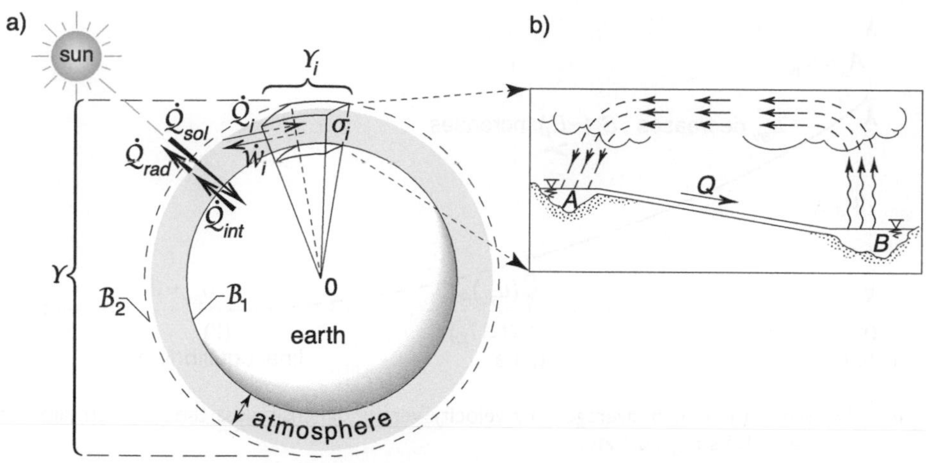

Figure 4.12 Example of a closed thermodynamic system and its constituent sub-systems. (a) System formed by the earth surface and atmosphere around it; (b) sub-system consisting of river, its in and out lakes A and B and atmospheric component of hydrologic cycle.

of the adjacent Germany). But if $\mathcal{E}_i$ of each Υ_i is on the long-range constant, then the entropy S_{*i} of Υ_i must increase with the time Θ.

A river-system, formed by its "in" and "out" lakes A and B, and its atmospheric component of hydrological cycle, can be considered as one of the above described sub-systems Υ_i (Figure 4.12b), and therefore the entropy S_{*AB} of the river AB itself, as well as that of any of its components, must be an increasing function of Θ. Identifying the conditions of the channel R under study with those of the river just mentioned, we can write for the entropy S_* of the part CV of R

$$\frac{dS_*}{d\Theta} > 0. \tag{4.46}$$

(iv) Let us now return to Eq. (4.45). The inequality (4.46) renders the first term of Eq. (4.45) positive. Consequently, its second term must be negative:

$$\frac{d}{d\Theta}\left(\rho V u_{av}^2\right) < 0. \tag{4.47}$$

Since $V = A_{av}(L/2)$ while $Q = A_{av}u_{av}$ ($= const$), we have $Vu_{av}^2 = Qu_{av}(L/2)$, which makes it possible to express Eq. (4.47) (treating ρ as constant) as

$$\frac{d}{d\Theta}\left(u_{av}L/2\right) = u_{av}\frac{d(L/2)}{d\Theta} + \frac{du_{av}}{d\Theta}(L/2) < 0. \tag{4.48}$$

Here the first term on the right-hand side of the equality sign is non-negative, for the channel length $L/2$ either increases with Θ (meandering) or it remains the same (degradation-aggradation). But this means that the second term must necessarily be negative:

$$\frac{du_{av}}{d\Theta} < 0. \tag{4.49}$$

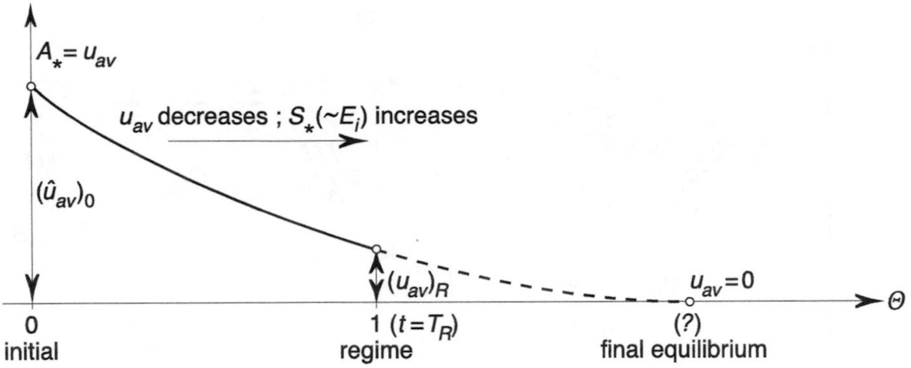

Figure 4.13 Schematic plot of average flow velocity versus dimensionless time as a stream evolves towards its regime state.

It follows that the increment of entropy is associated with the reduction of the flow kinetic energy $E_k = (\rho V u_{av}^2)/2$, and thus of its velocity u_{av}: the stream tends to alter its configuration (its slope, geometry, and effective roughness) so as to minimize its velocity u_{av}. Hence

$$A_* = u_{av}. \tag{4.50}$$

(v) If the flow "which is left to itself" (Wark 1971) were not subjected to any constraints, then the channel formation, and the time-decrement of u_{av} (time-increment of $S_* \sim E_i$) would continue until the state of the *final thermodynamic equilibrium*, viz $u_{av} = 0$, is achieved. In reality, however, the constraints (imposed by the sediment transport rate and/or bed forms) are always present, and they compel the channel formation to terminate much before the final equilibrium is reached. And it is this (earlier) state, where the channel formation factually stops (at $t = T_R$), which is commonly referred to as the *regime state* (and which is described by the *finite* characteristics $(u_{av})_R$, S_R, h_R, B_R, ..., etc.). The conditions described are illustrated schematically in Figure 4.13.

(vi) As is well known, the increment of entropy is associated with the increment of "disorder" or "randomness". In the present context, the randomness is at molecular level. The (macroscopic) internal energy e_i of each moving unit mass is the sum total of the kinetic energies of the randomly moving molecules comprising this mass and of the potential energies of their interaction. (The randomly directed paths of the molecules are straight, their random collisions are elastic).

Consider Eq. (4.43), where $E_i = \rho V e_{i,av}$ and $u_{av}^2/2 = e_{k,av}$. Ignoring the time-variation of V, we obtain from Eq. (4.43)

$$e_{av} = e_{k,av} + e_{i,av} = const. \tag{4.51}$$

This relation indicates that the progressive increment of $s_{*,av}$ (and thus of $e_{i,av}$) during T_R, i.e. the continual increment of the energy of *random* (microscopic) molecular motion, takes place at the cost of the decrement of the energy $e_{k,av} \sim u_{av}^2$ of the ordered (macroscopic) fluid motion (i.e. the time-decrement of flow velocity u_{av}, during T_R, means the increment of the molecular disorder).

It may also be noted here, that turbulent fluctuations, though random in their nature, form an "energy-input" to e_k, not to e_i. (Recall that the intensity of turbulence decreases when u_{av} decreases). [The explanations above rest on Landau and Kitaigorodskiy 1965, Landau et al. 1965, and Goldstein and Goldstein 1993.]

4.4.4 Froude number; comparison with experiment

(i) Eliminating h between the continuity and resistance equations, i.e. between $Q = u_{av}Bh$ and $Q = Bhc\sqrt{gSh}$, and taking into account that c^2S implies the Froude number $Fr = u_{av}^2/gh = Q^2/(Bh)^2gh$, one obtains

$$\frac{u_{av}^3}{Fr} = \frac{gQ}{B}. \tag{4.52}$$

At the later stages of regime development, B can be identified with $B_R = const$ (see Sub-section 4.4.1 (ii)), which gives the possibility to express the time-derivative of Eq. (4.52) as

$$3\frac{1}{u_{av}}\frac{du_{av}}{d\Theta} = \frac{1}{Fr}\frac{dFr}{d\Theta}. \tag{4.53}$$

This relation indicates that, at the later stages, the relative time-decrement of u_{av} can be considered to be related to that of Fr by a constant proportion, and thus that u_{av} and Fr achieve their minima together ($dFr/d\Theta = 0$ means $du_{av}/d\Theta = 0$). Hence, in the following, the (familiar) Froude number will be adopted as the dimensionless counterpart of A_* (as has already been done in Jia 1990, Yalin 1992, Yalin and da Silva 1997):

$$\Pi_{A_*} = Fr \tag{4.54}$$

(If, however, the physical meaning is disregarded, and the non-dimensionalization is viewed from the mathematical standpoint alone, then, of course, any dimensionless combination which monotonously increases with u_{av}, and whose Θ-derivative becomes zero when $du_{av}/d\Theta$ becomes zero (such as e.g. u_{av}/υ_{*cr}, $u_{av}D^2/Q$, ..., etc.) can be taken as Π_{A_*}).

(ii) Consider the function $Fr = \phi_{Fr}(\Xi, \eta_*, N)$ introduced in Sub-section 3.1.5. For a specified granular material and fluid, Ξ is a constant, and Fr reduces into the curve-family $Fr = \psi_{Fr}(\eta_*, N)$, say. Figure 4.14 shows schematically the geometric nature of this curve-family: η_* is the abcissa, each "Fr-curve" corresponds to a constant value of N – the larger is N $(= Q/(BD\upsilon_{*cr}) = Z(u_{av}/\upsilon_{*cr}))$, the lower is its Fr-curve.

The Fr-curves situated below the "boundary curve" Fr_* will be referred to as "sand-like" Fr-curves. These curves exhibit "dips", i.e. they possess the proper minima (in the sense $\partial Fr/\partial\eta_* = 0$) at some points P_s having $\eta_* \gg 1$ (and forming a line ("valley") L_s). These dips are solely due to bed forms (ripples and/or dunes), and therefore they become less and less prominent when $Z = h/D$, and thus N decreases. The Fr-curves

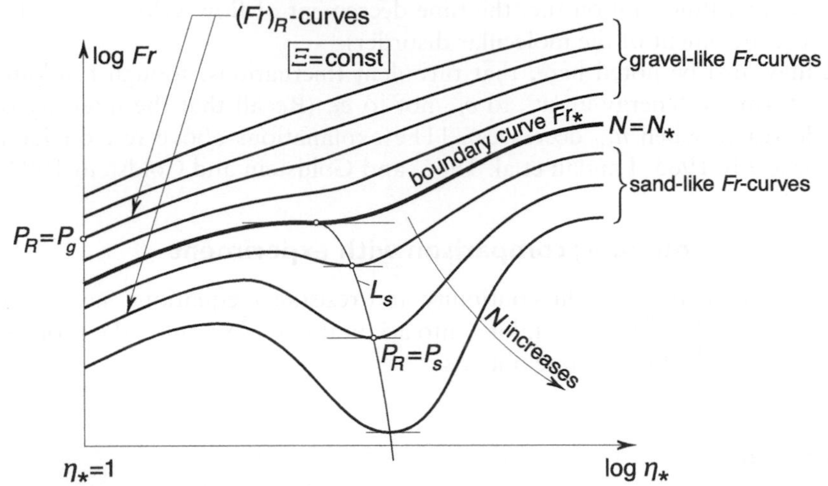

Figure 4.14 Schematic representation of sand- and gravel-like *Fr*-curves on the $(Fr; \eta_*)$-plane.

which are situated above the boundary curve Fr_*, and which thus do not have dips, will be referred to as "gravel-like" *Fr*-curves. The smallest ordinates *Fr* of the gravel-like curves are at $\eta_* = 1$ (points P_g).

A regime channel can be represented on the $(Fr; \eta_*)$-plane by one "regime point" P_R. The *Fr*-curve passing through the point P_R is "its" $(Fr)_R$-curve – which can be sand-like or gravel-like. The regime point P_R *must* be in coincidence either with a point P_s or with a point P_g:

1 If the regime channel formation is sand-like, then $P_R = P_s$. In this case $(\eta_*)_R \gg 1$ and thus $(q_s)_R \gg 0$ (transport is present: "live-bed" regime channel).
2 If the regime channel formation is gravel-like, then $P_R = P_g$. In this case $(\eta_*)_R = 1$ and thus $(q_s)_R = 0$ (no transport: "threshold" regime channel).

The reason for the "coincidences" $P_R = P_s$ and $P_R = P_g$ is explained, in association with the content of Sub-section 4.4.3 (iv), in the next chapter (Sub-section 5.1.3 (i)).

Figures 4.15a,b show the *Fr*-curves corresponding to $\Xi = 7.6$ (fine sand) and $\Xi = 1265$ (gravel), together with the regime data of all the sources available to the authors (Refs. [1d] to [20d] in Appendix D). Note that in both cases the regime data converge indeed to the locations of the smallest *Fr* (to the "valley" L_s in Figure 4.15a, and to $\eta_* \to 1$ in Figure 4.15b).

It should be pointed out here that, in addition to all the usual reasons for the scatter in sediment related plots, the regime-plots involve also the subjectivity in judgment on whether the stream at hand is or is not in its regime state. More often than not, it is simply assumed that if the channel appears to be no longer deforming, during the "sufficiently long period of time" (selected by the observer), then it is in regime. Hence the regime-data stemming from an individual source cannot be considered as reliable,

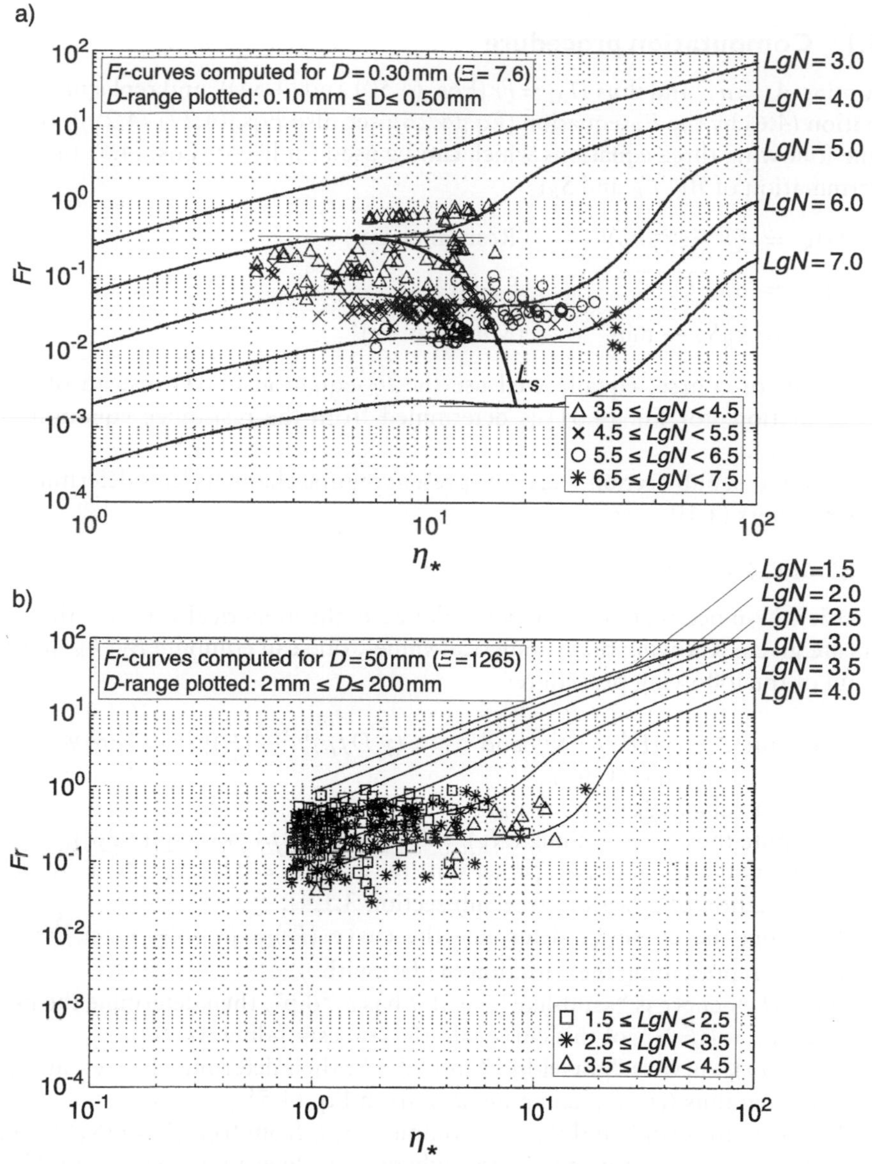

Figure 4.15 Graphs of *Fr*-curve families (using the dimensionless specific flow rate *N* as curve-classifying parameter) plotted together with the data, for two different values of the material number, namely: (a) 7.6 (fine sand); 1265 (gravel).

and it is only the average behaviour of the points of a very large number of sources and streams, i.e. the average behaviour of the "cloud of points", which can be relied upon. This explains why one should use *all the available* regime-data to test a regime hypothesis or formula.

4.5 COMPUTATION OF REGIME CHANNELS R

4.5.1 Computation procedure

Adopting $A_* = u_{av}$ and thus $\Pi_{A_*} = Fr$ (Eqs. (4.50) and (4.54)), and replacing the $(Q_s)_R$-equation (4.6) by the B_R-equation (Eq. (4.26); see also Eqs. (4.27)-(4.29)), one arrives at the following three equations which can be used (in lieu of Eqs. (4.5)-(4.7)) for the determination of B_R, h_R and S_R:

$$(Fr)_R = Q^2/(gB_R^2 h_R^3) \qquad \text{(Resistance Eq.)} \qquad (4.55)$$

$$B_R = \alpha_B \sqrt{Q/v_{*cr}} \qquad (B_R\text{-Eq.}) \qquad (4.56)$$

$$(Fr)_R = c_R^2 S_R \to \min \qquad \text{(Minimum } Fr\text{)} \qquad (4.57)$$

(We could, of course, equally well use the h_R-equation (4.31) instead of the resistance equation, for Eq. (4.31) is determined from the resistance equation and the B_R-equation).

As stated in Sub-section 4.2.2, the present regime channel (R) is determined by the six parameters (4.10), viz

$$Q,\ \rho,\ \nu,\ \gamma_s,\ D,\ g. \qquad (4.58)$$

Hence it is assumed that, for each particular case, the numerical values of these parameters are given. Knowing these numerical values, one can compute B_R, h_R and S_R with the aid of the following procedure:

1 Compute $\Xi = (\gamma_s D^3/\rho\nu^2)^{1/3}$ and $v_{*cr} = (\gamma_s D/\rho)^{1/2}[\Psi(\Xi)]^{1/2}$ (where $\Psi(\Xi)$ is given by Eqs. (1.33) and (1.34)).
2 Adopt a value, $(B_R)_j$ say, as B_R.
3 Compute the corresponding $(N_R)_j = Q/((B_R)_j D v_{*cr})$ (see Eq. (3.36)).
4 Knowing thus Ξ and $(N_R)_j$, determine the curve $c_j = \phi_c(\Xi, \eta_*, (N_R)_j)$ representing the variation of c_j with η_* (see Sub-section 3.1.4).
5 Knowing the curve c_j, determine the curve $(Fr)_j = (\alpha/(N_R)_j)(c_j^2\eta_*)^{3/2}$ (see Eq. (3.43)) representing the variation of $(Fr)_j$ with η_*.
6 If the $(Fr)_j$-curve is "sand-like", i.e. if it has a "dip", then determine the minimum $((Fr)_j)_{min} = ((Fr)_R)_j$ of this curve and the corresponding $((\eta_*)_R)_j$.
7 If the $(Fr)_j$-curve is "gravel-like" (no "dip"), then determine $((Fr)_R)_j$ at $(\eta_*)_R = 1$.
8 Knowing thus $((Fr)_R)_j$, compute $(h_R)_j$ from Eq. (4.55).
9 Knowing thus $(h_R)_j$ and $((\eta_*)_R)_j$, compute $(S_R)_j$ from $((\eta_*)_R)_j = g(S_R)_j(h_R)_j/v_{*cr}^2$.
10 Compute $(B_R)_{j+1}$ from Eq. (4.56), where α_B is determined by Eqs. (4.27)-(4.29). The values of c_f and c which appear in Eqs. (4.27)-(4.29) are given by Eqs. (3.4) and (3.12), respectively.
11 If $(B_R)_{j+1} = (B_R)_j$, then the problem is solved: $B_R = (B_R)_j$ (and thus $h_R = (h_R)_j$ and $S_R = (S_R)_j$).
12 Otherwise repeat the procedure until $(B_R)_{j+1} = (B_R)_j$ is achieved.

The Fortran and MATLAB versions of the computer program BHS-STABLE, which computes the regime characteristics as described above, can be downloaded from the CRC Press website.

Table 4.1 Examples of results of computation of regime channel characteristics

	B_R (m)	h_R (m)	S_R ($\times 1000$)	N_R	$(Fr)_R$	$(\eta_*)_R$	c_R
Example 1: $Q=1669.7m^3/s, D=0.18mm$ (Bhagirathi River)							
Computed:	234.8	6.37	0.086	$10^{6.5}$	0.020	28.6	15.2
Reported:	218.1	5.95	0.058	$10^{6.5}$	0.028	18.6	22.1
Example 2: $Q=15.6m^3/s, D=0.33mm$							
Computed:	18.29	1.25	0.325	$10^{5.3}$	0.038	19.0	10.8
Reported:	18.10	1.24	0.200	$10^{5.3}$	0.039	12.0	14.1
Example 3: $Q=141.4m^3/s, D=0.5mm$ (Beaver River)							
Computed:	66.16	2.57	0.195	$10^{5.4}$	0.0276	20.0	11.9
Reported:	54.90	2.74	0.240	$10^{5.5}$	0.0329	25.5	11.7
Example 4: $Q=13252.0m^3/s, D=0.656mm$ (Mississippi River)							
Computed:	724.7	15.6	0.049	$10^{6.2}$	0.0089	22.6	13.5
Reported:	532.2	15.2	0.077	$10^{6.3}$	0.0180	31.8	15.3
Example 5: $Q=848.9m^3/s, D=0.8mm$ (Savannah River)							
Computed:	155.6	4.55	0.198	$10^{5.5}$	0.0322	20.2	12.8
Reported:	106.7	5.18	0.110	$10^{5.6}$	0.0500	11.6	20.5
Example 6: $Q=4386.0m^3/s, D=31mm$ (North Saskatchewan River)							
Computed:	242.6	6.79	0.340	$10^{3.6}$	0.1070	1.0	17.7
Reported:	244.0	7.62	0.350	$10^{3.6}$	0.0740	1.12	14.6

Table 4.1 (corresponding to $\gamma_s/\gamma=1.65$, $\nu=10^{-6}m^2/s$) shows some examples of computed results.[8] The results of the present method of determination of regime channel characteristics are further illustrated by applying the method to the totality of the available regime data (see Sub-section 4.1.1 for a brief description of the data). As is the case for the examples in Table 4.1, here too the computations were carried out with the program BHS-STABLE. The resulting plots of computed versus measured values of regime flow width B_R, regime flow depth h_R and regime slope S_R are shown in Figures 4.16a-c, respectively. Each of these plots contains the points ranging from fine sand to gravel. Like in Figure 4.3, the larger is the number of the point-symbol, the larger is the grain size. As follows from Figures 4.16a-c, the present method yields a satisfactory agreement between the computed regime channel characteristics and their measured counterparts.

4.5.2 Special case: gravel-like regime channel formation

The complexity in the computation of B_R, h_R and S_R is entirely due to the sand-like behaviour where $(\eta_*)_R$ and c_R are determined by the transcendental expressions (and

[8]The examples in Table 4.1 are from works listed in Appendix D, namely: [6d] (Examples 1, 3 and 5); [5d] (Example 2); [4d] (Example 4); and [12d] (Example 6).

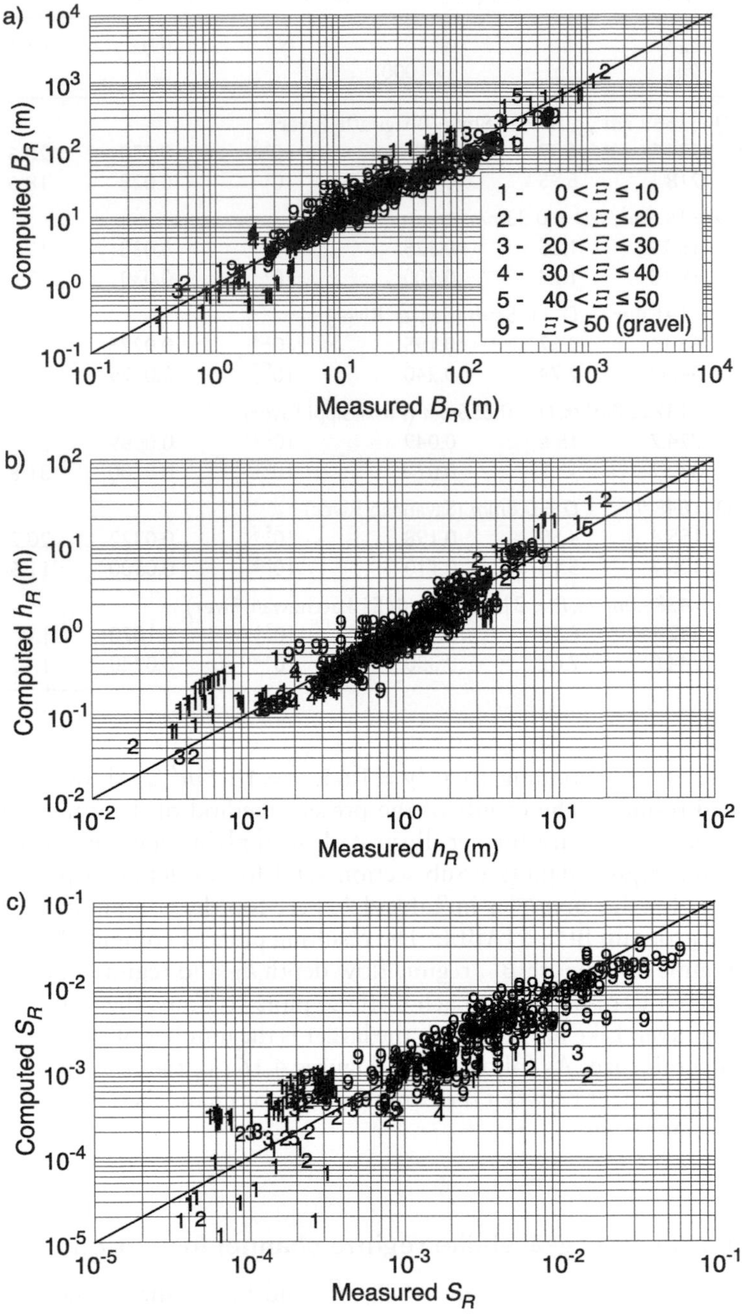

Figure 4.16 Plots of computed versus measured values of regime channel characteristics using the present method of computation of regime channels (Sub-section 4.5.1). (a) Flow width; (b) flow depth; (c) slope.

one has to resort to "trial and error"). No difficulty of this kind is present in the case of the gravel-like behaviour where we have

$$(\eta_*)_R = 1 \quad \text{i.e.} \quad S_R = \frac{\upsilon_{*cr}^2}{gh_R} = \frac{\gamma_s}{\gamma}\frac{D}{h_R}Y_{cr}, \tag{4.59}$$

and where the resistance due to (very flat) bed forms often can be ignored:

$$c_R \approx (c_f)_R \approx 7.66\left(\frac{h_R}{k_s}\right)^{1/6} \approx 6.82\left(\frac{h_R}{D}\right)^{1/6} \tag{4.60}$$

(where the second and third steps are with the aid of Eqs. (1.13) and (1.4), respectively). Moreover, in gravel channels the viscous influence is negligible, while $(c_f)_R <\approx 20$ is invariably valid. Hence Eqs. (4.28) and (4.29) give

$$\phi_1(\Xi) = 1.42; \quad \phi_2(c_R, (c_f)_R) = c_R/(c_f)_R, \tag{4.61}$$

where $(c_f)_R \approx c_R$ (by virtue of Eq. (4.60)). Using these relations in Eq. (4.27), we obtain $\alpha_B \approx 1.42$ ($\approx const$), and consequently

$$B_R \approx 1.42\sqrt{\frac{Q}{\upsilon_{*cr}}}. \tag{4.62}$$

Substituting the values of B_R, c_R and S_R given by Eqs. (4.62), (4.60) and (4.59) in the resistance equation (4.30), viz

$$Q = B_R h_R c_R \sqrt{g S_R h_R}, \tag{4.63}$$

(and replacing $\approx$ by $=$), we obtain for h_R the dimensionally homogeneous expression

$$h_R = \frac{D^{1/7}}{7.0}\left(\frac{Q}{\upsilon_{*cr}}\right)^{3/7} \quad \text{(where } 7.0 = (1.42 \times 6.82)^{6/7}\text{)}, \tag{4.64}$$

which yields for S_R, with the aid of (4.59),

$$S_R = \frac{\upsilon_{*cr}^2}{g} \cdot \frac{(7.0)}{D^{1/7}}\left(\frac{\upsilon_{*cr}}{Q}\right)^{3/7}. \tag{4.65}$$

Observe that the Q-exponents of h_R and S_R are $3/7 \approx 0.43$ and -0.43, respectively; i.e. they are exactly as they should be for gravel regime channels (see Eqs. (4.3) and (4.4)).

It follows that if it is known that the conditions are fully rough ($Y_{cr} = 0.045$; $\upsilon_{*cr} = ((\gamma_s/\gamma)gDY_{cr})^{1/2} = (1.65 \times 9.81 \times 0.045 \times D)^{1/2} = 0.853D^{1/2}$) and that the steepness of bed forms is negligible ($c_R \approx (c_f)_R$), then B_R, h_R and S_R can indeed be computed directly (without trial and error) from Eqs. (4.62), (4.64) and (4.65). Unfortunately, more often than not, such a priori knowledge is not available, and one has no alternative but to use the computation procedure in Sub-section 4.5.1.

4.5.3 Special case: constant flow width

Consider now the simplest case of the regime formation, namely the case of constant flow width B (a very long flume with a mobile bed, alluvial stream with protected banks, etc.). In this case, only S and h vary as to acquire their regime values. In the present case of the regime channel R (determined by the relations (4.55)-(4.57)), Eq. (4.56), which gives the regime value of B, looses its meaning and we are left with the resistance equation (4.55) and the minimization of A_* (Eq. (4.57)) for the computation of S_R and h_R.

How would we go about this problem in the case of the regime channel R_1? Since, in this case, both Q and Q_s are the characteristic parameters (which possess some given constant values), the flow *must* adjust its S and h so as to convey these given Q and Q_s. But this means that S_{R_1} and h_{R_1} *must* be computed from the resistance and sediment transport equations (Eqs. (4.5) and (4.6)). The minimization aspect of the regime-formation *cannot be used* in the case of R_1, if B is kept constant.

It follows that if $B = const$, then R reduces into the "answer" to the question, "For which values of S and h will the flow be able to convey a specified Q with the smallest flow velocity u_{av} $(= A_*)$?". Yet, R_1 becomes the answer to "For which values of S and h will the flow be able to convey the specified Q and Q_s?".

4.6 OTHER RATIONAL METHODS OF COMPUTATION OF REGIME CHANNELS

Rational methods for the determination of regime channels not based on the minimization of an energy-related quantity are also available in the literature. Such methods rest on the belief that the stable channel geometry acquired by a stream can be explained entirely on the basis of the equations of fluid motion, and the interaction of the moving fluid with the movable boundary. However, as highlighted by the recent work by Wilkerson and Parker (2011), there is no agreement on the appropriate theoretical framework. Indeed, wile the resistance equation and an expression related to sediment transport are invariably invoked in such approaches, additional equations result from e.g. the consideration of bank stability (see e.g. Eaton et al. 2004), secondary flows (e.g. Julien and Wargadalam 1995, Lee and Julien 2006), etc. Unfortunately, at present it is not possible to compare the different methods on the basis of an extensive regime dataset – as information on sediment load, differences between the bed and bank materials, bank strength, vegetation, etc. (as required for the practical application of different methods) is available for only few cases. For this reason, in the following we consider exclusively the method by Julien and Wargadalam (1995).

According to these authors:

$$B_R = 0.512 Q^{(2m+1)/(3m+2)} D^{(-4m+1)/(6m+4)} Y^{(-2m-1)/(6m+4)} \tag{4.66}$$

$$h_R = 0.133 Q^{1/(3m+2)} D^{(6m-1)/(6m+4)} Y^{-1/(6m+4)} \tag{4.67}$$

$$S_R = 12.4 Q^{-1/(3m+2)} D^{5/(6m+4)} Y^{(6m+5)/(6m+4)} \tag{4.68}$$

where $m = 1/\ln(12.2h/D)$. These equations presuppose the *a priori* knowledge of Y, and thus cannot provide a solution to the problem if only Q and D are specified. Julien and Wargadalam (1995) suggest that: "Should the value of the Shields parameter (Y) be unknown, the hydraulic geometry of a stable channel design can be calculated after assuming the threshold of bed material...", i.e. by taking $Y = Y_{cr}$ (or $\eta_* = 1$) – a condition which is strictly valid for gravel streams. As is well-known, and as follows from the content of this chapter, in the case of sand streams Y differs considerably from Y_{cr}. It thus seems worthwhile to investigate the nature of the results produced by the equations by Julien and Wargadalam (1995) when Y is identified with Y_{cr} ($\eta_* = 1$) for the case of gravel streams, and with $10Y_{cr}$ ($\eta_* = 10$) for the case of sand streams. Here $\eta_* = 10$ is to be viewed as a rough estimate of the average value of Y in sand streams at the regime state (see Figure 4.15a). The resulting plots of computed versus measured values of B_R, h_R and S_R are shown in Figures 4.17a-c, respectively (the data in these plots are the same as plotted in Figures 4.16a-c; the symbols 1 to 9 also have the same meaning).

Figures 4.18a,b provide a comparison between the values of h_R calculated on the basis of the equations by Julien and Wargadalam (1995) and the present method of computation of regime channels. Figure 4.18a corresponds to sand streams; Figure 4.18b, to gravel streams. Figure 4.18c is a similar plot, but for the slope S_R and including the totality of the data (sand and gravel). As can be inferred from these figures, the two methods under consideration, in general, yield rather comparable results for the case of gravel streams. However, substantial differences occur for the case of sand streams. On the basis of Figures 4.17a-c and 4.18a-c, it is concluded that the method of Julien and Wargadalam (1995) produces reasonable results of B_R and h_R for the case of sand streams when Y is simply identified with $10Y_{cr}$. Yet, the same cannot be said with regard to the stream slope.

4.7 REPRESENTATIVE FLOW RATE

The purpose of this chapter is to indicate how to compute a regime channel for a given constant Q, and not to indicate how this (representative) Q is to be determined from the time-varying flow rate Q_t of a natural stream. However, the topic is of a great practical interest (see e.g. White et al. 1986, Radecki-Pawlik 2015), and the following is an outline of the authors' view on it.

Since $Q_t = f(t)$ emerges as a result of records carried out for many years (T), it can thus be viewed as a sample of a quasi-periodic random function which extends over $0 < t < T$.

Although following the prevalent trend, the authors would recommend to identify Q with the bankfull flow rate Q_{bf}, i.e. with that ordinate of $f(t)$ which yields $Q_t > Q_{bf}$ for the duration $\Delta t = 0.006T$ (Wolman and Leopold 1957, Nixon 1959), they do not think that this method forms an adequate answer to the question "What is the representative Q?". Indeed:

1 There is no evidence that the (practically instantaneous) bankfull flow rate of an alluvial channel, having a time-varying Q_t, would be able to produce the same channel if it were to flow (in an arbitrary initial channel) as a constant flow rate.

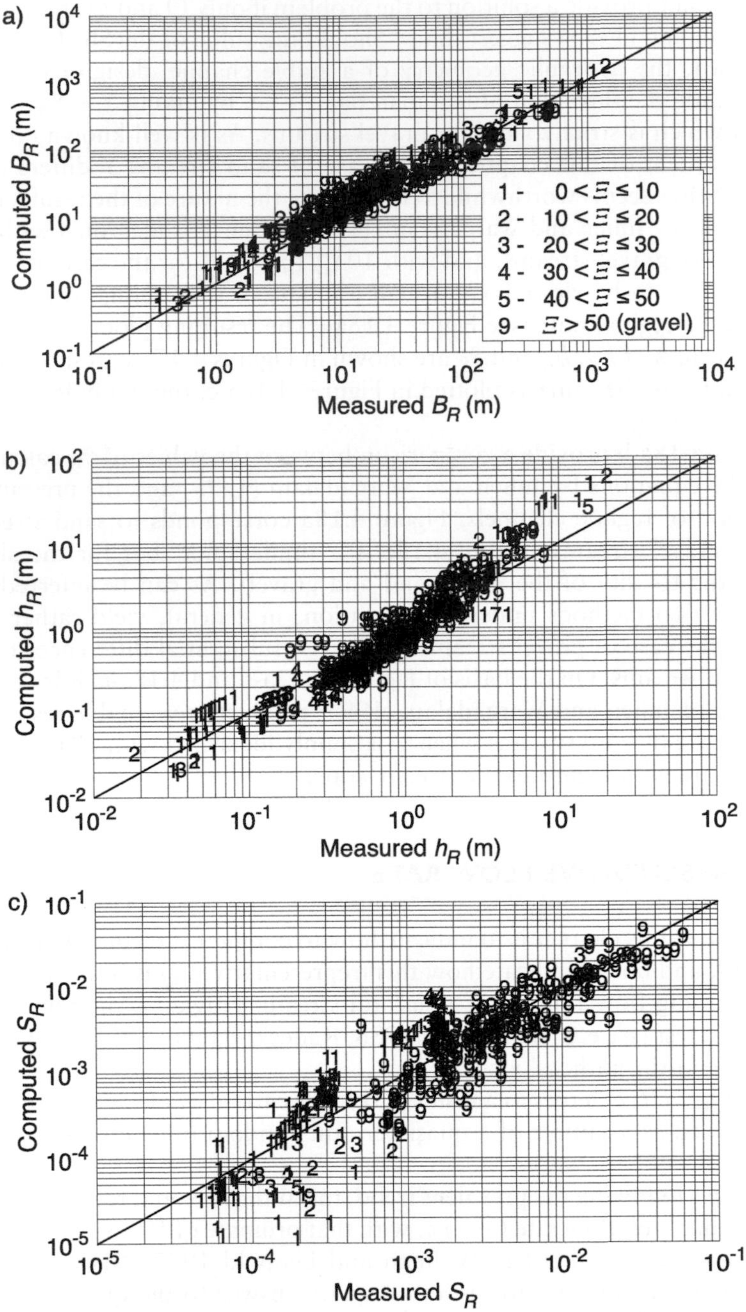

Figure 4.17 Plots of computed versus measured values of regime channel characteristics using the equations by Julien and Wargadalam (1995) (by taking $Y = Y_{cr}$ for the case of gravel streams, and $Y = 10Y_{cr}$ for the case of sand streams. (a) Flow width; (b) flow depth; (c) slope. Legend in (a) applies also to (b) and (c).

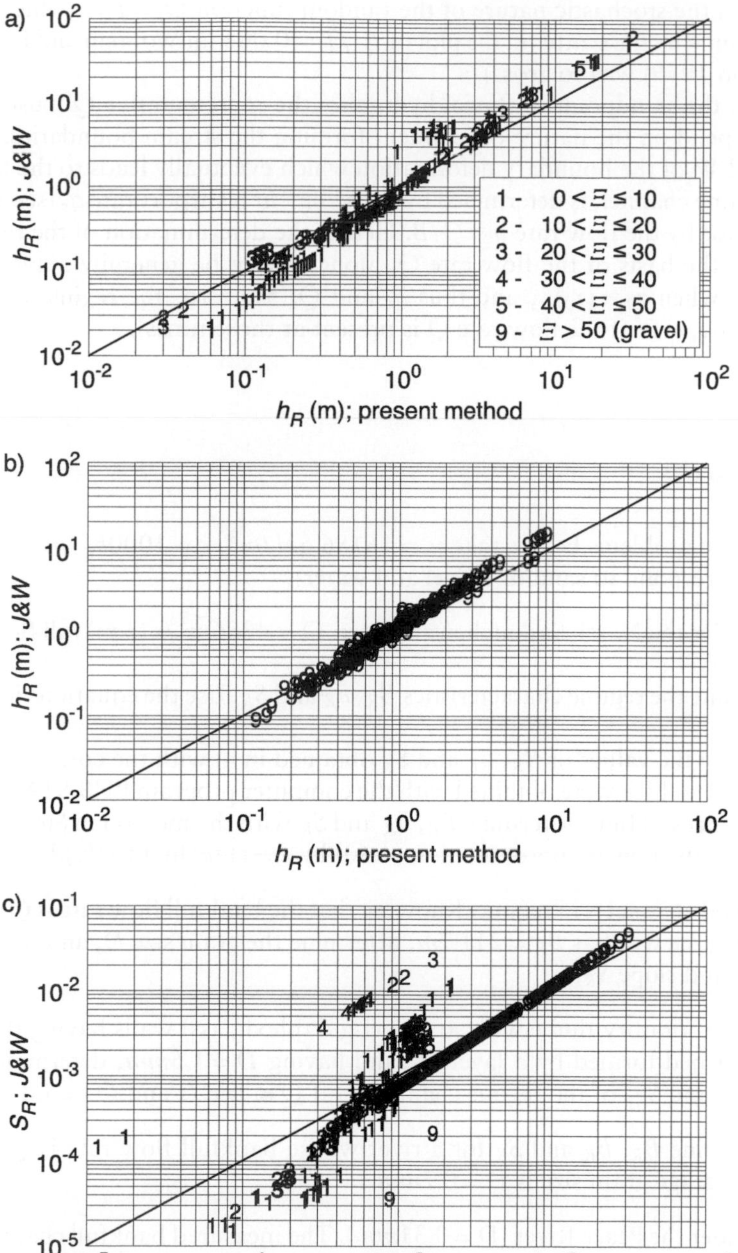

Figure 4.18 Plots of computed values of regime channel characteristics using the equations by Julien and Wargadalam (1995) (with $Y = Y_{cr}$ for the case of gravel streams, and $Y = 10Y_{cr}$ for the case of sand streams) versus those computed using the present method. (a) Flow depth (sand streams); (b) flow depth (gravel streams); (c) slope. Legend in (a) applies also to (b) and (c).

2 From a mathematical standpoint, the ratio $\Delta t/T$ must be expected to vary depending on the stochastic nature of the random function $Q_t = f(t)$ (which may partly explain why the scatter in the plot of $\Delta t/T = 0.006$ (in Wolman and Leopold 1957, Nixon 1959) was so gross).

3 From the standpoint of fluvial hydraulics the representative Q must be expected to depend on the nature of sediment forming the stream boundaries. Indeed, the speed W of the boundary deformation which eventually leads to the formation of a regime channel, is determined by the sediment transport rate q_s (see Section 1.7), and not by the flow rate $q \approx Q/B$. Hence, the determination of the representative Q on the basis of the flow rate Q_t alone cannot be generally applicable (e.g. at times when $\tau_0 < (\tau_0)_{cr}$, and thus q_s and Q_s are zero, the regime channel is not forming, even if the flow rate Q is present at those times).

PROBLEMS

To solve the problems below, take $\gamma_s = 16186.5 N/m^3$, $\rho = 1000 kg/m^3$, $\nu = 10^{-6} m^2/s$ (which correspond to sand or gravel and water).

4.1 Consider the North Saskatchewan River: $Q = 4386 m^3/s$ (bankfull flow rate), $D = 31 mm$ (gravel).
a) Determine the regime characteristics B_R, h_R and S_R. Use the equations in Sub-section 4.5.2.
b) Compare the values of B_R, h_R and S_R obtained in a) with the corresponding values in Table 4.1 (which were obtained with the computer program BHS-STABLE). Explain the differences. [Hint: Determine B_R, h_R and S_R with the method in Sub-section 4.5.2, but replace the power approximation (4.60) by $c = (1/\kappa)\ln(11h/k_s)$.]

4.2 A gravel channel is in regime. Knowing that the bankfull flow rate is $Q = 3000 m^3/s$ and the regime width is $B_R = 211.7 m$, determine the grain size D, and the regime flow depth h_R and slope S_R.

4.3 For a laboratory flume with straight rigid (plexiglass) walls having the width $B = 2m$, and a bed formed by a layer of sand having $D = 1.5 mm$, determine the regime values h_{R_1} and S_{R_1} of the regime channel R_1 if $Q = 1 m^3/s$ and $Q_s = 5.0 \times 10^{-5} m^3/s$.

4.4 Determine B_R, h_R and S_R for a river whose bankfull flow rate is $Q = 1500 m^3/s$ and the average grain size is $D = 0.7 mm$.

4.5 Consider the Peace River ($D = 0.31 mm$). The measured bankfull characteristics are $Q = 9905 m^3/s$, $B = 619.2 m$, $h = 9.33 m$, $S = 0.000084$ (see Ref. [12d] in Appendix D). Is this river in regime?

4.6 Consider a straight region of a stream flowing in a cohesionless alluvium ($D = 1 mm$). It is assumed that the stream is in regime. The total volumetric sediment transport rate (past the undulated bed) at the regime stage is $(Q_s)_R = 3.0 \times 10^{-3} m^3/s$. Determine the regime characteristics B_R, h_R, S_R, and the value of the bankfull flow rate Q_{bf}.

(Assume that the turbulent flow is fully rough and that the sediment is transported as bed-load only; take $\lambda_c = c/c_f = 0.7$).

4.7 Eq. (4.31) gives the regime flow depth h_R in terms of regime values of the Froude number. Show that in terms of the regime values of u_{av} and η_*, h_R is given as

$$h_R = \frac{1}{\alpha_B}\frac{Q^{1/2}v_{*cr}^{1/2}}{(u_{av})_R} \quad \text{and} \quad h_R = \frac{1}{\alpha_B}\frac{1}{c_R(\eta_*)_R^{1/2}}\frac{Q^{1/2}}{v_{*cr}^{1/2}},$$

respectively.

4.8 Consider a $N = const$ curve on the $(Fr; \eta_*)$-plane corresponding to a given value of Ξ, and let P be the point of this curve where c (which also varies along this curve when η_* varies) has its minimum value. What is the inclination of the tangent (implying $\partial Fr/\partial \eta_*$) at this point?

4.9 Write a computer program to draw the Fr-curves corresponding to $\log N = 2$, 3, 4, 5, and 6 on the $(Fr; \eta_*)$-plane corresponding to $\Xi = 10.12$ (i.e. $D = 0.4mm$); take $1 \leq \eta_* \leq 100$.

REFERENCES

Ackers, P. (1964). Experiments on small streams in alluvium. *Journal of the Hydraulics Division*, ASCE, 90(4), 1-37.

ASCE Task Committee on Relations between Morphology of Small Streams and Sediment Yield of the Committee on Sedimentation of the Hydraulics Division (1982). Relationships between morphology of small streams and sediment yield. *Journal of the Hydraulics Division*, ASCE, 108(11), 1328-1365.

Bettess, R. and White, W.R. (1987). Extremal hypothesis applied to river regime. In *Sediment Transport in Gravel-Bed Rivers*, edited by C.R. Thorne, J.C. Bathurst and R.D. Hey, John Wiley and Sons, 767-789.

Bettess, R. and White, W.R. (1983). Meandering and braiding of alluvial channels. *Proceedings of the Institution of Civil Engineers*, 75, Part 2, 525-538.

Blench, T. (1966). *Mobile-bed fluviology*. University of Alberta, Edmonton, Alta.

Cao, S. and Knight, D.W. (1997). Entropy-based design approach of threshold alluvial channels. *Journal of Hydraulic Research*, 35(4), 505-524.

Chang, H.H. (1988). *Fluvial processes in river engineering*. John Wiley and Sons, Inc.

Chang, H.H. (1980). Stable alluvial canal design. *Journal of the Hydraulics Division*, ASCE, 106(5), 873-891.

Chang, H.H. (1979). Minimum stream power and river channel patterns. *Journal of Hydrology*, 41, Issues 3-4, 303-327.

Chang, H.H. and Hill, J.C. (1977). Minimum stream power for rivers and deltas. *Journal of the Hydraulics Division*, ASCE, 103(12), 1375-1389.

Chitale, S.V. (1973). Theories and relationships of river channel patterns. *Journal of Hydrology*, 19(4), 285-308.

Chiu, C.L. and Said, C.A.A. (1995). Maximum and mean velocities and entropy in open-channel flow. *Journal of Hydraulic Engineering*, 121(1), 26-35.

Chiu, C.L. (1991). Application of entropy concept in open-channel flow study. *Journal of Hydraulic Engineering*, 117(5), 615-628.

da Silva, A.M.F. (2009). On the stable geometry of self-formed alluvial channels: theory and practical application. *Canadian Journal of Civil Engineering*, 36(10), 1667-1679.

Davies, T.H.R. and Sutherland, A.J. (1983). Extremal hypothesis for river behaviour. *Water Resources Research*, 19(1), 141-148.

Davies, T.H.R. and Sutherland, A.J. (1980). Resistance to flow past deformeable boundaries. *Earth Surface Processes*, Vol. 5, 175-179.

Eaton, B.C., Church, M. and Millar, R.G. (2004). Rational regime model of alluvial channel morphology and response. *Earth Surface Processes and Landforms*, 29, 511-529.

Farias, H.D. (1993). Morphology of regime alluvial channels: A review. In Advances in Hydro-Science and Engineering. *Proceedings of the 1st International Conference on Hydro-Science and Engineering*, Washington, DC, edited by S.S.Y. Wang., Vol. I, Part B, 1423-1428.

Garde, R.J. and Raju, K.G.R. (1985). *Mechanics of sediment transportation and alluvial stream problems*. 2nd edition, Wiley Eastern, New Delhi.

Goldstein, M. and Goldstein, I.F. (1993). *Understanding the laws of energy*. Harvard University Press, Cambridge.

Jia, Y. (1990). Minimum Froude number and the equilibrium of alluvial sand rivers. *Earth Surface Processes and Landforms*, 15, 199-209.

Julien, P. and Wargadalam, J. (1995). Alluvial channel geometry: theory and applications. *Journal of Hydraulic Engineering*, 121(4), 312-325.

Kirillin, V. A., Sychev, V.V. and Scheindlin, A.E. (1976). *Engineering thermodynamics*. Translated from Russian by S. Semyonov, Mir Publishers.

Landau, L.D. and Kitaigorodskiy, A.I. (1965). *Physics*. Publishing House NAUKA, Moscow.

Landau, L.D., Achiezer, A.I. and Lifshitz, E.M. (1965). *General physics (mechanics and molecular physics)*. Publishing House NAUKA, Moscow.

Lee, J. and Julien, P. (2006). Downstream hydraulic geometry of alluvial channels. *Journal of Hydraulic Engineering*, 132(12), 1347-1352.

Leopold, L.B. and Wolman, M.G. (1957). River channel patterns: braided, meandering and straight. *U.S. Geological Survey Professional Papers*, 282-B, 39-84.

Leopold, L.B. and Maddock, T. (1953). The hydraulic geometry of stream channels and some physiographic implications. *U.S. Geological Survey Professional Papers*, 252, 1-57.

Moran, M.J. and Shapiro, H.N. (1992). *Fundamentals of engineering thermodynamics*. 2nd edition, John Wiley and Sons.

Munson, B.R., Young, D.F. and Okiishi, T.H. (1994). *Fundamentals of fluid mechanics*. 2nd edition, John Wiley and Sons.

Nixon, M. (1959). A study of the bank-full discharges of rivers in England and Wales. *Proceedings of the Institution of Civil Engineers*, 12, 157-174.

Pefley, R.K. and Murray, R.I. (1966). *Thermofluid mechanics*. McGraw-Hill Book Company.

Radecki-Pawlik, A. (2015). Why do we need bankfull and dominant discharges. Chapter 20, in *Rivers – Physical, Fluvial and Environmental Processes*, edited by

P. Rowińsky and A. Radecki-Pawlik, GeoPlanet: Earth and Planetary Book Series, Springer International Publishing, Switzerland, 497-518.
Shu-you, C. and Knight, D.W. (2002). Review of regime theory of alluvial channels. *Journal of Hydrodynamics*, Ser. B, 3, 1-7.
Singh, V.P. (2003). On the theories of hydraulic geometry. *International Journal of Sediment Research*, 18(3), 196-218.
Singh, V.P., Yang, C.T. and Deng, Z.Q. (2003). Downstream hydraulic geometry relations. 1. Theoretical development. *Water Resources Research*, 39(12), 1337, doi:10.1029/2003WR002484.
Song, C.C.S. and Yang, C.T. (1982). Minimum stream power: theory. *Journal of the Hydraulics Division*, ASCE, 106(9), 1477-1487.
Soo, S.L. (1967). *Fluid dynamics of multiphase systems*. Blaisdall Publishing Co., Waltham, Massachussetts, Toronto, London.
Spalding, D.B. and Cole, E.H. (1973). *Engineering thermodynamics*. 3rd edition, Edward Arnold (Publishers) Ltd.
Van Wylen, G.J. and Sonntag, R.E. (1965). *Fundamentals of classical thermodynamics*. John Wiley and Sons.
Wark, K. (1971). *Thermodynamics*. 2nd edition, McGraw-Hill Book Company.
White, W.R., Bettess, R. and Shiqiang, W. (1986). *A study on river regime*. Report No. SR 89, Hydraulics Research Ltd., Wallingford, UK.
White, W.R., Bettess, R. and Paris, E. (1982). Analytical approach to river regime. *Journal of the Hydraulics Division*, ASCE, 108(10), 1179-1193.
White, W.R., Paris, E. and Bettess, R. (1981). *River regime based on sediment transport concepts*. Report IT 201, Hydraulics Research Ltd., Wallingford, UK.
Wilkerson, G. and Parker, G. (2011). Physical basis for quasi-universal relationships describing bankfull geometry of sand-bed rivers. *Journal of Hydraulic Engineering*, 137(7), 793-753.
Wolman, M.G. and Leopold, L.B. (1957). River flood plains: some observations on their formation. *U.S. Geological Survey Professional Papers*, 282-C, 87-107.
Yalin, M.S. and da Silva, A.M.F. (2000). Computation of regime channels on thermodynamic basis. *Journal of Hydraulic Research*, 38(1), 57-63.
Yalin, M.S. and da Silva, A.M.F. (1999). Regime channels in cohesionless alluvium. *Journal of Hydraulic Research*, 37(6), 725-742.
Yalin, M.S. and da Silva, A.M.F. (1997). On the computation of equilibrium channels in cohesionless alluvium. *Journal of Hydroscience and Hydraulic Engineering*, JSCE, 15(2), 1-13.
Yalin, M.S. (1992). *River mechanics*. Pergamon Press, Oxford.
Yalin, M.S. (1977). *Mechanics of sediment transport*. 2nd edition, Pergamon Press, Oxford, England.
Yang, C.T. (1994). Variational theories in hydrodynamics and hydraulics. *Journal of Hydraulic Engineering*, 120(6), 737-756.
Yang, C.T. (1992). Force, energy, entropy and energy dissipation rate. In *Entropy and energy dissipation in water resources*, edited by V.P. Singh and M. Fiorentino, Kluwer Academic Publishers, London, United Kingdom, 63-89.
Yang, C.T. (1987). Energy dissipation rate in river mechanics. In *Sediment transport in gravel-bed rivers*, edited by C.R. Thorne, J.C. Bathurst and R.D. Hey, John Wiley and Sons, 735-766.

Yang, C.T. (1984). Unit stream power equation for gravel. *Journal of Hydraulic Engineering*, 110(12), 1783-1797.
Yang, C.T. and Molinas, A. (1982). Sediment transport and unit stream power function. *Journal of the Hydraulics Division*, ASCE, 108(6), 774-793.
Yang, C.T., Song, C.C.S. and Woldenberg, M.J. (1981). Hydraulic geometry and minimum rate of energy dissipation. *Water Resources Research*, 17(4), 1014-1018.
Yang, C.T. and Song, C.C.S. (1979). Theory of minimum rate of energy dissipation. *Journal of the Hydraulics Division*, ASCE, 105(7), 769-784.
Yang, C.T. (1976). Minimum unit stream power and fluvial hydraulics. *Journal of Hydraulic Engineering*, 102(7), 919-934.

Chapter 5

Formation of regime channels; meandering and braiding

In Chapter 4 we were concerned with the *reason* for the formation of regime channels, and also with the computation of their pertinent characteristics. In this chapter we will study *how* a regime-formation process takes place. It will be shown that this process does not always proceed in the same way; in some cases the stream is meandering, in others it is braiding, and yet in some others the plan geometry of the stream remains as it is.[1]

5.1 MEANDERING AND REGIME DEVELOPMENT

5.1.1 Initiation and subsequent development of meandering

Relationship between meander wavelength and flow width

Since the distance between two crossovers O_i and O_{i+2}, i.e. the meander wavelength Λ_M is assumed to remain constant throughout the regime channel development (see Sub-section 4.4.1 (iii)), let us identify the length Λ_M with the length scale of one-row large-scale horizontal coherent structures (LSHCS's) that occur in the (still straight) channel at the end ($\hat{T}_0$) of the adjustment period. This yields $\Lambda_M = 6\hat{B}_0$. Since, however, the channel width varies during $T_R - \hat{T}_0$ within the narrow interval $\hat{B}_0 < B < B_R$ (Sub-section 4.4.1 (ii)), the error cannot be significant if $\hat{B}_0$ is simply replaced by any B from this interval. By doing so, one arrives at the relation

$$\Lambda_M \approx 6B \quad (= L_H). \tag{5.1}$$

This is generally in agreement with existing plots of Λ_M versus B, two examples of which, resulting from work by the authors and Soar and Thorne (2001), are shown in Figures 5.1a,b, respectively. Figure 5.1a was produced on the basis of the field and laboratory data previously used by Garde and Raju (1985) in their Figure 13.13, as well as data from Leopold et al. (1964). The data from Leopold et al. (1964) include not only alluvial streams, but also meltwater channels on ice and meanders of the Gulf Stream. Indeed, these authors appear to have been the first to realize that "the meander pattern of meltwater channels on the surface of glaciers have nearly identical geometry

[1]A stream whose plan geometry remains unchanged is often referred to (following Leopold and Wolman 1957) as a "straight" stream.

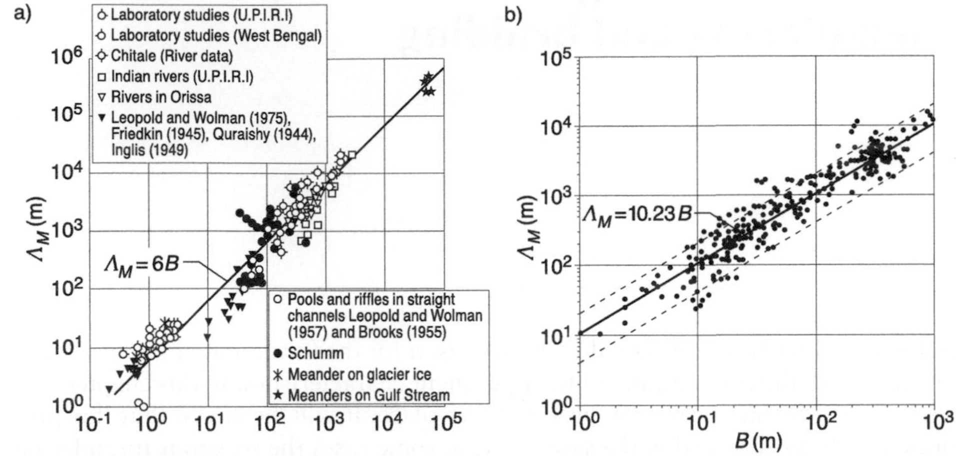

Figure 5.1 Examples of existing plots of meander wavelength versus flow width B; the solid lines are the best-fit to the data. (a) Adapted from Yalin and da Silva (2001); (b) adapted from Soar and Thorne (2001); the dashed lines mark $\Lambda_M = 4B$ and $20B$.

to the meander bends in rivers" and that "the geometry in plan view of meanders in the Gulf Stream is also similar to that of rivers". It should be noted here that, as pointed out by Leopold et al. (1964), p. 302, the "meandering channels on ice are formed without any sediment load or point-bar construction by sediment deposition", and that the meanders on the Gulf Stream too occur "... without debris load and, in this instance, without confining banks". Considering this, Yalin (1992), p. 161, defined meandering as a "self-induced plan deformation of a stream that is (ideally) periodic and anti-symmetrical with respect to an axis, x say, which may or may not be exactly straight" – where the term self-induced is used to imply that the deformation is induced by the stream itself, as opposed to being forced upon the stream by its environment.[2]

Eq. (5.1), however, deserves some further discussion. Indeed, on the basis of the content of Section 2.5, it is only natural to expect the proportionality factor in the relationship of Λ_M with B to vary within some range of values, in principle comparable to that of the proportionality coefficient in the relationship of Λ_a with B (namely, ≈ 4 to 14). That is, $\Lambda_M/B \approx 4$ to 14, where Λ_M is to be considered as the average meander length. Here, the term average is to be interpreted as the average of multiple individual realizations of Λ_M obtained under similar conditions, and not as the absolute average obtained by considering all streams, irrespective of stream size, flow conditions, physiographic settings, etc. In this sense, one cannot say that the just mentioned expectation is validated by plots such as those in Figures 5.1a,b – as they only convey that individual values of Λ_M obtained under a wide variety of conditions, as a rule, vary from $\approx 4B$ to $20B$, say. It would indeed be particularly worthwhile to further investigate the

[2]Clearly, the term "meandering" is used throughout this book in the sense implied by this definition.

nature of the proportionality coefficient in the meander wavelength equation, and in particular analyze whether a dependency on B/h and h/D similar to that of the proportionality coefficient in the alternate bar length equation is discernible – even if the "random element" in real meandering rivers is rather strong, and likely to obfuscate the matter. In the absence of such information, and in similarity to the approach adopted in Section 2.2, such matters are disregarded here, and the form (5.1) is adopted for purposes of subsequent discussion.

Information supplied by the data

Figure 5.2a is the extension of the plot in Figure 2.30 where all the available meandering data (M and m) and the data of braiding streams (B and b) are also plotted.[3] In the following, the points (M and m) and (B and b) will be referred to simply as points M and B. Figure 5.2b is included merely to facilitate the identification of the zones populated by the points mentioned.

(i) Observe from Figure 5.2a that the line $\mathcal{L}_{1,2}$ does not form the upper boundary only of the alternate-bar points A: it can be taken as a common upper boundary of the points A and M (i.e. $\mathcal{L}_{1,2} \equiv \mathcal{L}_{M,B}$). The fact that the points A and M have the same upper boundary implies that meanders, just like alternate bars, occur only in the streams having one-row ($n=1$) LSHCS's (Figure 2.11b). The time-average streamlines of such streams exhibit in plan view an "internal meandering" (Figure 5.3a); its longitudinal period is the same as the length scale $L_H \approx 6\hat{B}_0$. Consequently, the stream acquires a periodic along x non-uniformity which, in turn, causes the (deformable) banks of the stream to deform with the same periodicity. Thus the channel meandering of the wave length $\Lambda_M = L_H \approx 6\hat{B}_0$ initiates.

Even though the points A and M have a common upper boundary, their lower boundaries are different, and the data-points suggest that these lower boundaries can be characterized by the lines $\mathcal{L}_{0,1}$ and $\mathcal{L}_M$, respectively (Figure 5.2). The area between the lines $\mathcal{L}_M$ and $\mathcal{L}_{0,1}$ contains almost exclusively the points M – the points A are hardly present. This fact can be taken as an (experimental) indication that the (earlier occurring) alternate bars cannot be the *cause* of meandering, as has been thought by some authors (e.g. Kinoshita 1961, Sukegawa 1970, Ackers and Charlton 1970, Lewin 1976, Kondratiev et al. 1982).

Indeed, the following two cases can be present:

1 If B/h is small, then the LSHCS's are not rubbing the bed (see Figure 2.10b and related text), and therefore they cannot produce "their" bed forms, viz alternate bars. Yet, the sequence of these structures can still initiate meandering by their

[3]The data sources of point-symbols M, m, B, b in Figure 5.2 are given in Appendix E:
Point-symbol M: [3e], [5e], [6e], [7e], [8e], [9e], [10e], [14e], [16e], [18e], [21e], [22e], [23e], [26e], [29e], [32e];
Point-symbol m: [2e], [13e], [25e], [31e];
Point-symbol B: [1e], [7e], [8e], [9e], [15e], [16e], [17e], [20e], [24e], [27e], [28e], [30e], [32e];
Point-symbol b: [4e], [11e], [12e], [13e], [19e], [25e].
The point-symbols A, C and P are the same as in Figure 2.30 (see footnote 15 in Chapter 2 for a list of the data sources).

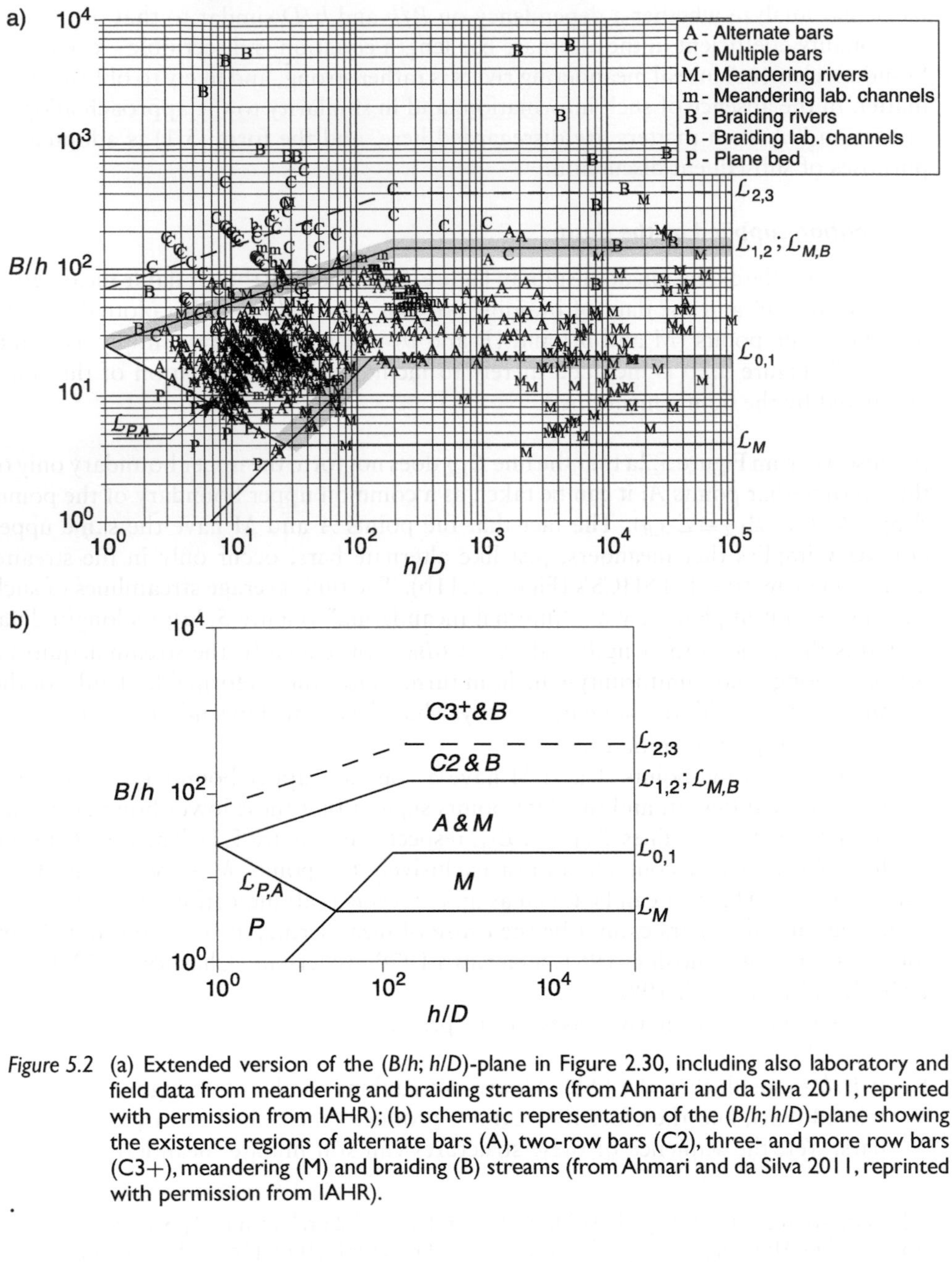

Figure 5.2 (a) Extended version of the (B/h; h/D)-plane in Figure 2.30, including also laboratory and field data from meandering and braiding streams (from Ahmari and da Silva 2011, reprinted with permission from IAHR); (b) schematic representation of the (B/h; h/D)-plane showing the existence regions of alternate bars (A), two-row bars (C2), three- and more row bars (C3+), meandering (M) and braiding (B) streams (from Ahmari and da Silva 2011, reprinted with permission from IAHR).

direct impact on the banks, as well as by the convective action of the "internal meandering" they generate. Thus the LSHCS's can "imprint" on the channel banks the length $L_H \approx 6\hat{B}_0$, in the form of the meander length $\Lambda_M = 6\hat{B}_0$, *without* alternate bars (Figure 5.3). This occurs in the zone between the lines $\mathcal{L}_M$ and $\mathcal{L}_{0,1}$.

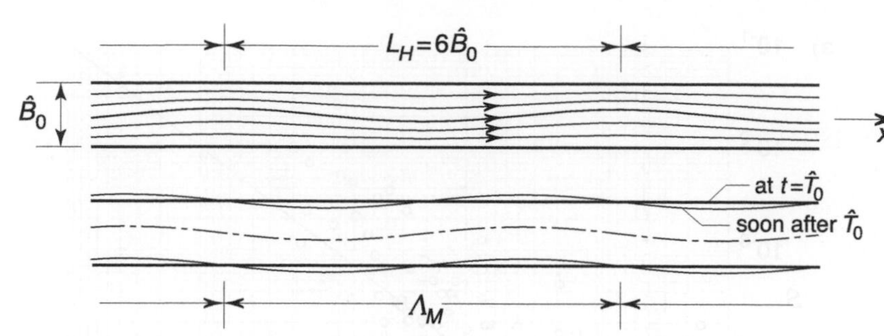

Figure 5.3 Initiation of meandering. (a) Internal meandering of the flow in a straight channel; (b) channel evolving from straight to meandering.

2 If B/h is larger than the ordinates of $\mathcal{L}_{0,1}$, then the LSHCS's *are* rubbing the bed (Figure 2.10c), and they produce first the alternate bars, which, acting as "guide-vanes", facilitate (accelerate) the bank deformation described in point 1 above. In this case the points A as well as M can be present in the same zone. This occurs between the lines $\mathcal{L}_{0,1}$ and $\mathcal{L}_{1,2}$.

(ii) As follows from the considerations so far in this book, in the authors' view a a self-induced (non-guided) meandering is but one of the means used by an alluvial stream to minimize its velocity u_{av} $(=A_*)$ as to achieve its regime state. The authors maintain that the channel meandering (which initiates as described in the previous paragraph) continually grows in its amplitude, i.e. the meander loops continually expand (and thus $L^{-1} \sim S \sim u_{av}^2$ progressively decreases), ultimately because of the regime trend. It should be mentioned here, however, that meandering is only one of the means available to an alluvial stream to achieve its regime state (it can resort also to aggradation, degradation, braiding, some combinations thereof, etc.). Yet, it appears that at least in the case of alluvial streams, the *expansion* of meandering is motivated solely by the minimization of flow velocity. Considering the aforementioned, one can assert that a meandering stream in an unlimited alluvium is either in its regime state ($S = S_R$; $\Theta \geq 1$), or it is in the process of achieving it ($S_0 > S > S_R$; $0 < \Theta < 1$): no meander growth is exhibited by a stream having $S = S_0 < S_R$. The first of these statements (that a meandering stream is either in its regime state or in the process of achieving it) was tested by plotting the actual slope of various sand and gravel meandering streams versus the regime slope S_R, as shown in Figures 5.4a,b, respectively.[4] Here S_R was calculated using the computer program BHS-STABLE (see Sub-section 4.5.1). Observe from Figures 5.4a,b that indeed, as a rule, the data plot above or around the line implying $S = S_R$. (The "diffusion" of the data to the region below the straight line implying $S = S_R$ is comparable to that in Figure 4.16 – and as such, is within the percentage of error inherent to the method of computation itself, as well as data uncertainies as discussed at the end of Sub-section 4.4.4).

[4]The data plotted in Figures 5.4a,b are derived from Refs. [3e], [7e], [19e], [26e], [32e].

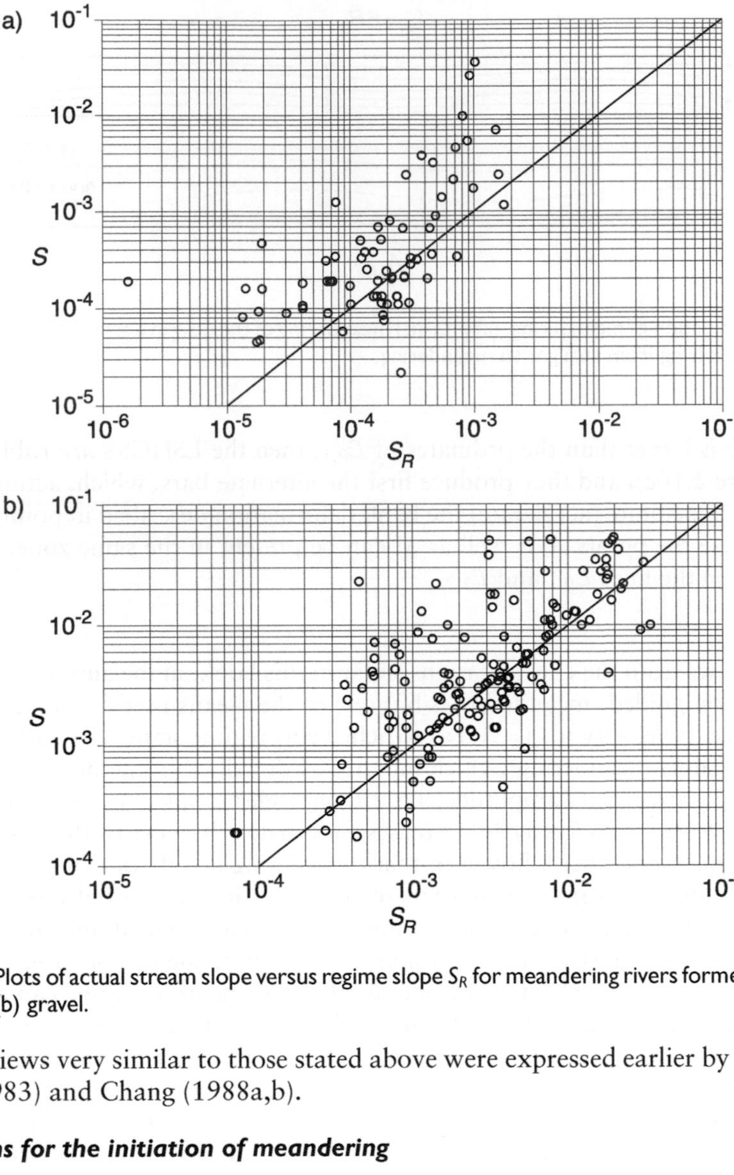

Figure 5.4 Plots of actual stream slope versus regime slope S_R for meandering rivers formed by: (a) sand; (b) gravel.

The views very similar to those stated above were expressed earlier by Bettess and White (1983) and Chang (1988a,b).

Conditions for the initiation of meandering

It should be clear that belonging to the meandering region of the $(B/h; h/D)$-plan is a necessary but not sufficient condition for streams to meander.[5]

From the aforementioned, it follows that the set of required conditions for a stream to meander are:

1 the stream transports the sediment;
2 the stream is turbulent;

[5]The same applies to braiding, but the topic is outside of the scope of this section.

3 the stream slope is larger than the regime slope;
4 the initial values of B and h (after the initial adjustment period $\hat{T}_0$) must be such that the corresponding data-point plots in the meandering region of the plan.

It must be emphatically pointed out that other considerations may need to be involved to address specific circumstances such as, for example, the fact that bed degradation occurring under limited sediment supply may limit or suppress the need for a stream to meander.

5.1.2 Tracing of meander development on the $(Fr; \eta_*)$- and $(B/h; h/D)$-planes

In this sub-section we will consider the regime development of the conventional regime channels, which have one-row LSHCS's, i.e. which are situated below the boundary $\mathcal{L}_{1,2}$ in Figure 5.2. During the regime development of such channels, S and u_{av} decrease, whereas B increases (Section 4.4). Since the relative intensity of the time-variation of S, viz $(\partial S/\partial\Theta)/S$, is much larger than $(\partial h/\partial\Theta)/h$ and $(\partial c^2/\partial\Theta)/c^2$ (see Problem 5.9), the time-decrement of S and u_{av} (during T_R) can be unambiguously reflected by the time-decrements of the dimensionless variables η_* ($\sim hS$) and Fr ($= u_{av}^2/(gh) = c^2 S$). Similarly, the time-increment of B can be characterized by the time-decrement of (the dimensionless) $N = Q/(BD\upsilon_{*cr})$. The decrement of S need not necessarily be due to meandering (as may be inferred from the title of this sub-section): the explanations and diagrams below are equally valid for the reduction of S by degradation-aggradation.

(i) Consider first the $(Fr; \eta_*)$-plane (Figure 5.5). Knowing $(\eta_*)_R$ and $(Fr)_R$ (from the regime channel computation), one can locate on the $(Fr; \eta_*)$-plane the "regime-point" P_R (which is the "lowest" point (P_g or P_s) of the corresponding $(Fr)_R$-curve – see Sub-section 4.4.4 (ii)). Similarly, knowing the initial channel (B_0, h_0, S_0), and having computed[6] the adjusted initial channel $(\hat{B}_0, \hat{h}_0, \hat{S}_0)$, one can determine their $(\eta_*)_0$; $(Fr)_0$ and $(\hat{\eta}_*)_0$; $(\hat{Fr})_0$; and locate the respective "initial points" P_0 and $\hat{P}_0$. [The fact that η_*, Fr and u_{av} must continually decrease during T_R, hints already that the regime-formation cannot initiate from *any* arbitrarily selected point P_0. A little reflection may help to realize that the point P_0 must be selected in the area remaining on the right of the respective $(Fr)_R$-curve, the lower boundary of this area being the horizontal line passing through the regime point (P_g or P_s)].

The time-development of a regime channel can be pictured by the "motion" of a point m on the $(Fr; \eta_*)$-plane (Yalin 1992). At $t = 0$ and $t = \hat{T}_0$, the point m is at P_0 and $\hat{P}_0$; at $t = T_R$, it is at P_R. What kind of a path l will the point m follow? Since (the average) $\partial B/\partial\Theta$ within $0 < t < \hat{T}_0$ is much larger than within $\hat{T}_0 < t < T_R$ (see Figure 4.5), the point m first must rapidly approach "its" $(Fr)_R$-curve (that is, by starting to move nearly horizontally to the left), and then closely follow this curve until

[6]See Sub-section 5.3.1 (iii) for the computation of adjusted initial channel.

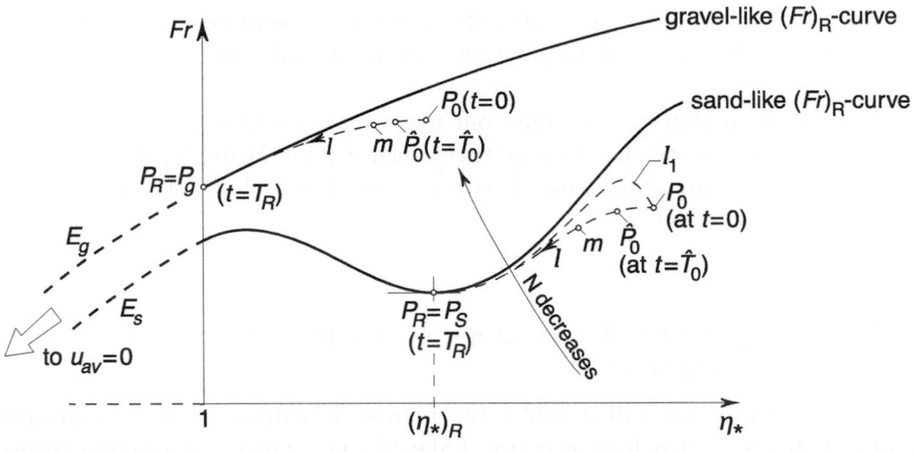

Figure 5.5 Tracing of meander development on the $(Fr; \eta_*)$-plane.

it merges into it at its lowest point P_R (paths l in Figure 5.5). (An approach such as e.g. l_1 cannot take place, for it is associated with the increment of Fr).[7]

From Sub-section 4.4.3 (v) it should be clear that both gravel-like and sand-like Fr-curves can be extended in fact up to the point $[\eta_* = 0;\ Fr = u_{av} = 0]$ which signifies the state of the *final* thermodynamic *equilibrium*. In the log-log system of Figure 5.5, this point is, of course, infinitely left-below – as indicated by the "extension curves" E_g and E_s.

Combining the aforementioned with the content of Sub-section 4.4.3 (v), one infers that the regime points P_g and P_s imply two different situations. Indeed:

1 *Gravel-like formation*. In this case the point m, moving along "its" gravel-like $(Fr)_R$-curve with the "intention" to reach the final equilibrium state, will actually stop when it reaches $\eta_* = 1$ (point $P_R = P_g$); for the channel slope S, and thus η_*, cannot continue to decrease if sediment is not transported any longer ($q_s \equiv 0$ when $\eta_* \leq 1$).
2 *Sand-like formation*. In this case the moving point m stops at the point $P_R = P_s$. This happens in spite of the fact that at this point, $(\eta_*)_R \gg 1$ and thus $(q_s)_R \gg 0$, meaning that the sediment is moving and, *in principle*, $S \sim \eta_*$ can continue to decrease further. The reason for the termination of the decrement of S, and thus of the channel formation, lies in the fact that at P_s we have $\partial(Fr)_R/\partial\eta_* = 0$, and consequently the (local) extrema $u_{av} = (u_{av})_{min}$ and $S_* = (S_*)_{max}$. Suppose that the point m moving along l has reached the point P_s, and yet it still continues to move along the $(Fr)_R$-curve (upwards-left). This would mean that u_{av}, which has decreased to its minimum, begins to increase again; while the entropy S_*, which has reached its maximum, begins to decrease again. Clearly, such u_{av}- and S_*-variations

[7]No research has been carried out so far with the intention to reveal the general form of the function $l = \phi(\eta_*; Fr)$ connecting the points P_0 and P_R. An effort on this score would certainly be worthwhile.

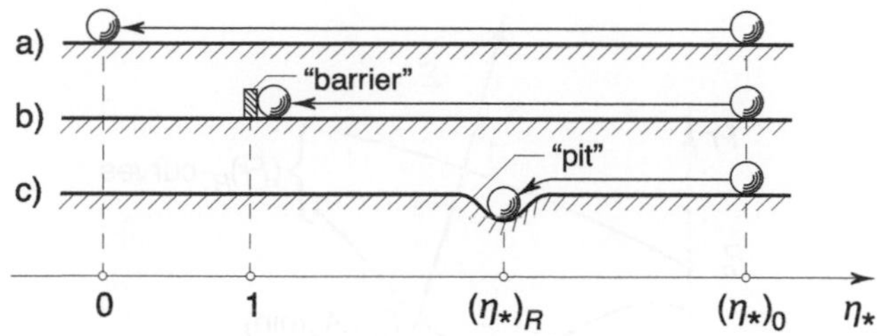

Figure 5.6 "Rolling ball" analogy to the time-development of a regime channel. (a) Ball rolling unimpeded; (b) ball stops at "barrier"; (c) ball falls into a "pit".

are physically impossible (they contradict the Second Law), and therefore the channel development *must* stop at P_s.

The above mentioned conditions are symbolized by the "rolling-ball" analogy shown in Figure 5.6:

a) The ball rolls on a rough horizontal plane starting at $\eta_* = (\eta_*)_0$ with an initial velocity $u = u_0$. Eventually the ball's kinetic energy becomes completely converted, by friction, into its internal energy (no heat transfer), and the ball stops: $u = 0$ at $\eta_* = 0$ (final equilibrium).

b) The ball cannot reach the "target-location" $\eta_* = 0$, for it is stopped by the "barrier" at $\eta_* = 1$ (q_s-constraint; *gravel-like development*).

c) The ball cannot reach the location $\eta_* = 0$, because it falls into the "pit" at $(\eta_*)_R$ ($\gg 1$) (K_s-constraint; *sand-like development*).

Although in this text we are dealing with the regime channels R, it may be of some interest to reveal how the regime channels R_1 are determined in the $(Fr; \eta_*)$-plane. The first equation of R_1 (see Eqs. (4.5) to (4.7)) is the same as that of R: it is the resistance equation $u_{av} = c\sqrt{gSh}$, i.e. $Fr = u_{av}^2/gh = c^2 S$. Hence, for a specified experiment, the graph of the first equation of R_1 is (also) a Fr-curve corresponding to a certain $N = Q/BDv_{*cr}$ (follow from Figure 5.7).

The second equation of R_1 is the sediment transport equation. Identifying it with the Bagnold's form (3.49), with $\overline{u} = u_{av}$, we obtain

$$Q_s = \beta' B u_{av}(\lambda_c^2 \tau_0 - (\tau_0)_{cr})/\gamma_s = \beta' B h u_{av} \frac{(\tau_0)_{cr}}{\gamma_s h}(\lambda_c^2 \eta_* - 1)$$

$$= \beta' Q \frac{\Psi(\Xi)}{Z}(\lambda_c^2 \eta_* - 1) \qquad (5.2)$$

i.e.

$$K_* = \left[\frac{Q_s}{Q}\frac{1}{\beta \Psi(\Xi)}\right] = \frac{\lambda_c^2 \eta_* - 1}{Z}, \qquad (5.3)$$

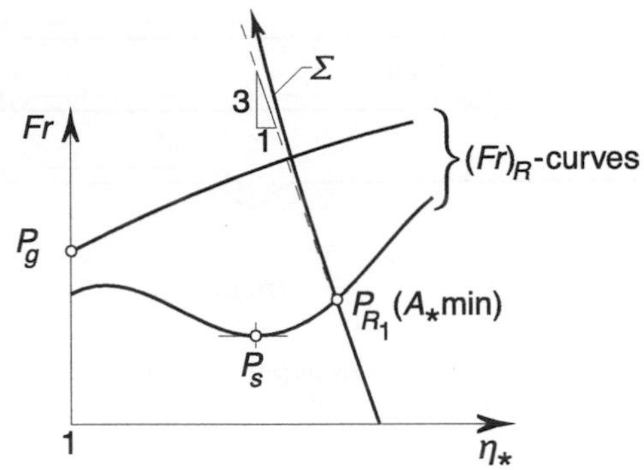

Figure 5.7 Intersection of line Σ with *Fr*-curves on the $(Fr; \eta_*)$-plane.

where use was made of $Bhu_{av} = Q$, $h/D = Z$, and $(\tau_0)_{cr} = \gamma_s D\Psi(\Xi)$. Moreover,

$$Fr = \frac{Q^2}{gB^2h^3} \quad \text{gives} \quad K_{**} = \left[\frac{Q^2}{gB^2D^3}\right] = Fr \cdot Z^3. \tag{5.4}$$

Eliminating Z between Eqs. (5.3) and (5.4), we arrive at

$$Fr = \frac{K}{(\lambda_c^2 \eta_* - 1)^3} \qquad (\text{with } K = K_*^3 K_{**}). \tag{5.5}$$

If B is given, then for a specified R_1-formation experiment, K is also given. In this case, Eq. (5.5) implies a curve, Σ say, in the $(Fr; \eta_*)$-plane (for the value of $\lambda_c^2 = (c/c_f)^2$ corresponding to any point of the $(Fr; \eta_*)$-plane can always be computed by using one of the methods presented in Chapter 3). In order to reveal the basic form of Σ, we consider the simplest case of $\lambda_c = 1$ (no bed forms). In this case (as should be clear from Eq. (5.5)), when $\eta_* \gg 1$, then Σ becomes indistinguishable from a 3/1-declining straight line, and when $\eta_* \to 1$, then it must approach a vertical asymptote (Figure 5.7). Naturally, if $\lambda_c < 1$ (bed forms), then some deviations from the basic form just described must be expected – however, the inclination of its asymptotes (viz $-3/1$ and ∞) will remain unchanged.

If $B = const_B$, as in the case of flume experiment described in Sub-section 4.5.3, then the regime point P_{R_1} is simply the intersection point of Σ with the *Fr*-curve corresponding to $N = Q/const_B D\upsilon_{*cr}$, and no minimization of any A_* is involved. If, however, $B = B_{R_1}$ is unknown (general case), then we have a multitude of *Fr*-curves (each of which corresponding to a constant $N_i \sim 1/B_i$) and a multitude of Σ-curves (each of which corresponding to a constant $K_j \sim 1/B_j^2$). Consequently, we have a multitude of the points P where the *Fr*- and Σ-curves corresponding to the *same* B $(B_i = B_j)$ intersect each other.

According to the R_1-approach, the regime point P_{R_1} must be identified with that P where the adopted A_* acquires its smallest value. (If $A_* = u_{av}$, and thus $\Pi_{A_*} = Fr$, as in the present R-approach, then R_1 can coincide with R only if the point P_{R_1} coincides

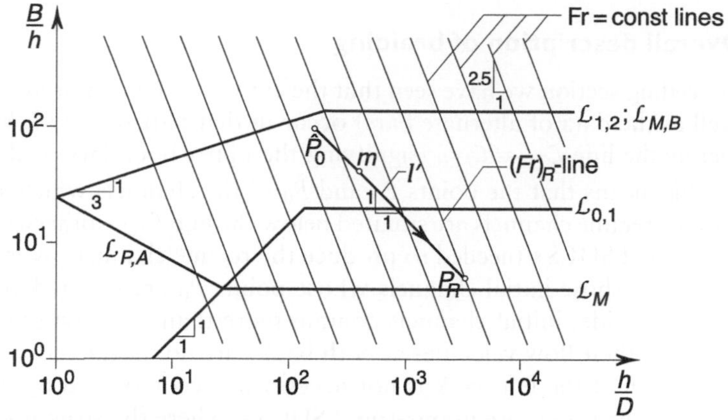

Figure 5.8 Tracing of meander development on the (*B*/*h*; *h*/*D*)-plane.

with the lowest point P_s ($= P_R$) of the *Fr*-curve: i.e. if Q_s (of R_1) is "selected" so as to be equal to $(Q_s)_R$ (of R)).

(ii) Consider now the $(B/h; h/D)$-plane in Figure 5.8. The resistance equation $Q = Bhc\sqrt{gSh}$, can be expressed as

$$\frac{B}{h} = \alpha \left(\frac{h}{D}\right)^{-2.5} \tag{5.6}$$

where

$$\alpha = \frac{Q}{D^{2.5} c\sqrt{gS}} = \frac{(const)_a}{\sqrt{Fr}} \qquad \left[\text{with } (const)_a = Q/\sqrt{gD^5}\right]. \tag{5.7}$$

These relations indicate that, for a given experiment (i.e. for given Q and D), each of the 2.5/1-declining straight lines on the log-log $(B/h; h/D)$-plane is associated with a certain value of α and thus of Fr ("$Fr = const$" lines in Figure 5.8).

Consider the regime development on the $(B/h; h/D)$-plane with the aid of the moving point m starting at the adjusted initial point $\hat{P}_0$. In this case, for any $t \in [\hat{T}_0; T_R]$, we can adopt $B \approx B_R$ and thus

$$\frac{B}{h} \approx \frac{B_R/D}{h/D} = \frac{(const)}{h/D}, \tag{5.8}$$

which is a 1/1-declining straight line l' (Figure 5.8): the point m is moving in the direction of arrow. The regime-formation stops when m "hits" that particular $Fr = const$ line which corresponds to $Fr = (Fr)_R$, i.e. when it reaches the regime point P_R.

If the regime-development is entirely by meandering (with no input of degradation-aggradation), then it can be accomplished only if the regime point P_R is above the lower boundary $\mathcal{L}_M$ of the meander region (Figure 5.8) – otherwise the meander development stops when m reaches $\mathcal{L}_M$. Similarly, P_R cannot be reached by m if the channel sinuosity $S_0/S_R = \sigma$ exceeds ≈ 8.5 (where the meander loops begin to touch each other (Figure 6.3)).

5.2 BRAIDING AND REGIME DEVELOPMENT

5.2.1 Overall description of braiding

(i) In the preceding section we have seen that the conventional regime and meandering data (as well as the data of alternate bars) occur in that part of the $(B/h; h/D)$-plane which is *below* the line $\mathcal{L}_{1,2} \equiv \mathcal{L}_{M,B}$ (signifying the upper boundary of the single-row LSHCS's). This means that the points P_0 and $\hat{P}_0$ of the channels which subsequently develop into the regime channels are situated below the line $\mathcal{L}_{1,2}$ – otherwise they could not have one-row LSHCS's (needed to produce the regime channels by meandering).

Consider now those initial channels whose points P_0 are situated *above* the line $\mathcal{L}_{1,2}$.[8] Such (very wide) initial channels contain more than one-row LSHCS's ($n > 1$, Figure 2.12), and their flow velocities at both banks are not negatively correlated (i.e. the increment of u at the left bank is not necessarily accompanied by its decrement at the right bank, as is the case in one-row LSHCS's, where the structures themselves extend throughout the flow width). The time-average streamlines of the initial channel mentioned will be e.g. as those shown in Figure 5.9a (which corresponds to $n = 2$); and they will generate on the bed surface, as has been explained in Chapter 2, "their" n-row bars ($n = 2$ in Figure 5.9a). The points B and C in Figure 5.2 are due to the channels of this type.

(ii) If the banks were rigid, as in the case of flume experiments, then we would have indeed the *submerged* multiple bars as the outcome of the experiment. However, the banks of a natural stream are erodible and the flow, which is impeded by the bars, erodes them. The bar near a bank "forces the water into the flanking channels, which, to carry the flow, deepen and cut laterally into the original banks. Such deepening locally lowers the water surface and the central bar emerges as an island" (p. 39, Leopold and Wolman 1957). In short, "when bank erosion occurs (and the channel widens), the water level (and thus the flow depth) decreases, and the bars emerge" (p. 1761, Schumm and Khan 1972).[9] In other words, the cross-sectional geometry develops as illustrated in Figure 5.9b. The initial stream ($k = 1$) widens and a sequence (along x) of islands I_1 is being created ("island-braiding"). With the passage of time, the elevation of the free surface continually decreases, while the size of the islands (in plan view) increases. Thus the flow around islands I_1 splits into a pair of secondary streams ($k = 2$). Owing to the same formation-mechanism, the secondary streams produce "their own" islands I_2, and their own pair of secondary streams ($k = 4$), ..., and so on (Figure 5.9c). This consecutive branching process, referred to as *braiding*, perpetuates itself until the equilibrium or the regime state is reached. (Figure 5.9d shows another possible scheme of channel splitting: k increases as an arithmetic sequence, but the flow rate conveyed by the central channels is larger).

---

[8] Braiding and the associated regime development start with the continual channel-widening (at $t = 0$). Hence the "adjustment" of the initial channel by its widening (in the sense of Sub-section 4.4.1 (ii)) has no meaning in the present context, and therefore no initial point $\hat{P}_0$ is present in the study of braiding below.

[9] The original past tense has been altered to the present tense by the authors; brackets added for clarification.

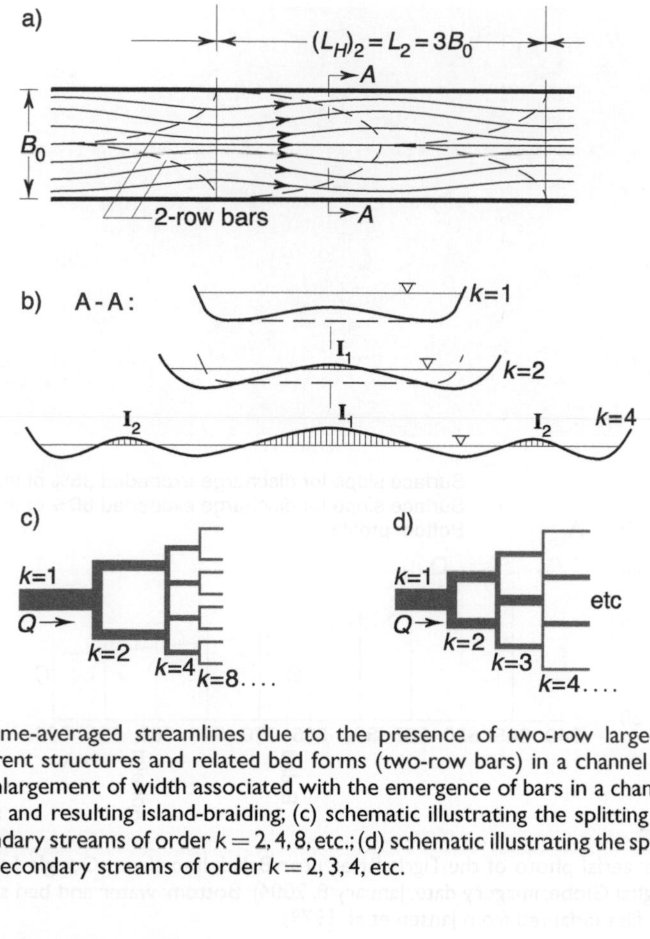

Figure 5.9 (a) Time-averaged streamlines due to the presence of two-row large-scale horizontal coherent structures and related bed forms (two-row bars) in a channel with rigid banks; (b) enlargement of width associated with the emergence of bars in a channel with erosible banks and resulting island-braiding; (c) schematic illustrating the splitting of a stream into secondary streams of order $k = 2, 4, 8$, etc.; (d) schematic illustrating the splitting of a stream into secondary streams of order $k = 2, 3, 4$, etc.

5.2.2 The role of the slope

(i) The occurrence of braiding is, as a rule, associated with "large" values of the slope S. This association has been revealed first by Leopold and Wolman (1957) (see also Leopold et al. 1964) by means of their well known S versus Q plot (which can be found also e.g. in Jansen et al. 1979, Chang 1988a). However, it is a misconception to think, as is frequently done, that braiding *causes* the stream slope to increase in the course of its development (as e.g. meandering causes it to decrease). On the contrary, it is the elevated slope of the terrain which promotes braiding. A natural single-channel stream starts to braid, and remains braiding, in those stretches where the original topographic slope of the terrain is steeper than usual. In the regions where this slope flattens, braiding disappears: separate channels merge again into a single channel (see e.g. the river regions AB (braiding) and BC (meandering) shown in Figure 5.10). The reason for the aforementioned can be explained as follows.

(ii) Consider a terrain which can be idealized by two parallel planes, AB and CD, and by a steeper plane BC between them (Figure 5.11). The characteristics of AB

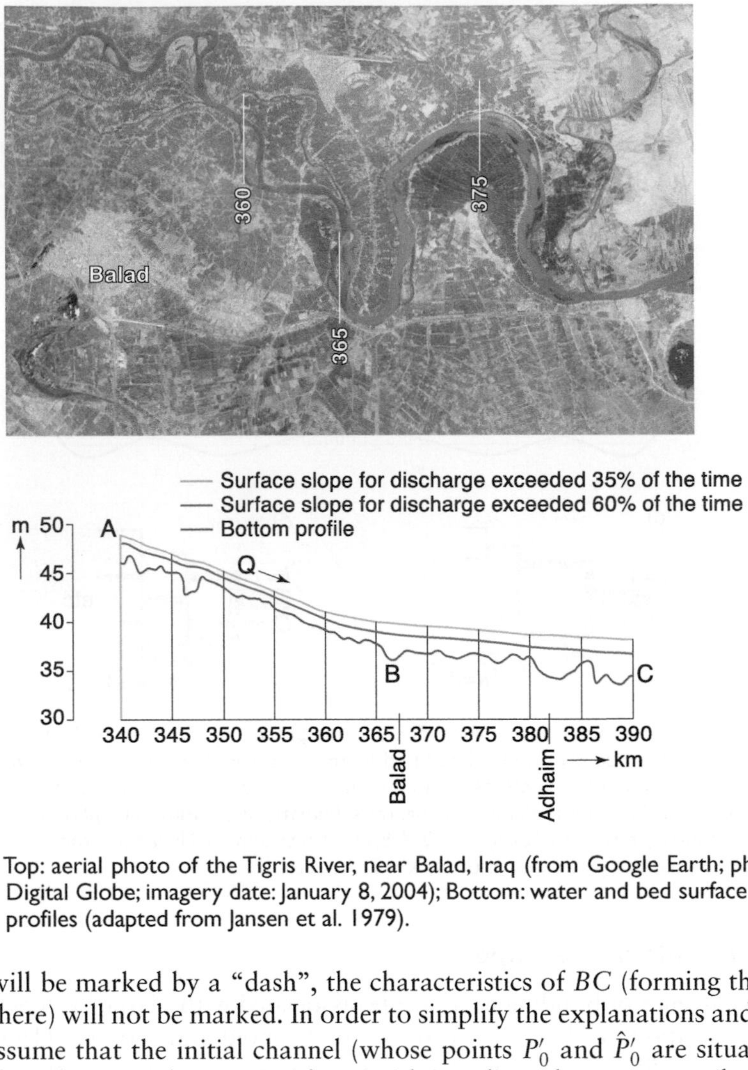

Figure 5.10 Top: aerial photo of the Tigris River, near Balad, Iraq (from Google Earth; photo source: Digital Globe; imagery date: January 8, 2004); Bottom: water and bed surface longitudinal profiles (adapted from Jansen et al. 1979).

and CD will be marked by a "dash", the characteristics of BC (forming the focus of attention here) will not be marked. In order to simplify the explanations and notation, we will assume that the initial channel (whose points P'_0 and $\hat{P}'_0$ are situated below $\mathcal{L}_{1,2}$) is selected so as to be in coincidence with its adjusted counterpart (hence, in the following any $\hat{a}$ will be denoted simply as a).

Suppose that a constant-width initial channel is excavated throughout $ABCD$ (Figure 5.11). We have

$$S_0 = \lambda S'_0 \quad \text{and} \quad B_0 = B'_0 \tag{5.9}$$

where $\lambda \gg 1$. The channels convey the same Q ($= const$), and therefore

$$B_0 h_0^{3/2} c_0 (gS_0)^{1/2} = B'_0 h_0'^{3/2} c'_0 (gS'_0)^{1/2} \quad (= Q), \tag{5.10}$$

which yields, ignoring the possible difference between c_0 and c'_0,

$$\frac{h_0}{h'_0} = \frac{1}{\lambda^{1/3}}, \quad \frac{B_0/h_0}{B'_0/h'_0} = \lambda^{1/3}, \tag{5.11}$$

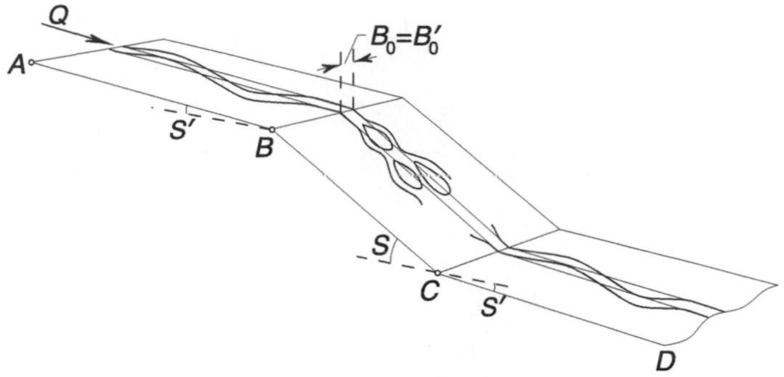

Figure 5.11 Occurrence of braiding in the steep river region BC.

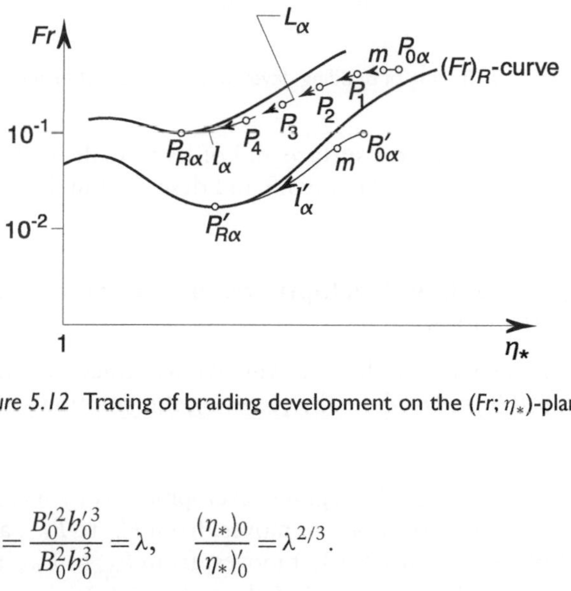

Figure 5.12 Tracing of braiding development on the $(Fr; \eta_*)$-plane.

and

$$\frac{(Fr)_0}{(Fr)'_0} = \frac{c_0^2 S_0}{c_0'^2 S'_0} = \frac{B_0'^2 h_0'^3}{B_0^2 h_0^3} = \lambda, \qquad \frac{(\eta_*)_0}{(\eta_*)'_0} = \lambda^{2/3}. \tag{5.12}$$

(e.g. if $\lambda = 8$, then $h_0 = h'_0/2$ and $(Fr)_0 = 8(Fr)'_0$).

Let $P_{0\alpha}$ and $P'_{0\alpha}$ be the "initial points" of the channels BC and AB (or CD) in the $(Fr; \eta_*)$-plane (Figure 5.12), while $P_{0\beta}$ and $P'_{0\beta}$ are those in the $(B/h; h/D)$-plane (Figure 5.13). From Eqs. (5.11) and (5.12) it is clear that the ordinates of the points $P_{0\alpha}$ and $P_{0\beta}$ are *larger* than those of their counterparts $P'_{0\alpha}$ and $P'_{0\beta}$, and Figures 5.12 and 5.13 are drawn accordingly. But this means that, in some cases, $P'_{0\beta}$ is *below* while $P_{0\beta}$ is *above* the line $\mathcal{L}_{1,2}$ forming the boundary between the zones of *single-row* and *multiple-row* LSHCS's (as shown in Figure 5.13).

(iii) Suppose now that $P'_{0\beta}$ is indeed below the line $\mathcal{L}_{1,2}$ (i.e. it is in the zone of the single-row LSHCS's), while $P_{0\beta}$ is above it (i.e. it is in the zone of the multiple-row LSHCS's). In this case, after the beginning of experiment the flow in the initial channel AB (or CD) will start to meander, while that in BC will start to widen, and it will start

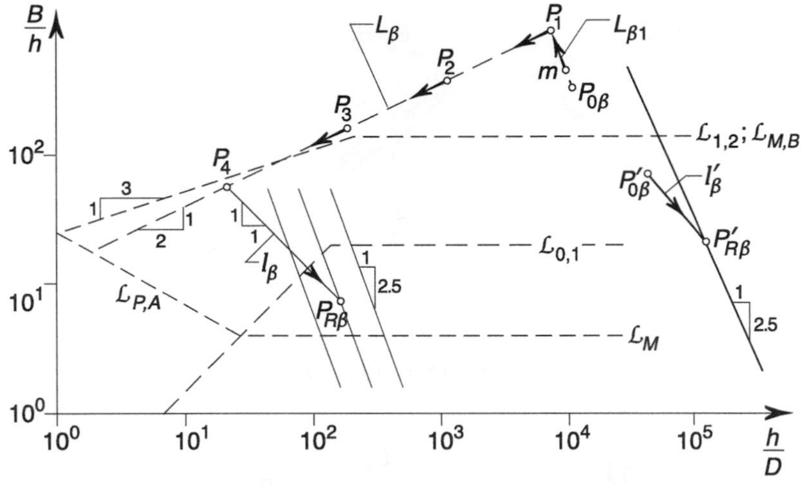

Figure 5.13 Tracing of braiding development on the $(B/h; h/D)$-plane.

to reduce the elevation of its free surface and of its flow depth h_0 (i.e. it will start to exhibit braiding as illustrated in Figure 5.9 and described in Sub-section 5.2.1 (ii)).

5.2.3 Tracing of braiding development on the $(Fr; \eta_*)$- and $(B/h; h/D)$-planes

As done in Sub-section 5.1.3, in the following we will trace the braiding development on the $(Fr; \eta_*)$ and $(B/h; h/D)$-planes by picturing the "motion" of a point m on these planes.

(i) In the channels AB and CD the regime development is by meandering. And in the $(Fr; \eta_*)$-plane of Figure 5.12 the point m moves, from $P'_{0\alpha}$ to $P'_{R\alpha}$, along a line l'_α, while in the $(B/h; h/D)$-plane of Figure 5.13, it moves, from $P'_{0\beta}$ to $P'_{R\beta}$, along a line l'_β. Both of these cases were considered earlier (in Sub-section 5.1.2).

(ii) Consider now the channel BC whose development forms the topic of this sub-section. As previously mentioned, after the beginning of experiment this initial channel widens while its depth reduces, i.e. B increases while h decreases. Consequently, in the $(B/h; h/D)$-plane, the point m moves starting from $P_{0\beta}$, along a declining-line $L_{\beta 1}$ in the direction of arrow (Figure 5.13). This motion of m continues until the progressively lowering free surface touches the highest points of the bars. As a result of this "touching", the (single-channel) initial flow degenerates into a multitude of interconnected flows whose boundaries in the flow-plan are islands and banks. No reliable formulation of the regime-development can be produced at the present for this early stage of "island-braiding" – which can equally well be viewed as a single-flow obstructed by islands or as a multitude of interconnected irregular flows. Hence, currently, we have no alternative but to confine the formulations to those comparatively later stages where the presence of the "stream-braiding" (in the sense of Figure 5.9c) can be assumed.

Since the regime development of the channel width takes place relatively fast, we postulate (as has been done already in Yalin 1992) that at the stage k, each of the k-streams conveys on average Q/k and that the flow width B_k of each individual stream can be identified (on average) with its regime value given by Eq. (4.26):

$$B_k = (\alpha_B)_k \sqrt{\frac{Q/k}{\upsilon_{*cr}}}. \tag{5.13}$$

Here, $(\alpha_B)_k$ is a known function of Ξ, c_k and $(c_f)_k$, which is determined by the relations (4.27)-(4.29). Using Eq. (5.13) in the expression of the dimensionless variable $N = Q/(BD\upsilon_{*cr})$ (see Eq. (3.36)), we determine for the stage k

$$N_k = \frac{\lambda_\alpha Q/k}{(B_1/\sqrt{k})D\upsilon_{*cr}} = \frac{Q}{B_1 D\upsilon_{*cr}} \cdot \frac{\lambda_\alpha}{\sqrt{k}} \qquad (\text{with } \lambda_\alpha = (\alpha_B)_1/(\alpha_B)_k). \tag{5.14}$$

The branching streams corresponding to any k are very shallow, their sinuosity usually is insignificant ($1 < \sigma <\approx 1.5$, say), and they are all spread on the platform BC whose slope is S. Hence it would be reasonable to adopt, for any k,

$$S_k = S \quad (= const). \tag{5.15}$$

Substituting Eqs. (5.13) and (5.15) in the resistance equation

$$\frac{Q}{k} = B_k h_k^{3/2} c_k \sqrt{g S_k}, \tag{5.16}$$

one determines

$$h_k = \frac{1}{k^{1/3}} \left(\frac{Q\upsilon_{*cr}}{g(\alpha_B)_k^2} \right)^{1/3} \frac{1}{(Fr)_k^{1/3}}, \tag{5.17}$$

which is but Eq. (4.31) expressed for the stage k. Here

$$(Fr)_k = c_k^2 S. \tag{5.18}$$

[Note that in the case of braiding the decrement of the Froude number is (mainly) due to the decrement of the resistance factor (c_k) – and not due to the decrement of both c and S as in the case of meandering].

Since $B_k \sim 1/k^{1/2}$ while $h_k \sim 1/k^{1/3}$, we have $B_k/h_k \sim 1/k^{1/6}$ and $h_k/D \sim 1/k^{1/3}$. Hence

$$(B/h)_k \sim (h/D)_k^{1/2}, \tag{5.19}$$

which indicates that if $(\alpha_B)_k$ and c_k *would not* vary with the stage k, then the points $P_1, P_2, ..., P_k$ representing the individual channels at the stages 1, 2, ..., k in the $(B/h; h/D)$-plane would form a (broken) 1/2-declining line L_β (Figure 5.13). And the point m would move (or rather "jump") from one point P_k to the next as the integer k increases. In reality $(\alpha_B)_k$ and c_k are, of course, not constant and L_β indicates merely the general direction of the curve connecting the factual points P_k.

Since the inclination 1/2 of the line L_β is larger than the inclination 1/3 of the line $\mathcal{L}_{1,2}$, the latter line will eventually be intersected by the former, i.e. one of the points P_k (which in the case of Figure 5.13 is P_4) will enter the region of the single-row LSHCS's. Hence, starting from this point (P_4), the regime formation of individual channels will continue by meandering – along a continuous 1/1-declining line l_β, until the regime point $P_{R\beta}$ is reached.

Note from Eq. (5.14) that the value of N_k, just like that of Fr and η_*, decreases with the increment of k. Hence, the development of an individual channel in the $(Fr; \eta_*)$-plane (which must, of course, also be by "jumps" of m from one point P_k to the next) should proceed in such a manner that all three, N_k, $(Fr)_k$, $(\eta_*)_k$, decrease with the increment of k (broken line L_α in Figure 5.12). The last stage $P_4 \to P_{R\alpha}$, shown by a continuous line l_α, signifies the final development by meandering.

5.2.4 Additional remarks on braiding

(i) Although the conditions in the $(Fr; \eta_*)$-plane are described above in terms of one of the individual k-streams, they can be taken so as to correspond to the braiding stream system *as a whole*. Indeed, since the bank friction is neglected, the values of the (intensive) characteristics h, S, c, u_{av}, Q/B $(=q)$, and thus of Fr, η_*, N $(=q/Dv_{*cr})$, do not depend on the channel width. Consequently, at any given division stage k, they have the same values for an individual channel (B_k, h_k, S), as well as for the whole system of channels (kB_k, h_k, S).

(ii) In most of the practical cases the regime state $P_{R\beta}$ can be reached only in seldom cases, for the braiding development depicted in Figures 5.12 and 5.13 becomes interrupted whenever the (usually irregular) individual streams happen to "touch" each other (in analogy to the interruption of meander development by the meander loops "touching" each other). And this is why a braiding stream system usually is unstable.

(iii) In the case of meandering, the regime state is approached (and whenever possible is achieved) because of the continual decrement of the channel slope S (from S_0 to S_R), and thus the continual decrement of the channel Froude number $Fr = c^2 S$ (from $(Fr)_0$ to $(Fr)_R$) – the channel and its Q maintaining their integrity in the process. In the case of braiding, it is the opposite: S remains constant, the channel is branching; the number of branches progressively increases, their flow rate decreases: $Q/2$, $Q/3$, ..., Q/k, ..., etc. The smaller is Q/k, the larger is "its" regime slope $(S_R)_k \sim (Q/k)^{n_S}$ (where the exponent is *negative* (see Eq. (4.4)). Hence, the progressive decrement of the flow rate ($Q/2$, $Q/3$, ..., Q/k) causes the progressive increment of $(S_R)_k$, and the regime state of the k-stream (and thus of the braiding system as a whole) is reached when k, k/Q, and $(S_R)_k$ become as sufficiently large as to render $(S_R)_k$ equal to the platform slope S.

In summary, one can say that if to achieve its regime state the stream resorts to meandering, then its regime state is achieved when the initial (valley) slope S_0 decreases down to S_R (Figure 5.14a). And if it resorts to braiding, then the regime state is achieved when the regime slope $(S_R)_k$ of the k-streams becomes as large as the platform slope S (Figure 5.14b).

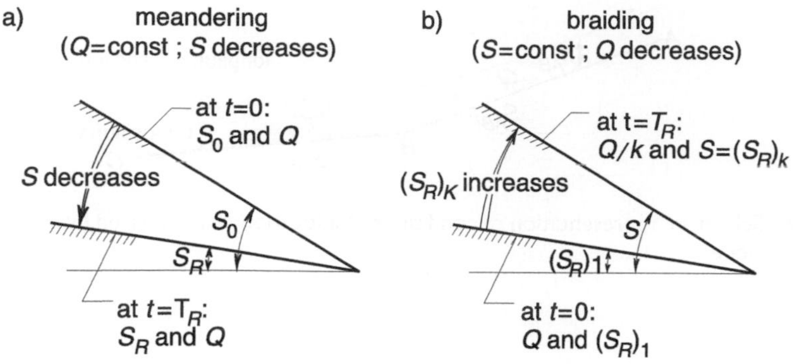

Figure 5.14 Changes in stream slope during the regime development. (a) Decrease in stream slope from S_0 to S_R when the stream resorts to meandering; (b) increase in slope of the k-streams when the stream resorts to braiding.

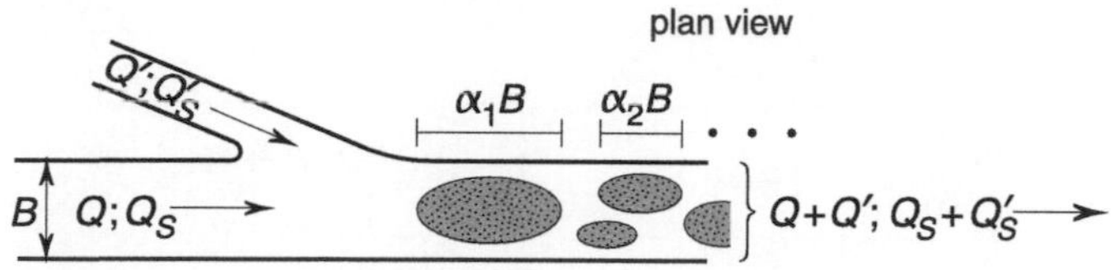

Figure 5.15 Schematic representation of conditions at a confluence.

In the case of meandering, the decrement of S_0 means the increment of the channel sinuosity σ; in the case of braiding, the increment of $(S_R)_k$ implies the decrement of Q/k. i.e. σ and k have analogous effects, as far as the regime development is concerned.

(iv) In order to keep a certain system of presentation going, it has been assumed throughout this section that braiding originates from a single initial channel whose width-to-depth ratio B_0/h_0 is as sufficiently large as to yield the multiple-row bars on its bed surface. However, this kind of a (formal) braiding-initiation is not the *only* possible braiding initiation. The formation of "islands" and the consequent stream-splitting can take place in an alluvial stream whenever its bed becomes covered by the deposition zones (scaling with the flow width) and its banks are erodible. For example, if the ratio Q'_s/Q' of a tributary is larger than the Q_s/Q-value of the main stream, then $(Q'_s + Q_s)/(Q' + Q) > Q_s/Q$, which means that the stream region downstream of the confluence is likely to acquire the deposition zones as well as the consequent braiding (Figure 5.15). On the other hand, the occurrence of depositions is only a necessary (but not sufficient) condition for the occurrence of braiding. Indeed, consider e.g. the conditions sketched in Figure 5.16, where $S_{AB} > S_{BC}$, while $h_{AB} < h_{BC}$. The flow rate Q which transports Q_s along AB will not be able to transport it further along BC, whose slope is flatter. Hence BC will inevitably become covered by depositions. Yet, these depositions may not emerge as "islands", for the flow depth along (the flatter) BC is larger than that along AB (in contrast to Figure 5.11, where the flow depth along the steeper BC is smaller than that along AB).

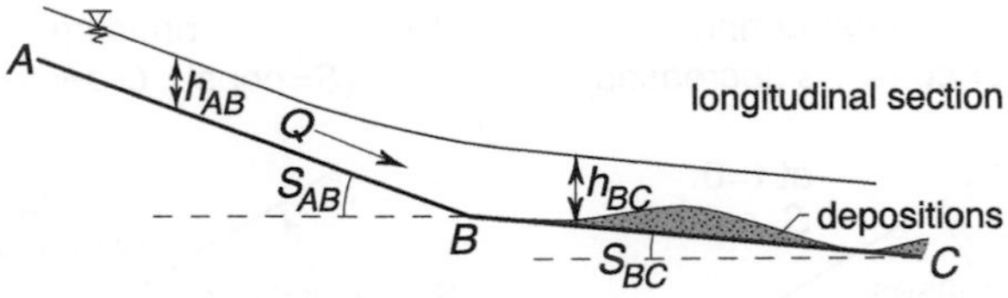

Figure 5.16 Schematic representation of conditions at a location characterized by a sudden change in slope.

Figure 5.17 Inland delta of the Rio Negro, Brasil.

5.2.5 Delta formation

(i) If the braiding region of a river can be called the "inland delta" (e.g. Rio Negro, Figure 5.17), then a delta can be referred to as a "stream-end braiding". Indeed, consider Figure 5.18a which depicts schematically a prismatic initial channel entering a semi-infinite stationary fluid. As is well known from fluid mechanics, the vertically averaged streamlines s of such a channel flow can be assumed to be straight parallel lines only in the channel regions far upstream of the channel-end I-I. Near the channel-end the streamlines of the flow are neither straight nor parallel: in the central region of the channel the streamlines *diverge* from each other; in the region near the banks, they *converge* to each other and to the banks. Thus in the bank-region we have $\partial u/\partial x > 0$, and thus $\partial q_s'/\partial x > 0$ (where q_s' typifies the transport rate in the bank region), whereas in the central region, $\partial u/\partial x < 0$, and thus $\partial q_s/\partial x < 0$ (where q_s is the transport rate

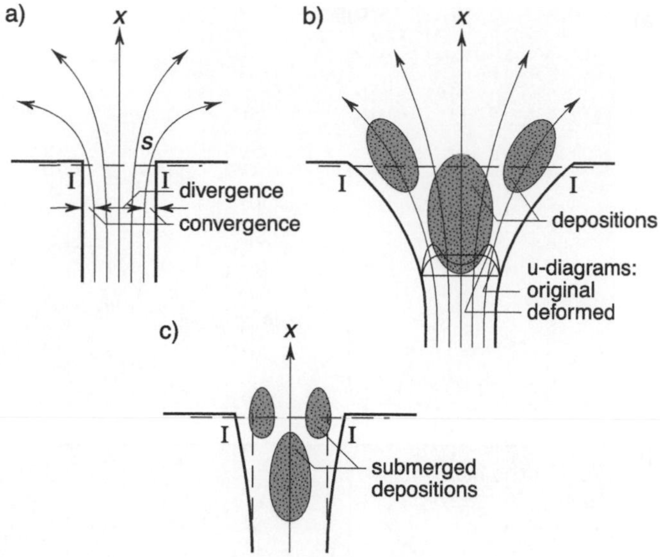

Figure 5.18 Prismatic initial channel entering a stationary fluid. (a) Vertically-averaged streamlines; (b) example of extensive bank divergence and prominent depositions occurring as a result of flow deceleration; (c) example of moderate bank divergence and associated depositions.

in the central region). Consequently, with the passage of time, the banks of the initial channel must be eroded, while the bed of the central region must be subjected to deposition – the intensity of each of these trends must increase when the section I-I is approached. This means that with the passage of time the banks must acquire a divergent (∇-like) shape, while the central part of the bed must become covered by depositions scaling with B_0. Clearly, the exact shape of the ∇-like divergence of banks, at any given instant, as well as the prominence of depositions, is determined largely by the nature of channel flow and sediment. e.g. if Q_s/Q and B_0/h_0 are "large", and the banks are easily erodible, then the ∇-like divergence of banks can be extensive, and the depositions can be prominent. Such depositions are likely to emerge as "islands", and consequently to induce an island-braiding (Figure 5.18b) or even a stream-braiding. Conversely, if Q_s/Q and B_0/h_0 are "small", and the banks are not easily erodible, then the ∇-like divergence of banks may not be substantial, while the depositions may not be as prominent as to protrude the free surface and thus induce braiding. In this case the stream will terminate in the form of a somewhat widening single channel. Such a stream-end is usually referred to as the "river-mouth" (Figure 5.18c).[10]

Acting as submerged weirs with limited width, the emerging depositions (or "islands") deform the original flow velocity field so as to promote a further deposition of material and thus a further erosion of banks (see the u-diagrams in Figure 5.18b). Consequently, the plan area of the delta-system as a whole increases, the individual streams comparatively narrowing in the process.

[10] Some authors prefer to use the term "estuary" for this purpose (Mangelsdorf et al. 1990).

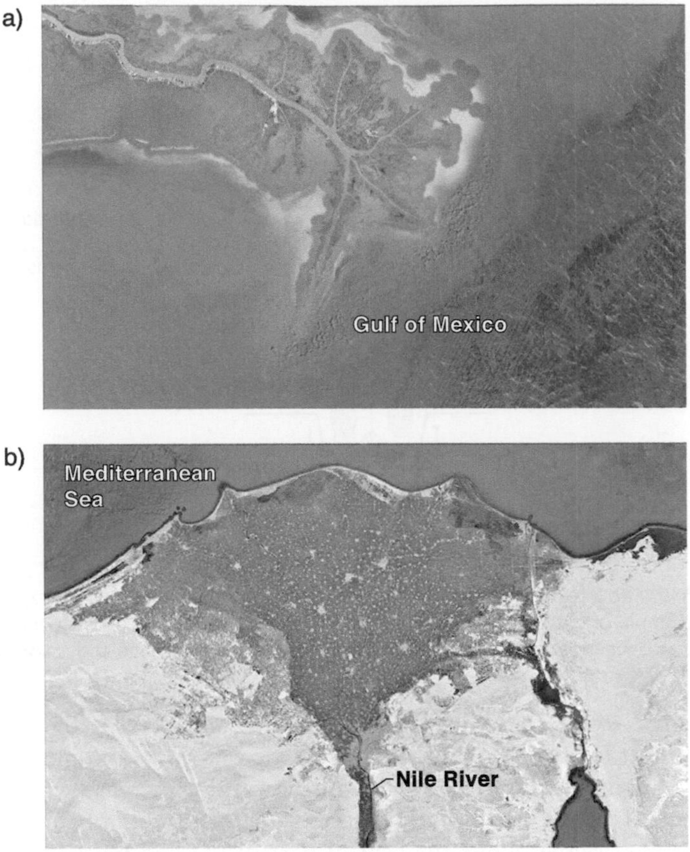

Figure 5.19 Examples of river deltas (satellite images). (a) Mississippi River delta (from Google Earth; photo source: TerraMetrics (image by NOAA)); (b) Nile River delta (from Google Earth; photo source: Landsat; imagery date: December 12, 2015).

(**ii**) It has been assumed that the alluvial stream enters a semi-infinite static fluid. Yet, most of the large natural rivers enter a fluid mass (lake, sea, ocean) which is not static: currents, short waves, long (tidal) waves, etc. can be present. Clearly, these "sea-born" factors must contribute also, and in fact substantially, to the delta formation: "Deltas are consequences of the conflict between rivers and the sea" (Wright and Coleman 1976). It has been found that the plan-geometry exhibited by deltas varies considerably depending on the relative "strength" of fluvial- and marine-forces. The more the delta formation is "fluvial-dominated", the more is the irregularity and the larger is the number of protrusions in the shoreline – as e.g. in the case of the Mississippi River delta shown in Figure 5.19a ("although the sea actively strives to mold the alien sediments to its own design" ... "the sediments are deposited at the coast more rapidly than they can be reworked by the waves" (Wright and Coleman 1976)). By contrast, the more the delta formation is marine, i.e. wave-dominated, the more regular is the shoreline ("the waves quickly mold the river-born sediment into a (simple) marine-coastline

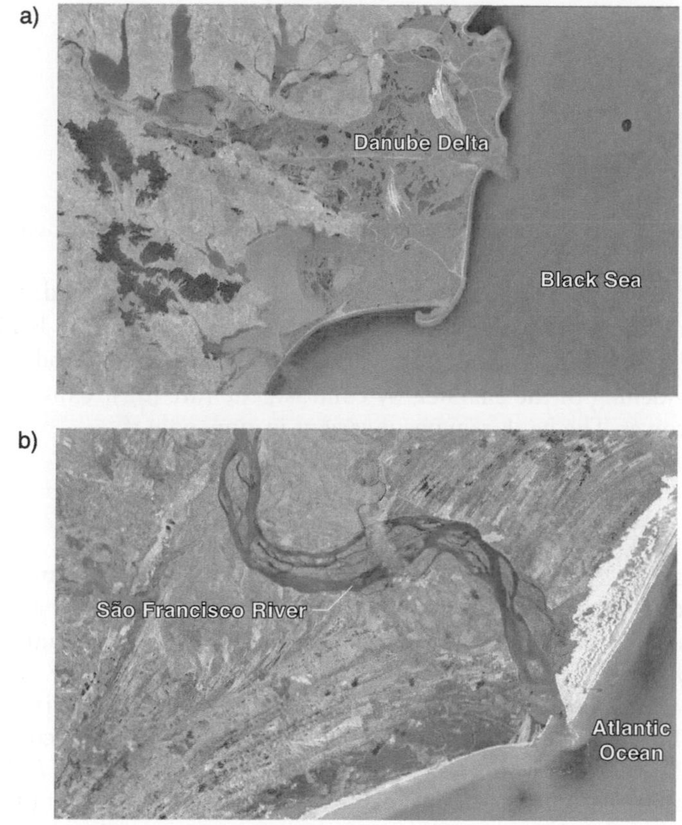

Figure 5.20 Further examples of river deltas (satellite images). (a) Danube River delta (from Google Earth; photo source: Landsat/Copernicus; imagery date: December 13, 2015); (b) São Francisco River delta (from Google Earth; photo source: CNES/Astrium and TerraMetrics; imagery date: June 25, 2016).

configuration" (Wright and Coleman 1976)). The delta of the Nile River shown in Figure 5.19b can serve as example for this case. The shoreline of the Danube River delta in Figure 5.20a, with a pronounced stream-braiding, typifies the intermediate case. The São Francisco River (Brasil) in Figure 5.20b can be shown as an example for a "river mouth", with island braiding. (See AAPG Reprint Series 1976, Bates 1976, Coleman 1976, Wright and Coleman 1976, Nemec 1990, Postma 1990, for more information on deltas).

5.3 TIME DEVELOPMENT OF A SINGLE-CHANNEL STREAM

5.3.1 Adjustments of channel geometry

(i) Consider the cross-section of a sediment transporting straight alluvial stream (Figure 5.21): Q and S have certain constant values. Two different transport rates are crossing (per unit flow length) the imaginary vertical section σ_b passing through the

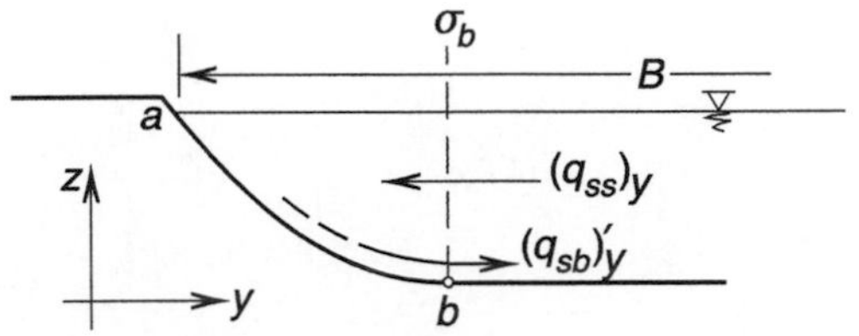

Figure 5.21 Lateral sediment transport rates in the bank region of a cross-section of flow.

bank-end point b. One is the y-component $(q_{sb})'_y$ of the bank-load, the other is the suspended-load $(q_{ss})_y$ (which migrates from the bed-region to the bank-region, and eventually deposits on the bank ab). The transport rate $(q_{sb})'_y$ erodes the bank and causes the widening of the channel; by contrast, the rate $(q_{ss})_y$ causes the channel to become narrower. The bank remains invariant in time only if

$$(q_{sb})'_y - (q_{ss})_y = 0, \tag{5.20}$$

where both $(q_{sb})'_y$ and $(q_{ss})_y$ are treated as positive quantities.

In the case of wide channels under study, the mechanical structure of flow at the banks, and thus the value of $(q_{sb})'_y$ (which is generated by the flow at the banks) can also be considered to be practically independent of B. The same cannot be said with regard to $(q_{ss})_y$, which stems from the central region of the flow. The larger is B, the larger is the extent of lateral fluctuations of the large-scale horizontal turbulence (large-scale horizontal eddies and coherent structures) and of the consequent lateral turbulent diffusion. Hence $(q_{ss})_y$ must be an increasing function of B. But this means that a continual increment of B must cause a continual increment of $(q_{ss})_y$, with $(q_{sb})'_y$ remaining relatively unaffected. The time-increment of B can take place only if $(q_{sb})'_y > (q_{ss})_y$. Yet, from the above mentioned, it is clear that the intensity of this inequality must progressively decrease as B increases. Eventually B must reach such a value that would render the inequality $(q_{sb})'_y > (q_{ss})_y$ to become the equality (5.20). In this case, the channel widening stops: the *equilibrium* flow width is achieved.

(ii) Consider now a regime channel formation experiment. Let us assume, for a moment, that this experiment is conducted so that the slope S is kept constant, at a value S_k, say (e.g. by a continual tilting of the alluvial platform). In this case the channel widening, and in fact the channel formation experiment itself, terminates when the continually increasing B reaches (at T_k, say) the value of the equilibrium flow width B_k. The value of B_k depends on the value of the slope S_k (hence the subscript k in B_k). The conditions described are depicted in Figure 5.22, which is but a special case of Figure 4.5. Note that the equilibrium width B_k is but the regime width of this experiment which corresponds to $S_k = const$. The analogous is valid in the case of an actual regime channel formation process (experiment), where the slope S is not constant, but continually decreases during T_R.

The mechanism of the actual experiment can perhaps be better explained with the aid of the following "discrete model". Replace the continuous time-decrement of S by its step-like time-decrement S_0, ..., S_i, S_{i+1}, ..., S_R; the "steps" are separated from

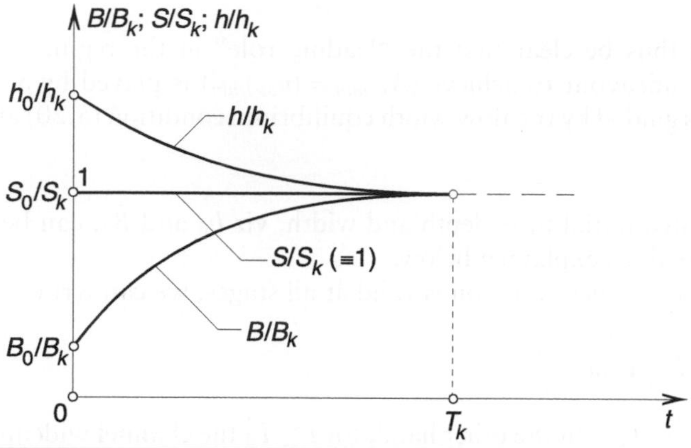

Figure 5.22 Schematic plot showing the variation with time of flow width B and flow depth h as a stream evolves towards its regime state, in the case of an experiment conducted with constant slope S.

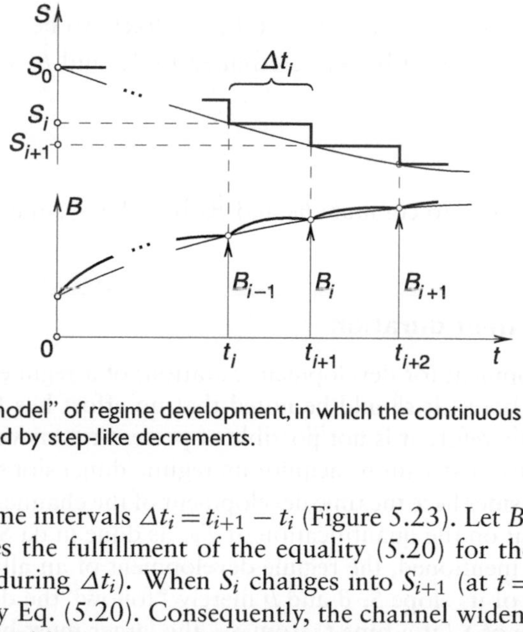

Figure 5.23 "Discrete model" of regime development, in which the continuous decrement of the slope S is replaced by step-like decrements.

each other by the time intervals $\Delta t_i = t_{i+1} - t_i$ (Figure 5.23). Let B_i be the equilibrium width which ensures the fulfillment of the equality (5.20) for the slope S_i (which is assumed to persist during Δt_i). When S_i changes into S_{i+1} (at $t = t_{i+1}$), the width B_i can no longer satisfy Eq. (5.20). Consequently, the channel widens further, until such "new" equilibrium width B_{i+1} is reached which satisfies Eq. (5.20) for S_{i+1}. Then S_{i+1} changes into S_{i+2}, ..., and so on. The actual continuous S- and B-curves are but the "smoothened" counterparts of the step-like S_i- and B_i-diagrams just described. One can say that the actual B-diagram is formed by the pseudo-equilibrium flow widths B_i which incessantly adjust themselves as to fit the continually decreasing S.[11] Only

[11] "Pseudo-equilibrium" because, strictly speaking, Eq. (5.20) cannot be satisfied if S, B and h vary (no matter how slowly) during T_R.

when S reduces into S_R, is the final equilibrium flow width, i.e. the regime width B_R, established.

It should thus be clear that the "leading role" in the regime development (i.e. in the flow's endeavour to achieve $(A_*)_{min} = (u_{av})_{min}$) is played by S: B and h merely "follow" S (as guided by the flow width equilibrium condition (5.20) and the resistance equation).

(iii) The adjusted initial flow depth and width, viz $\hat{h}_0$ and $\hat{B}_0$, can be determined (or, rather, estimated) as explained below.

Since the resistance equation is valid at all stages, we can write

$$Q = \hat{B}_0 \hat{h}_0 \hat{c}_0 \sqrt{g S_0 \hat{h}_0}, \tag{5.21}$$

for $\hat{S}_0 \approx S_0$ at $t = \hat{T}_0$. On the other hand, for $t \geq \hat{T}_0$ the channel widening is only feeble, and so must be the difference $(q_{sb})'_y - (q_{ss})_y$ (which is exactly zero only when the channel is in regime). This suggests that the regime-width formula (4.26), viz

$$B_R = \alpha_B \sqrt{Q/v_{*cr}}, \tag{5.22}$$

where α_B is determined by S_R and h_R (at T_R), is likely to be applicable also at $\hat{T}_0$ – provided that α_B is replaced by $\hat{\alpha}_B$ determined by S_0 and $\hat{h}_0$ (at $\hat{T}_0$). Adopting this approach, we arrive at

$$\hat{B}_0 = \hat{\alpha}_B \sqrt{Q/v_{*cr}}, \tag{5.23}$$

which makes it possible to compute $\hat{h}_0$ and $\hat{B}_0$ by solving simultaneously Eqs. (5.21) and (5.23).

5.3.2 Development duration

The time of development (or development duration) of a regime channel, T_R, is often mentioned in this book. It should be noted that no effort has been made to date to quantify T_R, and therefore it is not possible at present to answer the question "How long does it take for a stream to acquire its regime dimensions?" It therefore seems appropriate to consider here the time development of the channel slope S in an attempt to throw some light on the quantification of T_R, as done in da Silva (2009).

As previously mentioned, the regime development of an alluvial stream is by the time-development of its slope S; B and h merely "follow" the development of S. The farther is the value of S (at a time t) from S_R, the larger must be the "S-development speed". Hence the following relation can be adopted:

$$\frac{dS}{dt} = -\alpha(S - S_R), \tag{5.24}$$

in which α, which has the dimensions [1/time], must be, among others, an increasing function of the channel's sediment transport capacity – to be characterized by a representative sediment transport rate, $\tilde{q}_s$, say). Note that if e.g. $\tilde{q}_s = 0$, then α, and thus dS/dt, must also be zero.

By integrating Eq. (5.24) from $t = 0$ (when $S = S_0$) to t, one obtains

$$\frac{S}{S_R} = 1 + \left(\frac{S_0}{S_R} - 1\right) \cdot e^{-\alpha t}. \quad (5.25)$$

[It is assumed in the present considerations that $\hat{T}_0 \ll T_R$, and therefore the integration of Eq. (5.24), which strictly speaking should be carried out using $t = \hat{T}_0$ as lower limit of integration, was instead carried out using $t = 0$ (as lower limit of integration).]

Eq. (5.25) correctly gives $S = S_0$ when $t = 0$, and it indicates that $S = S_R$ is achieved theoretically only when $t = \infty$. Hence the practical duration T_R of the S-development (and thus of the development of the regime channel) must be obtained by using the methods analogous to the determination of the boundary-layer thickness δ, terminal velocity w_s, etc. Accordingly, T_R can be defined as the value of t when S reduces to $S_R \cdot (1 + \epsilon)$, where ϵ is "small" but different from zero (e.g. $\epsilon = 0.01$). Substituting $T = T_R$ and $S = S_R \cdot (1 + \epsilon)$ in Eq. (5.25), one obtains

$$\epsilon = \left(\frac{S_0}{S_R} - 1\right) \cdot e^{-\alpha T_R}, \quad (5.26)$$

which makes it possible to determine T_R if α is known and ϵ is selected. A special future research directed towards the determination of α, and thus the quantification of T_R, would seem particularly worthwhile.

Finally, note that Eq. (5.25) gives $S = S_R$ at $t \to \infty$, and $S = S_R \cdot (1 + \epsilon)$ at $t = T_R$. If it is desired that $S = S_R$ occurs when $t = T_R$, then every αt in Eq. (5.25) should be replaced by $\alpha t/(T_R - t)$:

$$\frac{S}{S_R} = 1 + \left(\frac{S_0}{S_R} - 1\right) \cdot e^{-\alpha t/(T_R - t)}. \quad (5.27)$$

The graph of S/S_R in Figure 4.5 is drawn in agreement with Eq. (5.27).

5.4 LARGE-SCALE BED FORMS AND REGIME DEVELOPMENT

(i) Consider first dunes, which initiate because of the large-scale vertical turbulence. In Sub-section 2.2.1 it has been explained that dunes do not originate in their full length Λ_d. Rather, they begin to emerge, just after the commencement of experiment (at $t = 0$), by having a length $(\Lambda_d)_0$ that is much smaller than Λ_d. The so-emerging dunes incessantly *coalesce*, and as a consequence their length step-by-step increases. This increment of dune length takes place until its largest possible extent Λ_d is reached (at $t = (T_\Delta)_d$): $(\Lambda_d)_0 < (\Lambda_d)_1 < ... < (\Lambda_d)_k ... < \Lambda_d \approx 6h$. The dunes grow during $(T_\Delta)_d$ remaining almost geometrically similar – their steepness $(\delta_d)_k$ $(= (\Delta_d)_k/(\Lambda_d)_k)$ does not vary (increase) significantly in the process.

But why should dunes grow at all? Why does the flow not "accept" the initial dune size $(\Lambda_d)_0$ as it is? The answer to this question cannot be given by the mechanics of the phenomenon, for the stream can flow uniformly past *any* undulated bed surface. Yet, the answer follows naturally from the thermodynamic minimization of $A_* = u_{av}$ (Chapter 4). Indeed, the growth of $(\Lambda_d)_k$ (for hardly varying $(\delta_d)_k$) means the increment of the bed resistance; and thus the decrement of the resistance factor c_k and the flow velocity $(u_{av})_k$ (Eq. (3.11) where $c=u_{av}/v_*$). In other words, the dunes grow because the flow (which "intends" to minimize its velocity) endeavours to render its bed as "obstructive" (to its motion) as possible.

(ii) All what has been said above for dunes (having $\Lambda_d \approx 6h$) is valid *mutatis mutandis* for bars (which initiate because of the large-scale horizontal turbulence, and which thus have $\Lambda_a \sim B$). That is, bars too grow step-by-step (by coalescence) during the period $(T_\Delta)_a$ of their development. In summary, the time-development of large-scale bed forms is also a manifestation of the regime trend $A_* = u_{av} \to \min$ (or $\Pi_{A_*} = Fr = c^2 S \to \min$). In this case it is the reduction of c which "leads" to the minimization process.

PROBLEMS

To solve the problems below, take $\gamma_s = 16186.5N/m^3$, $\rho = 1000kg/m^3$, $\nu = 10^{-6}m^2/s$ (which correspond to sand or gravel and water).

5.1 Consider the inclined (by 1/3) part of the boundary $\mathcal{L}_{1,2}$ between one-row and multiple-row bars in Figure 5.2a: the ordinate B/h of this part of the boundary varies as a function of h/D. Show that this ordinate can be expressed also as a function of (only) the dimensionless combination

$$\mathcal{M} = \frac{Q^2}{gSD^5},$$

and determine this function. (Treat the flow as two-dimensional; use $c_f = 7.66(h/2D)^{1/6}$ (Eq. (1.13)).

5.2 Determine whether the rivers below are meandering or braiding (the characteristics given, which are derived from Ref. [7e] in Appendix E, correspond to bankfull flow rate):
a) Mississippi River: $Q = 42450.0m^3/s$, $D = 0.5mm$, $B = 1382.0m$, $h = 20.13m$, $S = 0.000047$;
b) Yamuna River: $Q = 2122.5m^3/s$, $D = 0.15mm$, $B = 205.9m$, $h = 6.40m$, $S = 0.000328$;
c) Savannah River: $Q = 849.0m^3/s$, $D = 0.8mm$, $B = 106.7m$, $h = 5.18m$, $S = 0.00011$;
d) Brahmaputra River: $Q = 24762.5m^3/s$, $D = 0.3mm$, $B = 9455.0m$, $h = 1.52m$, $S = 0.000252$.

5.3 The initial alluvial channel which has a flat bed and which can be treated as two-dimensional, is determined by $D = 1.1mm$, $Q_s/Q = 2 \times 10^{-5}$, $B_0/h_0 = 100$. Determine the range of S-values for which the flow developing in this initial channel will certainly tend to (and perhaps will) exhibit braiding – not meandering. Use Bagnold's formula; take $\beta' = 0.3$.

5.4 The regime development of a gravel channel takes place by meandering: $Q = 230m^3/s$, $D = 20mm$. Determine the interrelation between its B and S near the end of the regime development.

5.5 If Q, D and c are treated as constants, then various constant values of S imply (in the $(B/h;h/D)$-plane) the straight lines inclined by -2.5. What would be the inclination of these lines if we had a (flat) gravel-bed and only Q and D were treated as constants?

5.6 Consider the development of a regime channel in the interval $0<t<T_R$: the (constant) flow rate is $Q=500m^3/s$; the average grain size of the cohesionless alluvium is $D=0.9mm$.
a) The initial channel (at $t=0$), which has a flat bed, has $B_0=70m$ and $S_0=1/500$. Determine its flow depth h_0, and plot the corresponding point P_0 on the $(B/h;h/D)$-plane in Figure 5.2a.
b) The adjusted initial channel (at $t=\hat{T}_0$) has $\hat{B}_0=105m$. Determine its flow depth $\hat{h}_0$, and plot the corresponding point $\hat{P}_0$ on the $(B/h;h/D)$-plane in Figure 5.2a. (Take $\hat{S}_0=S_0$).
c) Determine the regime characteristics B_R and h_R, and plot the corresponding point P_R on the $(B/h;h/D)$-plane in Figure 5.2a.
d) What is the inclination of the straight line connecting the points P_0 and $\hat{P}_0$? What is the inclination of the straight line connecting the points $\hat{P}_0$ and P_R?

5.7 The adjusted initial alluvial stream has $Q=140m^3/s$, $\hat{S}_0=0.001$, $D=0.5mm$, $\hat{h}_0=1.30m$. Will such a stream be meandering? (Hint: determine $\hat{c}_0$ first).

5.8 Prove that the paths l in Figure 5.5 *must* merge into the respective $(Fr)_R$-curves when approaching the regime points $P_R=P_g$ and $P_R=P_s$ (the paths l cannot intersect a $(Fr)_R$-curve at the points mentioned with a finite angle).

5.9 How many times is the magnitude of the relative time-variation of S, viz $(\partial S/\partial\Theta)/S$, larger than that of $(\partial h/\partial\Theta)/h$? How many times is it larger than $(\partial c^2/\partial\Theta)/c^2$? Treat Q and B as constants; assume that the bed is flat.

REFERENCES

AAPG Reprint Series (1976). *Modern deltas*. Selected Papers, No. 18.

Ackers, P. and Charlton, F.G. (1970). The geometry of small meandering streams. *Proceedings of the Institution of Civil Engineers*, Supplement, 12, 289-317.

Bates, C.C. (1976). Rational theory of delta formation. In *Modern Deltas*, AAPG Reprint Series No. 18, American Association of Petroleum Geologists, July.

Bettess, R. and White, W.R. (1983). Meandering and braiding of alluvial channels. *Proceedings of the Institution of Civil Engineers*, 75, Part 2, 525-538.

Chang, H.H. (1988a). *Fluvial processes in river engineering*. John Wiley and Sons, Inc.

Chang, H.H. (1988b). On the cause of river meandering. In *Proceedings of the International Conference on River Regime*, edited by W.R. White, published on behalf of Hydraulics Research Ltd., Wallingford, John Wiley and Sons, 83-93.

Coleman, J.M. (1976). *Deltas: processes of deposition and models for exploration*. Continuing Education Publication Company, Inc., U.S.A.

da Silva, A.M.F. (2009). On the stable geometry of self-formed alluvial channels: theory and practical application. *Canadian Journal of Civil Engineering*, 36(10), 1667-1679.

Garde, R.J. and Raju, K.G.R. (1985). *Mechanics of sediment transportation and alluvial stream problems*. 2nd edition, Wiley Eastern Ltd., New Delhi.

Ikeda, H. (1983). *Experiments on bed load transport, bed forms and sedimentary structures using fine gravel in the 4-meter-wide flume*. Environmental Research Center Papers, No. 2, The University of Tsukuba, Japan.

Jansen, P.Ph., van Bendegom, L., van den Berg, J., de Vries, M. and Zanen, A. (1979). *Principles of river engineering: the non-tidal alluvial river*.

Kinoshita, R. (1961). *Investigation of the channel deformation of the Ishikari River*. (In Japanese) Memorandum No. 36, Science and Technology Agency, Bureau of Resources, Japan.

Kondratiev, N., Popov, I. and Snishchenko, B. (1982). *Foundations of hydromorphological theory of fluvial processes*. (In Russian) Gidrometeoizdat, Leningrad.

Leopold, L.B., Wolman, M.G. and Miller, J.P. (1964). *Fluvial processes in geomorphology*. W.H. Freeman, San Francisco, U.S.A.

Leopold, L.B. and Wolman, M.G. (1957). River channel patterns: braided, meandering and straight. *U.S. Geological Survey Professional Papers*, 282-B, 39-84.

Lewin, J. (1976). Initiation of bed forms and meanders in coarse-grained sediment. *Geological Society of America Bulletin*, 87(2), 281-285.

Mangelsdorf, J., Scheurmann, K. and Weiss, F.-H. (1990). *River morphology. A guide for Geoscientists and Engineers*. Springer Series in Physical Environment, Springer-Verlag, Berlin.

Nemec, W. (1990). Deltas – remarks on terminology and classification. In *Coarse-grained deltas*, edited by A. Colella and D.B. Prior, Special Publication No. 10, International Association of Sedimentologists, Blackwell Scientific Publications, Oxford, 3-12.

Postma, G. (1990). Depositional architecture and facies of river and fan deltas: a synthesis. In *Coarse-grained deltas*, edited by A. Colella and D.B. Prior, Special Publication No. 10, International Association of Sedimentologists, Blackwell Scientific Publications, Oxford, 13-27.

Schumm, S.A. and Khan, H.R. (1972). Experimental study of channel patterns. *Geological Society of America Bulletin*, 83(6), 1755-1770.

Soar, P.J. and Thorne, C.R. (2001). *Channel restoration design for meandering rivers*. Report ERDC/CHL CR-01-1, ERDC, U.S. Army Corps of Engineers, Vicksburg, Mississippi, U.S.A.

Sukegawa, N. (1970). Condition for the occurrence of river meanders. *Journal of the Faculty of Engineering*, University of Tokyo, Japan, Vol. 30, 289-236.

Wright, L.D. and Coleman, J.M. (1976). Variations in morphology of major river deltas as functions of ocean waves and river discharge regimes. In *Modern Deltas*, AAPG Reprint Series No. 18, American Association of Petroleum Geologists, July.

Yalin, M.S. and da Silva, A.M.F. (2001). *Fluvial processes*. IAHR Monograph, IAHR, Delft, The Netherlands.

Yalin, M.S. (1992). *River mechanics*. Pergamon Press, Oxford.

Chapter 6

Geometry and mechanics of meandering streams

Since the duration (T_R) of regime development of a meandering stream (by the expansion of its loops) is usually by orders of magnitude larger than the duration T_b of its bed development, it became a convention to study the mechanics of meandering flows for virtually rigid banks – and so it will be done in the following. In this and the next chapter, the meandering flow will be idealized, as in Chapter 4, in the form of a sequence of anti-symmetrically identical loops conveying anti-symmetrically identical flows. Thus the transport rates entering and leaving a meander loop are equal, and the bed deforms (during T_b) around the unchanging average bed surface, which is the same as the flat initial bed surface at $t=0$.

6.1 CHANNEL-FITTED COORDINATES

In agreement with the prevailing trend (Smith and McLean 1984, Struiksma et al. 1985, Nelson and Smith 1989a,b, Shimizu and Itakura 1989, Shimizu 1991, etc.), the "channel-fitted" system of coordinates will be used in the following.

Consider the plan view of a meandering channel region in Figure 6.1. The distance l_c measured along the channel centreline, from a crossover-section O_i downstream, specifies the location of a flow cross-section (l_c-section). The centreline curve possesses at any of its l_c-sections an osculating circle. The radius of this circle (which is the curvature radius of the channel at l_c) will be denoted by R; its centre (which is the curvature centre of the channel at l_c) will be denoted by C. Both R and the location of C are thus functions of l_c; and the specification of l_c means the (unique) specification of these functions.

Let P be a space point. The elevation of P will be specified by the distance z measured from an adopted datum upwards; its position in the flow plan, by any of the coordinate-pairs such as

$$\phi \text{ and } r; \quad l_c \text{ and } n; \quad l_c \text{ and } r; \quad \ldots \text{ etc.} \tag{6.1}$$

In this text, the position of P will be specified mostly by

$$l_c,\ n \text{ and } z, \tag{6.2}$$

or by their dimensionless counterparts,

$$\xi_c=\frac{l_c}{L};\quad \eta=\frac{n}{B};\quad \zeta=\frac{z}{h_{av}}, \tag{6.3}$$

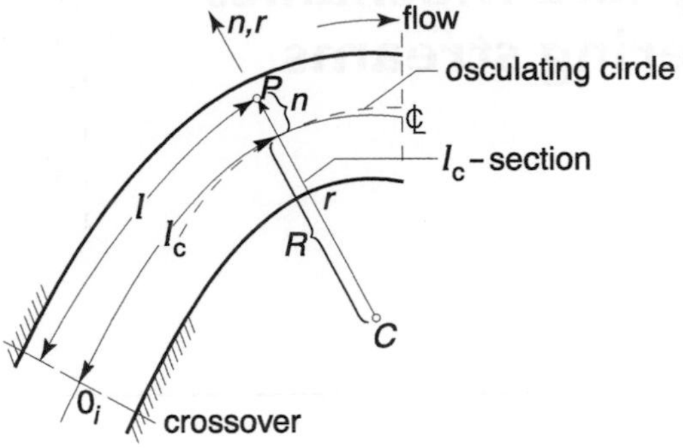

Figure 6.1 Definition sketch: channel-fitted coordinates.

where h_{av} is the channel average flow depth (see Sub-section 6.3.1 for averaging procedures).

1 For future reference, note that the following relations are valid

$$r = R + n; \quad dl = d\phi \cdot r; \quad dl_c = d\phi \cdot R; \quad dl = \left(1 + \frac{n}{R}\right) dl_c. \tag{6.4}$$

2 If the channel bed is flat and its slope along l_c is S_c, then its slope S along a coordinate line l (at n) is given by

$$S = S_c \frac{R}{R+n}. \tag{6.5}$$

3 Note also that for any function f we have

$$\frac{\partial f}{\partial r} = \frac{\partial f}{\partial n}. \tag{6.6}$$

6.2 SINE-GENERATED CHANNELS

6.2.1 Definition; geometric properties

(i) Following Langbein and Leopold (1966), Leopold and Langbein (1966), it appears to be generally accepted that the centreline of an idealized meandering stream can be represented best by the sine-generated function:

$$\theta = \theta_0 \cos\left(2\pi \frac{l_c}{L}\right), \tag{6.7}$$

where θ_0 and θ are the *deflection angles* at $l_c = 0$ and at (any) l_c, respectively (Figure 6.2).

Eq. (6.7) was introduced by the aforementioned authors as an approximation to the closed form integral resulting from the probabilistic derivation of meander path by von Schelling (1951). This rested on the postulate that for a given length between two

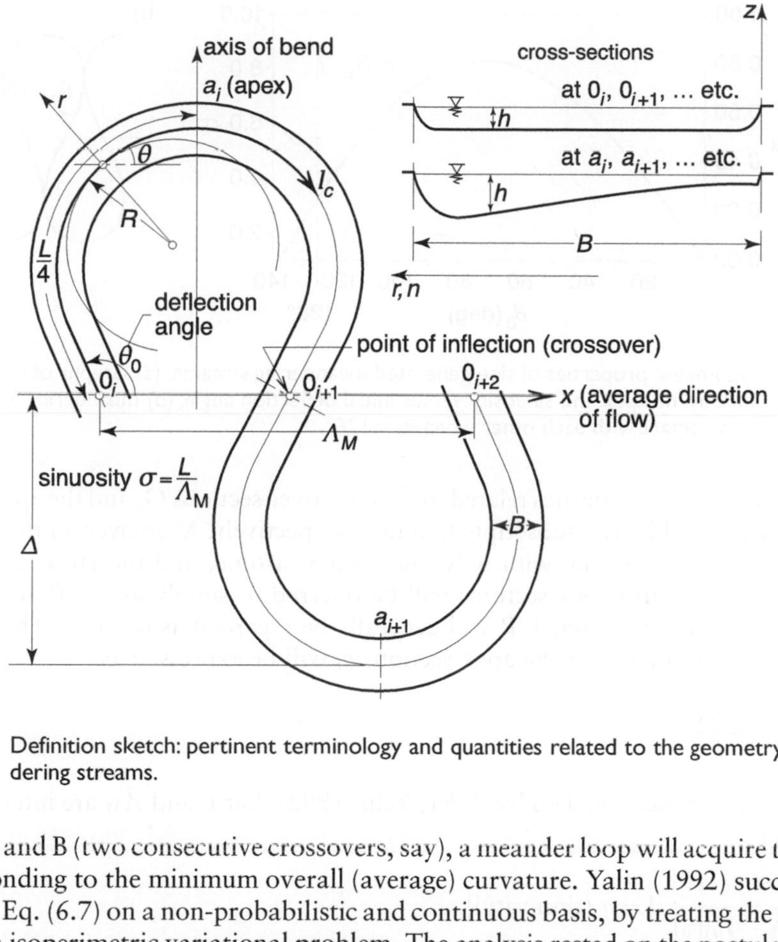

Figure 6.2 Definition sketch: pertinent terminology and quantities related to the geometry of meandering streams.

points A and B (two consecutive crossovers, say), a meander loop will acquire the shape corresponding to the minimum overall (average) curvature. Yalin (1992) succeeded in deriving Eq. (6.7) on a non-probabilistic and continuous basis, by treating the meander path as a isoperimetric variational problem. The analysis rested on the postulate that if a river is to turn in a meander loop having a given average curvature (square), then this loop must be such that the average rate of change (square) of its curvature is minimum. More recently, classical calculus of variations was also used by Movshovitz-Hadar and Shmukler (2006) to formulate the meander path. These authors, however, based their derivation on von Schelling's assumption of minimum overall curvature. Taking these works into consideration, it thus would seem that the sine-generated curve is such that it connects two points by simultaneously providing minimum overall (average) curvature and minimum average downstream rate of change in curvature.

(ii) The curvature $1/R$ of a sine-generated centreline, at a section l_c, is given by

$$\frac{1}{R} = -\frac{d\theta}{dl_c} = \frac{2\pi\theta_0}{L}\sin\left(2\pi\frac{l_c}{L}\right). \tag{6.8}$$

Eq. (6.8) indicates that when $l_c = 0$, $L/2$, L, ... etc. (at crossovers O_i), then $|1/R| = 0$; and when $l_c = L/4$, $3L/4$, $5L/4$, ... etc. (at apexes a_i), then $|1/R|$ is maximum.

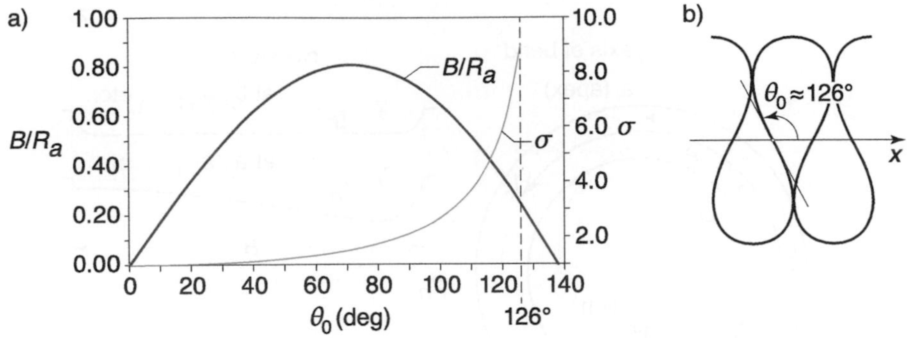

Figure 6.3 Geometric properties of sine-generated meandering streams. (a) Graphs of relative curvature at the apex and sinuosity versus initial deflection angle; (b) meandering loops coming into contact with each other when $\theta_0 \approx 126°$.

In this text, the quantities related to the crossover-sections O_i and the apex-sections a_i will be marked by the subscripts O and a, respectively. Moreover, in the following we will be mainly dealing with only one loop at a time, and therefore the l_c-values at the crossover- and apex-sections will be referred to simply as $l_c = 0$ and $l_c = L/4$. Owing to the same reason, $1/R$ will generally be regarded as positive. Thus, e.g. the maximum value of $1/R$ at the apex-sections a_i will be expressed as

$$\frac{1}{R_a} = \frac{2\pi\theta_0}{L}. \tag{6.9}$$

It can be shown (see e.g. da Silva 1991, Yalin 1992) that L and Λ_M are interrelated by θ_0 alone:

$$\frac{L}{\Lambda_M} = \frac{1}{J_0(\theta_0)} \quad [=\sigma \text{ (sinuosity)}]. \tag{6.10}$$

Here $J_0(\theta_0)$ is the Bessel function of the first kind and zero-th order (of the variable θ_0) – its graph is shown in Figure 6.3a; its polynomial approximation is given in Problem 6.1. Observe that when $\theta_0 \approx 138°$, then $J_0(\theta_0) = 0$, and $L; \sigma \to \infty$. However, this can never occur, for when θ_0 reaches the value $\approx 126°$ the meander loops come into contact with each other (Figure 6.3b) and the meandering flow pattern is destroyed. Hence $\theta_0 \approx 126°$ gives the largest practically possible sinuosity of sine-generated channels, viz $\sigma \approx 8.5$. The sinuosity of natural rivers usually varies, as a rule, within $1 < \sigma <\approx 5$.

Figure 6.4 gives an example for the agreement of the sine-generated function (6.7) with the plan view of a natural stream. Figure 6.5 shows how the sinuosity σ computed on the sine-generated basis, i.e. from Eq. (6.10), compares with the θ_0-values measured in large Russian rivers.

The expression (6.8) of the curvature of a sine-generated meandering stream can be expressed in dimensionless form, with the aid of the flow width B, as follows

$$\frac{B}{R} = \theta_0 \frac{2\pi B}{L} \sin\left(2\pi \frac{l_c}{L}\right). \tag{6.11}$$

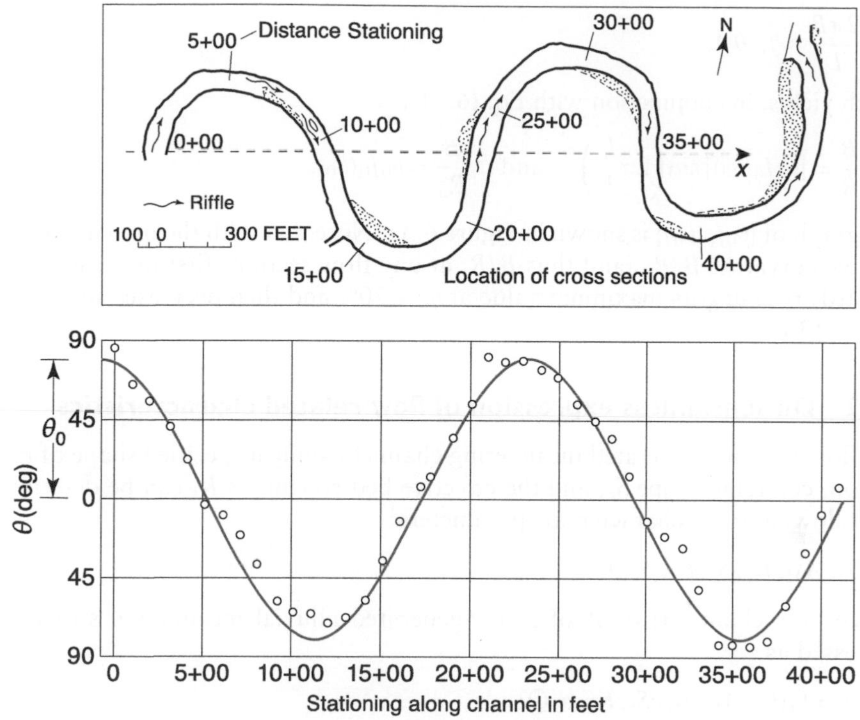

Figure 6.4 Top: plan view of the F. Popo Agie River near Hudson, Wyoming (adapted from Leopold and Langbein 1966); Bottom: corresponding plot of measured values of deflection angle at each hundred feet distance along the channel (circles) and graph of Eq. (6.7) (solid line).

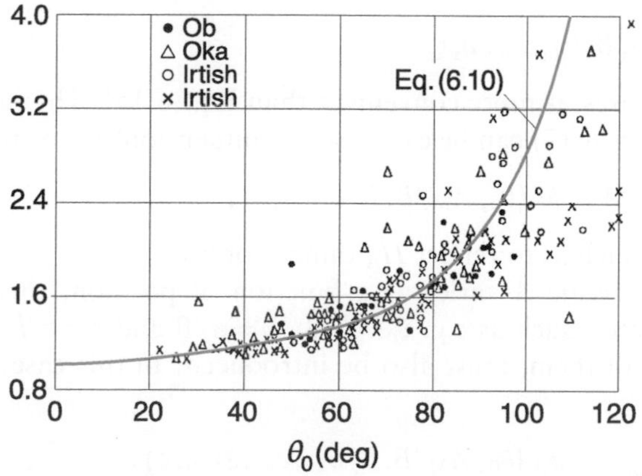

Figure 6.5 Plot of measured and calculated sinuosity versus initial deflection angle (adapted from Kondratiev et al. 1982).

Using Eq. (5.1), viz $\Lambda_M \approx 6B \approx 2\pi B$, in Eq. (6.10), one determines

$$\frac{2\pi B}{L} \approx J_0(\theta_0), \tag{6.12}$$

which yields, in conjunction with Eq. (6.11),

$$\frac{B}{R} = [\theta_0 J_0(\theta_0)] \sin\left(2\pi \frac{l_c}{L}\right) \quad \text{and} \quad \frac{B}{R_a} = \theta_0 J_0(\theta_0). \tag{6.13}$$

The graph of $[\theta_0 J_0(\theta_0)]$ is shown in Figure 6.3a. Note that with the increment of θ_0, the relative curvature B/R_a (and thus B/R, at any flow section) first increases from zero onwards reaching its maximum value at $\theta_0 \approx 70°$, and then decreases so as to vanish at $\theta_0 = 138°$.

6.2.2 Dimensionless expression of flow related characteristics

The flow in a sine-generated meandering channel having a specified shape of its cross-section, centreline slope S_c, and the effective bed roughness K_s can be determined by the following $n = 7$ characteristic parameters[1]

$$\theta_0, \Lambda_M, B, gS_c, K_s, \rho, Q. \tag{6.14}$$

Hence, any characteristic A of a sine-generated alluvial meandering stream can be expressed as

$$A = f_A(\theta_0, \Lambda_M, B, gS_c, K_s, \rho, Q). \tag{6.15}$$

Identifying A with the channel average flow depth h_{av} and the channel average resistance factor c_{av}, we can write

$$h_{av} = f_h(\theta_0, \Lambda_M, B, gS_c, K_s, \rho, Q) \quad \text{and} \quad c_{av} = f_c(\theta_0, \Lambda_M, B, gS_c, K_s, \rho, Q). \tag{6.16}$$

Eliminating two quantities, gS_c and K_s, from the three equations above, viz from Eq. (6.15) and two Eqs. (6.16), we determine for any A

$$A = f_A(\theta_0, \Lambda_M, B, h_{av}, c_{av}, \rho, Q) \tag{6.17}$$

which, in many cases, is more convenient than Eq. (6.15). The (usually dimensional) functional relation (6.17) can be expressed in dimensionless form as

$$\Pi_A = \rho^x Q^y h_{av}^z A = \phi_A\left(\theta_0, \Lambda_M/B, B/h_{av}, c_{av}\right), \tag{6.18}$$

where x, y, z are such as to render Π_A dimensionless.

If the characteristic A varies as a function of position, then the dimensionless position-coordinates, such as e.g. $\xi_c = l_c/L$, $\eta = n/B$ and $\zeta = z/h_{av}$ (see Eq. (6.3)), or only one or two of them, must also be introduced. In this case Eq. (6.18) becomes augmented into

$$\Pi_A = \rho^x Q^y h_{av}^z A = \phi_A\left(\theta_0, \Lambda_M/B, B/h_{av}, c_{av}, \xi_c, \eta, \zeta\right). \tag{6.19}$$

[1] If the geometry of the flow cross-section varies along l_c in a known manner, then l_c is an additional parameter which must be included into the set (6.14) (and, of course, also into the sets which follow from (6.14)).

It may be mentioned that there is no need for Π_A to be formed necessarily by Q and h_{av}: any two flow characteristics which are known to be dependent (among others) on Q and h_{av} can be used.[2] e.g. let $\mathcal{V}$ and $\mathcal{L}$ be *any* "velocity" and "length" which are dependent on Q and h_{av}. The resulting relation

$$\Pi_A = \rho^x \mathcal{V}^y \mathcal{L}^z A = \phi_A \left(\theta_0, \Lambda_M/B, B/h_{av}, c_{av}, \xi_c, \eta, \zeta\right). \quad (6.20)$$

is as good as Eq. (6.19).

6.3 MECHANICS OF MEANDERING FLOWS

6.3.1 Introductory considerations: averaging procedures

In this and the next chapter, we will often invoke different average values of quantities related to meandering streams. For the sake of clarity, these are defined below.

(i) The *vertically-averaged* value of a characteristic $A = f_A(l_c, r, z, ...)$, say, implies

$$\overline{A} = \frac{1}{h}\int_0^h A dh = \overline{f}_A(l_c, r, ...). \quad (6.21)$$

It should be noted here that when B/h is "large", a meandering stream can be formulated (and frequently is formulated nowadays) on a vertically-averaged basis (de Vriend 1977, Kalkwijk and de Vriend 1980, Smith and McLean 1984, Struiksma et al. 1985, Nelson and Smith 1989a, Shimizu and Itakura 1989, Struiksma and Crosato 1989, Jia and Wang 1999, etc.). In this case, e.g. the flow velocity vector

$$\mathbf{U} = u\mathbf{i}_l + v\mathbf{i}_r + w\mathbf{i}_z \quad (6.22)$$

is treated as

$$\overline{\mathbf{U}} = \overline{u}\mathbf{i}_l + \overline{v}\mathbf{i}_r, \quad (6.23)$$

where $\overline{u}$ and $\overline{v}$ are functions of l_c and r only (for $\overline{w} \to 0$ in wide channels). The vertical averaging reduces the three equations of motion into two, and makes it possible to characterize the flow by a single set of the streamlines in the $(l_c; n)$-plane.

The *cross-sectional average* value A_m of a characteristic A implies

$$A_m = \frac{1}{B}\int_{-B/2}^{B/2} \overline{A} dn = f_{A_m}(l_c, ...). \quad (6.24)$$

The *channel average value* A_{av} of A implies

$$A_{av} = \frac{2}{L}\int_0^{L/2} A_m dl_c = f_{A_{av}}(...) = const. \quad (6.25)$$

[2]See Sedov (1960), Yalin (1970, 1992) for the replacements of this kind.

(ii) The following should be mentioned:

1 In many cases, meandering affects the values of some characteristics (A) merely by "shifting" (or "redistributing") them in the radial direction ($n \sim \eta$) – without altering their cross-sectional average values along $l_c \sim \xi_c$ (da Silva 1995). In such cases, we have

$$A_{av} = A_m. \tag{6.26}$$

For example, the characteristics $\overline{u}$, h, c belong to this category – their cross-sectional average values are the same as their channel average values. This cannot be said e.g. for the free surface elevation z_f, internal energy e_i, entropy S_*.

2 Since B is assumed to be constant along $l_c \sim \xi_c$ (see Sub-section 4.4.1 (iii)), the term "B/h-ratio", used throughout the present text, is to be interpreted as "$B/h_{av} = B/h_m$-ratio".

3 The correlation coefficients (K_{AB}) encountered in the studies of meandering streams are usually "small", and they are thus invariably neglected (Smith and McLean 1984, Nelson and Smith 1989a, Struiksma and Crosato 1989, da Silva 1995). Considering this, the vertically-averaged value of the product of two quantities (A and B, say) will be expressed in this text as the product of their vertically-averaged values:[3]

$$\overline{AB} = \overline{A} \cdot \overline{B} + K_{AB} \approx \overline{A} \cdot \overline{B}. \tag{6.27}$$

4 In the case of a vertically-averaged flow in a sine-generated meandering channel, the relation (6.20) reduces into

$$\Pi_A = \rho^x \mathcal{V}^y \mathcal{L}^z A = \phi_A(\theta_0, \Lambda_M/B, B/h_{av}, c_{av}, \xi_c, \eta) \tag{6.28}$$

(with $\xi_c = l_c/L$ and $\eta = n/B$).

As is well known (Sedov 1960, Yalin 1970, 1992), not all of the dimensionless variables on the right-hand side of Eq. (6.28) must be expected to be present in the expression Π_A of every A.

6.3.2 Structure of meandering flow

In order to describe the internal structure of meandering flows, and following Yalin (1992) and da Silva (1995) (see also da Silva and Ebrahimi 2017), we will consider the flow by treating separately the effects on it of stream curvature and streamwise variation in curvature. So as not to encumber the explanations below with the consideration of additional effects, we will assume that the meandering streambed is flat.[4] If the

[3]Eq. (6.27) is used in the course of derivation of the vertically-averaged equations (7.1) to (7.6) in Sub-section 7.1.1 (i).

[4]It should be clear that the term 'flat bed' is used to imply a bed whose slope along the stream centreline l_c is the channel slope S_c, and whose slope in radial direction is zero (i.e., the bed cross-sectional profiles are horizontal).

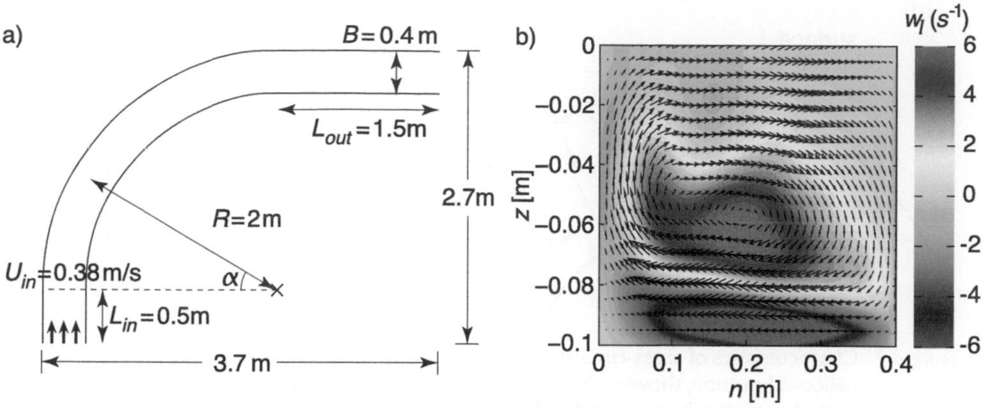

Figure 6.6 Results of numerical simulations by Davidsen (2007) in a 90° circular bend; flow depth = 0.1 m. (a) Computational domain (adapted from Davidsen 2007); (b) Cross-circulation velocity vectors and contour-map (colour) of downstream vorticity in the middle of the bend at $\alpha = 45°$ (from Davidsen 2007, reprinted with permission from Dr. T.S. Davidsen).

streambed is movable, then the flat bed is to be viewed as representing the conditions at the beginning of an experiment (i.e. the "flat initial bed" at time $t = 0$).

(i) Stream curvature invariably results in the emergence of a cross-circulatory motion (Γ), in which the fluid in the upper layers moves towards the outer bank, and that in the lower layers moves towards the inner bank (a motion that in contemporary literature is often referred to as centre region cell). Such motion is illustrated in Figure 6.6. The combination of (Γ) and u results in a *spiral* or *helicoidal* motion along the channel (see the schematic Figure 6.7a). Cross-circulation produces radial velocity diagrams (v_Γ-diagrams) along z consisting of one negative part and one positive part (Figure 6.7b). The radial velocities $v_{\Gamma,1}$ (at the free surface) and $v_{\Gamma,2}$ (near the bed) are in the opposite r-directions, and therefore the deviation angles ω_1 and ω_2 of its streamlines s_1 and s_2 are also of opposite sign. As already pointed out in Section 1.5, the negative and positive parts of the v_Γ-diagrams have equal areas, and therefore the vertically-averaged value of the cross-circulation velocity $\overline{v}_\Gamma$, and thus of Γ ($\sim \overline{v}_\Gamma$), vanishes:

$$\overline{v}_\Gamma = \frac{1}{h}\int_0^h v_\Gamma dz \equiv 0 \quad (\sim \Gamma). \tag{6.29}$$

(ii) The streamwise variation in curvature, on the other hand, makes the fluid to shift in all its "thickness" (flow depth) towards one of the banks (left or right). In the case of a (periodic) sequence of meander loops, the shifting of the fluid towards one or the other bank periodically alternates along the channel (Demuren and Rodi 1986, Yalin 1992, da Silva 1995). This is illustrated in Figure 6.8, where s_1 and s_2 mark the streamlines at the free surface and near the bed, respectively, and ω_1 and ω_2 are the local deviation angles between these streamlines and the longitudinal direction l. The amplitude of the lateral oscillation of streamlines (due to $d(1/R)/dl_c$) decreases as the bed is approached (hence $\omega_1 > \omega_2$). As a result, a vertical straight line segment 1-2 formed by fluid particles

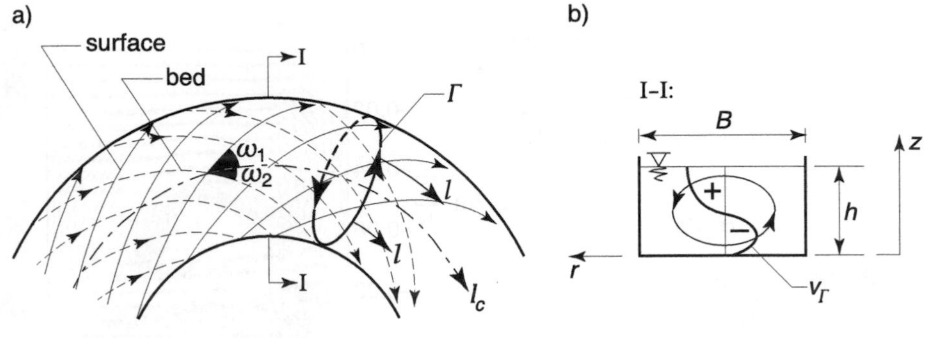

Figure 6.7 Characteristics of cross-circulatory motion (adapted from Yalin 1992). (a) Resulting spiral or helicoidal motion, showing deviation angles of the streamlines at the free surface (subscript 1) and near the bed (subscript 2); (b) cross-sectional view illustrating the related radial velocity diagrams.

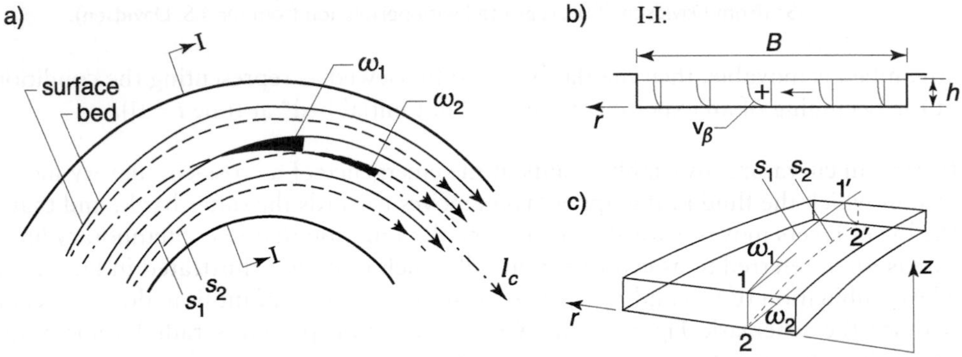

Figure 6.8 Effect of streamwise variation in curvature (adapted from Yalin 1992). (a) Shifting of the fluid towards one of the banks (left bank in this example); (b) related radial velocity diagrams; (c) streamlines and local deviation angles near the free surface (subscript 1) and near the bed (subscript 2).

at a certain instant, one second later, say, will be deformed into 1′-2′ as the fluid mass is shifted sideways. Clearly, this motion also produces radial velocities, in the following termed υ_β. At any given cross-section, the velocities υ_β along z are all in the same r-direction, and thus have the same sign (+ or −).

(iii) It follows that the three-dimensional flow in a stream exhibiting streamwise variation in curvature can be viewed as being formed by a "convective base" (due to the streamwise variation in curvature), upon which a cross-circulatory motion (due to the curvature itself) is superimposed. The resulting radial velocity diagram (υ-diagram) at any location is the sum of the cross-circulatory υ_Γ and the translatory υ_β radial velocities (Yalin 1992). As illustrated by the schematic Figure 6.9, depending on whether υ_β is positive or negative, and on the magnitudes of υ_Γ and υ_β in comparison to each other, the υ-diagrams of a meandering flow can acquire a wide variety of shapes,

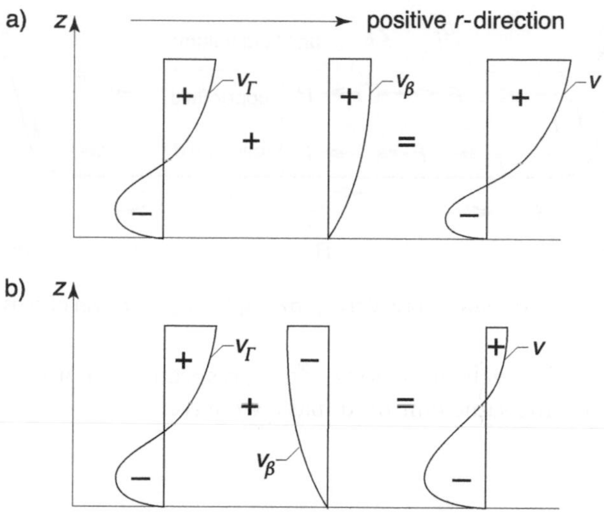

Figure 6.9 Schematic representation of diagrams of radial velocity v, and their cross-circulatory (v_Γ) and translatory (v_β) components (adapted from da Silva and Ebrahimi 2017). (a) Positive v_β; (b) negative v_β.

as demonstrated by numerous laboratory and field measurements as well as numerical simulations (see e.g. Demuren and Rodi 1986, Frothingham and Rhoads 2003, Blanckaert 2010, Termini and Piraino 2011). Note that since topographic steering affects the v-diagrams by locally increasing (or decreasing) the magnitudes of v_Γ and v_β, the statement just made is equally valid for the case of deformed beds – hence the reason why the just mentioned references cover also measurements over deformed beds.

6.3.3 Cross-circulation

In the following, we consider the cross-circulatory motion in greater detail. For this purpose, let us focus on the far-from-the-banks region B_c of a cross-section of a meandering flow (Figure 6.10). In order not to encumber the explanations below with details of secondary importance, and as done in the previous sub-section, we will assume that the bed is flat. Moreover, we will assume that the curvature radius r_s of the streamlines s at any location within B_c is nearly the same as the channel curvature r at that location. The free surface exhibits superelevation (i.e. the flow depth is somewhat larger at the outer bank than at the inner bank).

(i) Consider the z-axis selected within B_c (Figure 6.10): the cross-sectional location of this axis is r; $z = 0$ at the flow bed. The centrifugal inertia force acting along r on a unit fluid volume situated on this z-axis can be expressed (on the basis of aforementioned) as

$$F = \frac{\rho U^2}{r_s} \approx \frac{\rho u^2}{r}, \tag{6.30}$$

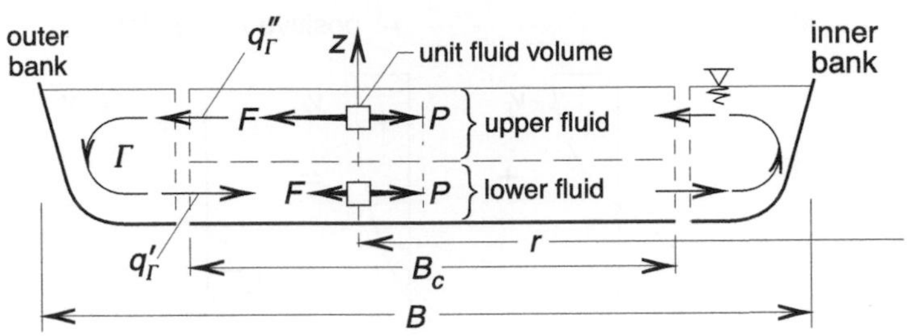

Figure 6.10 Cross-sectional view of meandering flow explaining the origin of cross-circulatory motion.

where U and u are flow velocities along s and l, respectively. For the net pressure force, acting along $-r$ on the same unit fluid mass, we have

$$P = -\gamma \frac{\partial h}{\partial r}, \tag{6.31}$$

where h is, in fact, the same as the elevation z_f of the free surface. As is well known, the vertically-averaged values of F and P are (approximately) equal. Yet, F progressively increases with z (for u does so), while P does not vary with z at all. Hence, in the upper part of the cross-section we have $F > P$, while in the lower, $F < P$. This causes the "upper fluid" to move toward the outer bank, while the "lower fluid", toward the inner bank (Figure 6.10). Consequently, the cross-sectional fluid exhibits the *cross-circulation* (Γ).

(ii) The generally non-equal forces F and P are brought into equilibrium by the z-derivative of the radial shear stresses τ_Γ. This fact can be reflected by the widely used differential equation (see e.g. Rozovskii 1957, Yen and Yen 1971, Kikkawa et al. 1976, Grishanin 1979, Falcon-Ascanio and Kennedy 1983, Chang 1988)

$$-\frac{\partial \tau_\Gamma}{\partial z} = F + P \tag{6.32}$$

which, with the aid of Eqs. (6.30) and (6.31), can be expressed as

$$-\frac{\partial \tau_\Gamma}{\partial z} = \frac{\rho u^2}{r} - \gamma \frac{\partial h}{\partial r}. \tag{6.33}$$

The z-distribution of $-\partial\tau_\Gamma/\partial z$, which is but the difference between the diagrams of F and P, is shown schematically in Figure 6.11a. As one can easily infer, the z-distribution of τ_Γ (which is zero at the free surface) must thus be as indicated in Figure 6.11b. At the level z_1 (where $\partial\tau_\Gamma/\partial z = 0$) we have $(\tau_\Gamma)_{max}$; $(\tau_\Gamma)_{min} = (\tau_0)_\Gamma$ being at the bed ($z = 0$). Since the radial shear stress τ_Γ (no matter what method is used for its determination) must be proportional to a power of the z-gradient of the cross-circulatory flow velocity v_Γ, one can also infer that the v_Γ-diagram should be as depicted in Figure 6.11c. The minimum v_Γ is at z_2 (where $\tau_\Gamma = 0$), the maximum, at the free surface. (Note e.g. from Rozovskii 1957, Chang 1988, that the most used v_Γ-distributions along z are indeed as depicted in Figure 6.11c).

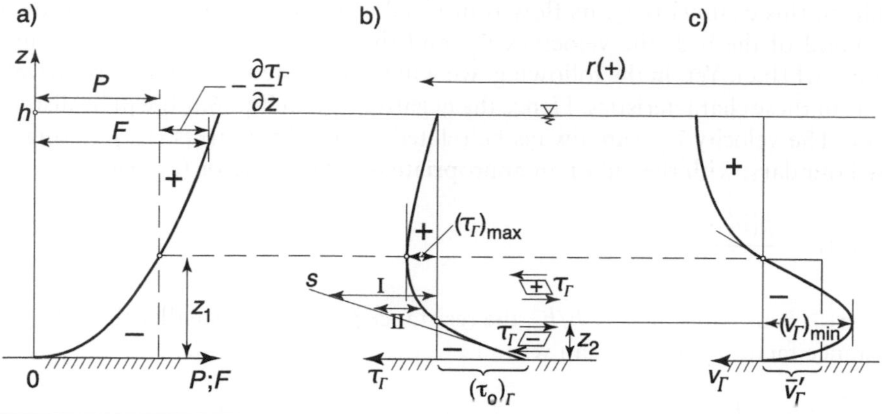

Figure 6.11 Cross-circulatory motion: vertical distribution of pertinent quantities. (a) Difference between the diagrams of F and P; (b) radial shear stress; (c) radial velocity.

Since at $z = z_1$ we have $\partial\tau_r/\partial z = 0$, one determines from Eq. (6.33)

$$\gamma\frac{\partial h}{\partial r} = \frac{\rho u_1^2}{r}, \tag{6.34}$$

where u_1 is the value of u at the level z_1. Substituting Eq. (6.43) in Eq. (6.33) and integrating along $[0; z]$, we get

$$\tau_r = \underbrace{(\tau_0)_r + \left(\frac{\rho u_1^2}{r}\right)z}_{\text{I}} - \underbrace{\frac{\rho}{r}\int_0^z u^2 dz}_{\text{II}}. \tag{6.35}$$

Here, for small z, the term II is negligible in comparison to I (as illustrated in Figure 6.11b, where the τ_r-curve becomes indistinguishable from the straight line s implying I). Approximating the τ_r-curve in the region $0 < z < z_2$ by the straight line s, we can write

$$\tau_r = (\tau_0)_r + \left(\frac{\rho u_1^2}{r}\right)z \quad (\text{if } z \in [0, z_2]), \tag{6.36}$$

which, considering that $\tau_r = 0$ when $z = z_2$, yields

$$(\tau_0)_r = -\left(\frac{\rho u_1^2}{r}\right)z_2. \tag{6.37}$$

The longitudinal flow velocity u_1 at the level z_1 can be expressed as a certain multiple of the vertically-averaged longitudinal flow velocity $\bar{u}$, while z_2 is but a fraction of the flow depth h:

$$u_1 = \alpha_1\bar{u}; \qquad z_2 = \alpha_2 h. \tag{6.38}$$

Using Eq. (6.38), one can write Eq. (6.37) as

$$(\tau_0)_r = -(\alpha_1^2\alpha_2)\rho\bar{u}^2\frac{h}{r}. \tag{6.39}$$

Let $\overline{v}'_\Gamma$ be the average cross-circulatory flow velocity in the lower "channel". (The height of this channel is z_1, its flow is directed toward the inner bank). In the neighbourhood of the bed, the velocities $\overline{v}'_\Gamma$ and the bed shear stress $(\tau_0)_\Gamma$ are negative (Figure 6.11b,c). Yet, in the following, we will be interested only in the (positive) magnitude of these characteristics. Hence the negative sign in Eq. (6.39) will be disregarded below. The velocity $\overline{v}'_\Gamma$ can always be related to the shear stress $(\tau_0)_\Gamma$ acting on the flow boundary, with the aid of an appropriate resistance factor (c_Γ, say):

$$(\tau_0)_\Gamma = \frac{\rho(\overline{v}'_\Gamma)^2}{c_\Gamma^2}. \tag{6.40}$$

Since $z_1 < h$ and thus $z_1/k_s < h/k_s$, the resistance factor c_Γ is smaller than the channel resistance factor c, and one can write

$$\frac{c_\Gamma}{c} = \alpha_c < 1. \tag{6.41}$$

From Eqs. (6.39), (6.40) and (6.41), one determines for the neighbourhood of the centreline (where r can be identified with R),

$$\left(\frac{\overline{v}'_\Gamma}{\overline{u}}\right)^2 = \alpha c^2 \frac{h}{R} \qquad \text{(with } \alpha = \alpha_c^2 \alpha_1^2 \alpha_2\text{)}, \tag{6.42}$$

where each of the (dimensionless) α_i is determined by the distribution of longitudinal flow velocities u along z. But since the shape of this distribution itself is determined mainly by c, one concludes that α is a so far unknown function of c (and perhaps of some other parameters).

The relations similar to Eq. (6.42), which indicate that the cross-circulatory velocity must increase with h/R, have already been derived in the past for fully developed flow in circular bends by various authors (Rozovskii 1957, Francis and Asfari 1971, Yen 1972, Engelund 1974, Kikkawa et al. 1976, Falcon-Ascanio and Kennedy 1983, etc.). Given that $h/R = (h/B) \cdot (B/R)$, such relations imply that the strength of cross-circulatory motion progressively decreases with increasing values of B/h and decreasing values of B/R. These earlier findings are well illustrated by the recent numerical simulations by Kashyap et al. (2012). It should, however, be noted here that the dependency of the strength (or intensity) of cross-circulation on B/h may be more complex than previously thought. Indeed, on the basis of numerical simulations in a 90° circular bend having $B/R = 0.2$, and in which B/h was varied from 2 to 8, Davidsen (2007) found that the intensity of cross-circulation at first progressively increased with B/h reaching a maximum for $B/h_{av} \approx 5$, and that it only progressively decreased with B/h_{av} for $B/h_{av} > \approx 5$. For $B/h_{av} = 8$, the intensity of cross-circulation had already been reduced to $\approx 75\%$ of that at $B/h_{av} = 5$ (see Figure 6.12). Here the intensity of cross-circulation is defined as the ratio between the cross-channel and along-channel kinetic energies, integrated over a cross-section.

The relations such as Eq. (6.42), however, merely convey that the importance of cross-circulation must increase with flow curvature; they do not incorporate any information peculiar to alluvial streams. For example, the relative curvature B/R of a natural meandering stream is not arbitrary; in the case of sine-generated meandering streams (which are regarded as the idealized counterparts of natural streams) the value

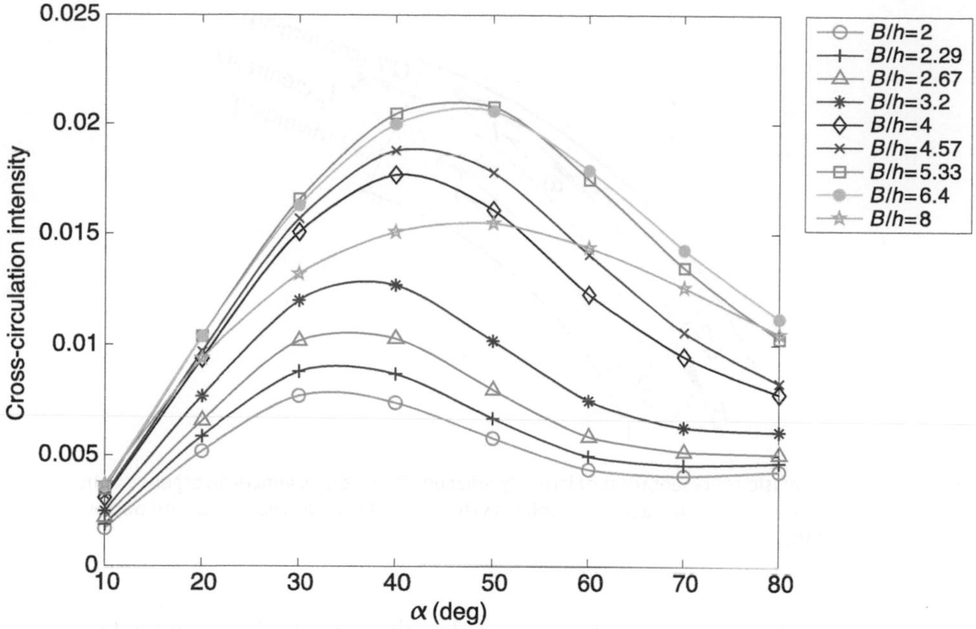

Figure 6.12 Plot of cross-circulation intensity as a function of position along the channel depicted in Figure 6.6a, for different values of the width-to-depth ratio (from Davidsen 2007, reprinted with permission from Dr. T.S. Davidsen).

of B/R *is determined* by Eq. (6.13). Using Eq. (6.13), one can express Eq. (6.42) as follows

$$\left(\frac{\overline{v}'_\Gamma}{\overline{u}}\right)^2 = \alpha c^2 [\theta_0 J_0(\theta_0)] \sin\left(2\pi \frac{l_c}{L}\right) \frac{h}{B}. \tag{6.43}$$

At the apex-section a (where $l_c = L/4$), the ratio $\overline{v}'_\Gamma/\overline{u}$ acquires its largest value

$$\left(\frac{\overline{v}'_\Gamma}{\overline{u}}\right)_a^2 = \alpha c^2 [\theta_0 J_0(\theta_0)] \frac{h}{B}. \tag{6.44}$$

This relation indicates that for a given flow in a sine-generated channel (for given $\overline{u}$, h, B and c), the value of $\overline{v}'_\Gamma$, and thus the importance of cross-circulation Γ ($\sim \overline{v}'_\Gamma$) varies with the channel's initial deflection angle θ_0; or, which is the same, with the channel's sinuosity $\sigma = (J_0(\theta_0))^{-1}$. From the graph of the function $[\theta_0 J_0(\theta_0)]$ ($= B/R_a$) in Figure 6.3a it is clear that this function acquires its maximum at $\theta_0 \approx 70°$. Thus the relevance of $\overline{v}'_\Gamma$, and thus Γ, decreases with the increment of the deviation of θ_0 from $\approx 70°$. Note that Eq. (6.43) further indicates that for any given sinuosity, or θ_0, the importance of $\overline{v}'_\Gamma$ and Γ progressively decreases with the increment of B/h (a statement which, on the basis of Davidsen 2007, should for the time being be taken as valid as long as B/h is not excessively small – which is never the case in real rivers).

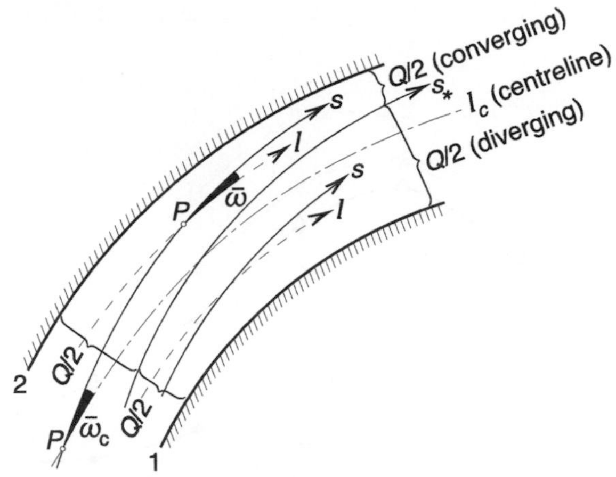

Figure 6.13 Schematic representation of laterally adjacent flow convergence-divergence zones formed by the vertically-averaged streamlines s (from da Silva 2006, reprinted with permission from IAHR).

Although the derivations above are only schematical, they nonetheless convey clearly that in the case of natural meandering streams the role of the cross-circulation (Γ) cannot be considered independently of B/h and θ_0. It may also be mentioned here, in passing, that in a distorted physical model, having B/h many times smaller than that of the prototype, the effect of Γ can be rather exaggerated – as has already been noticed long ago by Matthes (1941, 1948) (more on the topic in Section 6.6 (ii)).

6.3.4 The "convective base"

(i) The vertically-averaged streamlines of the laterally oscillating flow (due to the streamwise variation in curvature) are between the free surface streamlines s_1 and the near bed streamlines s_2 (see Figure 6.8c). Given that the deviation between these is small, it follows that the plan behaviour of the lateral oscillations of flow is very adequately reflected by the vertically-averaged counterpart of the actual three-dimensional flow.

It should be clear that in the case of a convective flow, the vertically-averaged streamlines s form, in flow plan, laterally adjacent convergence-divergence zones (in short, [CD]'s), as shown in Figure 6.13. In sine-generated streams, the pattern formed by the streamlines in plan view varies periodically along l_c. Hence the length of each convergence-divergence flow zone ([CD]) is equal to $L/2$.

Note that the direction of the (vertically-averaged) streamlines deviates from the direction of the coordinate lines l (Figure 6.13). This deviation, at a point P, will be characterized by the deviation angle $\overline{\omega}$: if P is on l_c, then $\overline{\omega}=\overline{\omega}_c$. Clearly, $\tan\overline{\omega}=\overline{v}/\overline{u}$. In a sine-generated channel, the largest cross-sectional $\overline{\omega}$ are in the neighbourhood of the centreline. Yet, $\overline{\omega}_c$ is usually less than $\approx 1/10$ say, and therefore

$$\overline{\omega}=\overline{v}/\overline{u} \tag{6.45}$$

can be adopted. At the flow boundaries 1 and 2 (banks), $\overline{\omega} \equiv 0$; $\overline{\omega}$ is also equal to zero at the cross-sections coinciding with the upstream-end and the downstream-end of a [CD]. The sign of $\overline{\omega}$ remains either positive or negative within each [CD]; and it alternates periodically along l_c between positive and negative.

Let s_* be the streamline separating the convectively accelerated and decelerated zones, i.e. converging and diverging flow zones (which periodically alternate along l_c). The streamline s_* divides the flow rate Q in two equal parts; each of the flows remaining to the left and to the right of s_*, conveys $Q/2$. The overall convective nature of a meandering flow can thus be reflected by the behaviour of s_* alone.

(ii) Taking into account the content of paragraph (i) above, let us then consider the vertically-averaged flow in sine-generated streams having a flat bed. The streams are assumed to be sufficiently wide so that the effect of Γ is of secondary importance only.

To the best knowledge of the authors', the first systematic experimental study to reveal the structure of such flow was conducted by da Silva (1995). This author used two channels having $\theta_0 = 30°$ and $110°$ and typifying "small" and "large" values of θ_0. In the experiments, $B = 0.40m$, $h_{av} \approx 0.03m$, $D = 2.2mm$; $\Lambda_M/B = 2\pi$, $(c_f)_{av} \approx 11$. The (flat) bed was immobilized. Each channel consisted of three consecutive meander loops. This work made it clear that streams having small and large values of θ_0 possess distinctly different locations (in flow plan) of their convergence-divergence zones – which means that the location of the [CD]'s in flow plan is not of a standard nature, but rather strongly dependent on (at least) θ_0. Considering this, da Silva et al. (2006) extended the previous measurements in da Silva (1995) by including also channels representing intermediate values of sinuosity ($\theta_0 = 50°$, $70°$ and $90°$) with the goal of revealing experimentally how the location in flow plan of the [CD]'s changes as θ_0 increases from "small" to "large". As examples, the vertically-averaged longitudinal velocity fields ($\overline{u}$-fields) measured by da Silva (1995) in the channels having $\theta_0 = 30°$ and $110°$, and that measured by da Silva et al. (2006) in the $70°$-channel, are depicted in Figure 6.14. It should be mentioned here that the measured velocity field in the $110°$ channel is in agreement with results of the study of free surface flow velocity conducted by Whiting and Dietrich (1993b) in a $100°$ channel as well as the measurements of the vertically-averaged flow carried out by Termini (1996) in a $110°$ channel.

From the just described measurements, it follows that the location in flow plan of the [CD]'s varies as shown in the schematic Figures 6.15a-c. In these figures, the meandering channels are "straightened" for the sake of simplicity; "CONV" and "DIV" indicate the zones of convergence and divergence of flow, respectively; and ξ_{c0} is the distance, normalized by L, measured along the stream centreline from the upstream crossover of a meander loop (with positive R) to the upstream end of the flow convergence-divergence zone partially or fully contained in that loop and exhibiting convergence at the left bank.

As depicted in Figures 6.15a-c:

1. *If* θ_0 *is "small"* ($\theta_0 <\approx 30°$, say), then a [CD] exhibiting $\overline{\omega} > 0$ is situated (approximately) between the apex sections a_i and a_{i+1}. In this case, the most intense convergence-divergence of flow, and thus the maximum value of $\overline{\omega}$, here termed $(\overline{\omega})_{max}$, occurs (approximately) at the crossover section O_{i+1}; while the maximum flow velocity occurs at the right bank (or its proximity) very near the apex section (as is the case in the

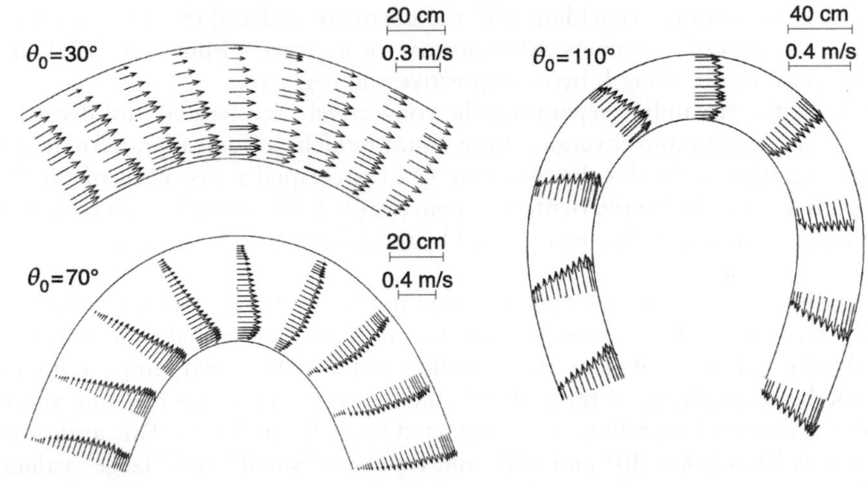

Figure 6.14 Measured $\bar{u}$-fields in three different sine-generated channels having $\theta_0 = 30°, 70°$ and $110°$.

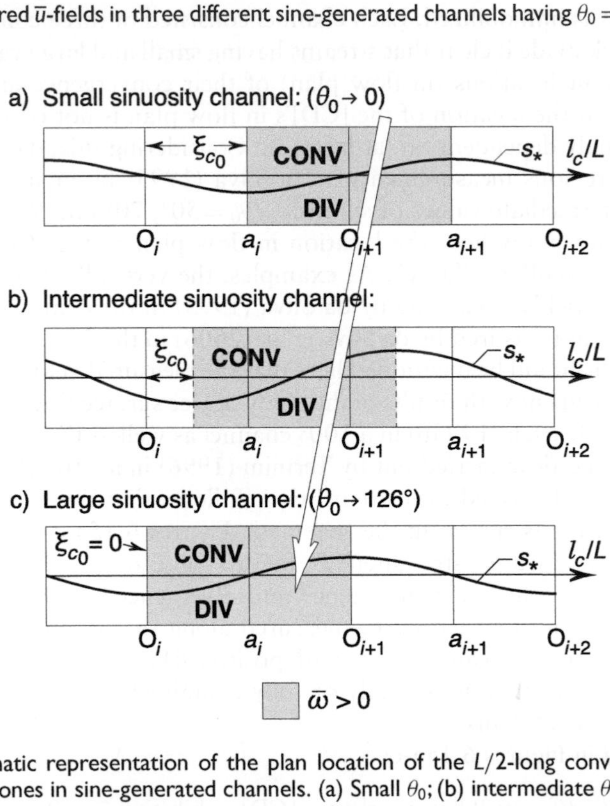

Figure 6.15 Schematic representation of the plan location of the $L/2$-long convergence-divergence flow zones in sine-generated channels. (a) Small θ_0; (b) intermediate θ_0; (c) large θ_0.

velocity field in the 30° channel in Figure 6.14). Following da Silva (1995, 1999), this type of convective flow which is *convergent* at the inner bank (approximately) between $a_{i-1}a_i$ and *divergent* (approximately) between a_i and a_{i+1}, will be referred to as *ingoing flow*.

2. *If θ_0 is "large"* ($\theta_0 > \approx 100°$, say), then a [CD] exhibiting $\overline{\omega} > 0$ is situated (approximately) between the crossover sections O_i and O_{i+1}. In this case, $\overline{\omega} = 0$ at (approximately) O_i, and the most intense convergence-divergence of flow occurs (approximately) at the apex section a_i; while the maximum flow velocity occurs at the inner bank (or its proximity) very near the crossover section O_i (as in the velocity field in the 110° channel in Figure 6.14). This type of flow, which is *divergent* at the inner bank (approximately) between O_iO_{i+1}, will be termed *outgoing flow* (da Silva 1995, 1999).

3. As θ_0 increases from "small" to "large", the location of the analogous [CD] exhibiting $\overline{\omega} > 0$ gradually shifts upstream, as implied by the arrow in Figure 6.15. The locus of maximum flow velocity thus shifts upstream as θ_0 increases, with intermediate values of θ_0 ($\theta_0 \approx 60°$ to 70°) exhibiting the largest flow velocity at the inner bank (or its proximity) somewhere between the sections O_i and a_i (as in the 70° channel in Figure 6.14).

Note that because of bank effect, in reality the largest flow velocity does not occur exactly at the inner bank – hence the reason for the inclusion of the words 'or its proximity' in the statements above.

The plan patterns of the streamlines of the vertically-averaged flow corresponding to small and large values of θ_0 are as shown in the schematic Figures 6.16a,b (diagrams on the left), respectively. [The flow rates between any two streamlines in Figures 6.16a,b are supposed to be equal. Hence the nearer to each other are the streamlines, the larger is the flow velocity $\overline{U}$ between them; the further they are from each other, the smaller is $\overline{U}$ (see the schematical $\overline{U}$-diagrams in the drawings on the right in Figures 6.16a,b).] The conditions described above can also be inferred from Figure 6.17, showing a plot of the measured values of ξ_{c0} versus θ_0 resulting from the aforementioned experiments.

The conceptual Figure 6.18 illustrates the distribution of $\overline{\omega}_c/(\overline{\omega}_c)_{max}$ along a [CD] of the length $L/2$ as shown in Figures 6.15a-c ($\overline{\omega} = 0$ and thus $\overline{\omega}_c/(\overline{\omega}_c)_{max} = 0$ signifying the ends of the accelerated/decelerated zones). Here $(\overline{\omega}_c)_{max}$ stands for the maximum value of $\overline{\omega}_c$ found along the channel centreline. There are reasons to believe that the upstream and downstream ends of the decelerated region (at the inner bank 1) of the "ingoing flow" coincide exactly with the apex sections a_i and a_{i+1} only in the limit $\theta_0 \to 0°$ ($\sigma \to 1$). Similarly, the upstream- and downstream-ends of the decelerated region (at the bank 1) of the "outgoing flow" are in coincidence with the crossover-sections O_i and O_{i+1} when $\theta_0 \to 138°$ ($\sigma \to \infty$). Accordingly, the curves C_0 and C_{138} in Figure 6.18 are the limiting curves corresponding to $\theta_0 = 0°$ and $\theta_0 = 138°$ – the curves C_{θ_0} corresponding to a θ_0 within the interval $0° < \theta_0 < 138°$ are situated between them, with the curve C_{30} starting slightly upstream of a_i, and the curve C_{110}, slightly downstream of O_i (hence the reason for using the word "approximately" in the statements 1 to 3 above). One can say that the C_{θ_0}-curves in Figure 6.18 "shift" to the left (from C_0 to C_{138}) as θ_0 increases. For future reference, note that Figure 6.18 also conveys that $(\overline{\omega}_c)_a$ is *always positive*: $(\overline{\omega}_c)_a \geq 0$ for any $\theta_0 \in [0°; 138°]$.

Figure 6.19 shows the graphs of the values of $(\overline{\omega}_c)_{max}$ versus θ_0 resulting from the above mentioned experiments by da Silva (1995) and da Silva et al. (2006). Note that for a given set of experimental conditions, as the deviation of θ_0 from $\approx 70°$ decreases

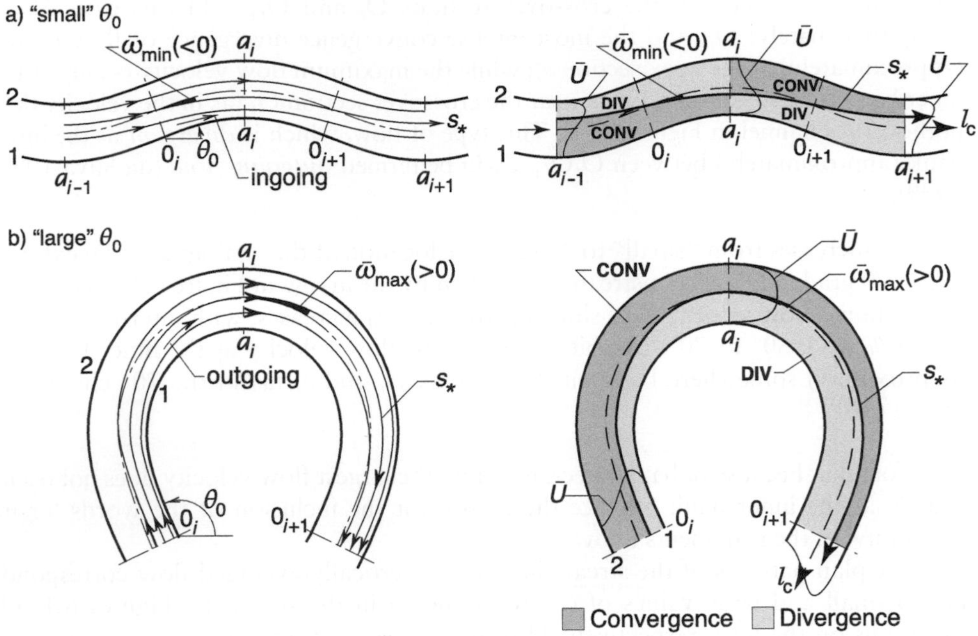

Figure 6.16 Plan patterns formed by the streamlines of the vertically-averaged flow over a flat bed (figures on the left) and related plan distribution of the convergence-divergence zones of flow (figures on the right). (a) Small θ_0; (b) Large θ_0.

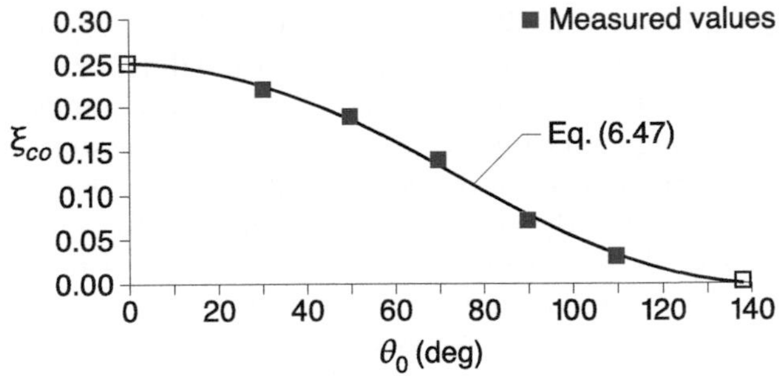

Figure 6.17 Plot of measured values of ξ_{c0} versus initial deflection angle and graph of Eq. (6.47) (adapted from da Silva et al. 2006).

and B/R_a $(=\theta_0 J_0(\theta_0))$ increases (see Figure 6.3a), the vertically-averaged flow must necessarily become "stronger" (in the sense that super-elevation increases, velocity gradients increase, the amplitude of oscillations of the streamlines s_* around the channel centreline increases, etc.). Therefore $(\overline{\omega}_c)_{max}$ must also increase. This explains why, in Figure 6.19, $(\overline{\omega}_c)_{max}$ reaches a maximum for $\theta_0 \approx 70° = 1.22 rad$.

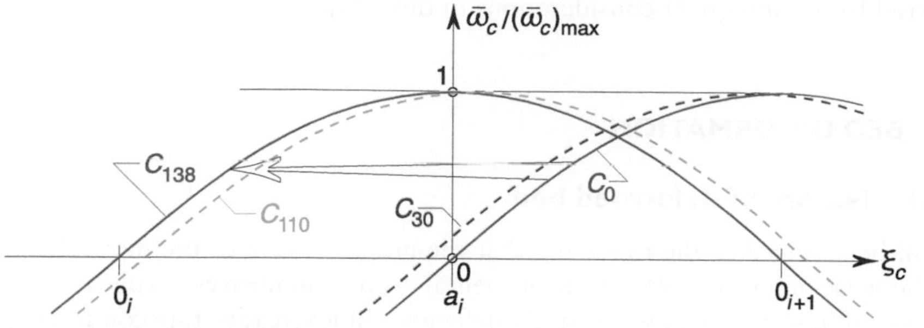

Figure 6.18 Conceptual illustration of the distribution of $\overline{\omega}_c/(\overline{\omega}_c)_{max}$ along the $L/2$-long convergence-divergence flow zone depicted in Figure 6.15.

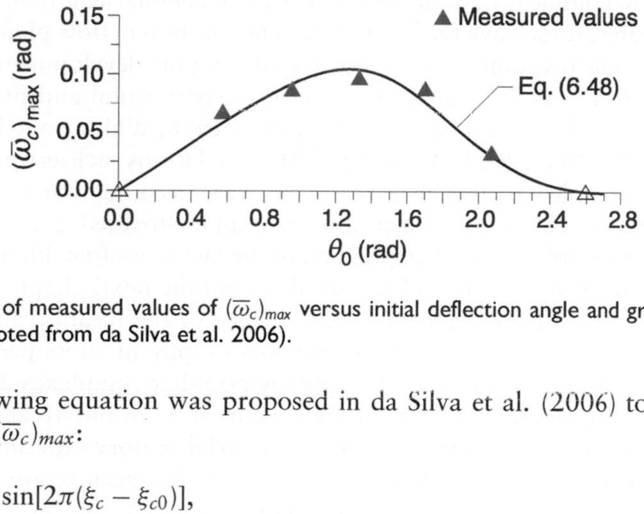

Figure 6.19 Plot of measured values of $(\overline{\omega}_c)_{max}$ versus initial deflection angle and graph of Eq. (6.48) (adapted from da Silva et al. 2006).

The following equation was proposed in da Silva et al. (2006) to represent the measured $\overline{\omega}_c/(\overline{\omega}_c)_{max}$:

$$\frac{\overline{\omega}_c}{(\overline{\omega}_c)_{max}} = \sin[2\pi(\xi_c - \xi_{c0})], \tag{6.46}$$

in which

$$\xi_{c0} = 0.0033\theta_0^4 + 0.0198\theta_0^3 - 0.11\theta_0^2 + 0.25 \tag{6.47}$$

and

$$(\overline{\omega}_c)_{max} = 0.115\theta_0 e^{-0.120\theta_0^{4.3}}. \tag{6.48}$$

Here θ_0 is in radians. The graphs of Eqs. (6.47) and (6.48) are shown as solid lines in Figures 6.17 and 6.19, respectively.

As follows from Sub-section 6.2.2, the location in flow plan of the [CD]'s (and consequently ξ_{c0}, $(\overline{\omega}_c)_{max}$, etc.) must be expected to depend on variables other than θ_0, and more specifically Λ_M/B, B/h_{av}, and c_{av}. However, no systematic studies have been carried out to date on the effect of these variables on the location of the [CD]'s. In spite of this, there is sufficient indication that the effect of c_{av} is minor when compared to that of θ_0 (for more on the topic, see da Silva 1995, Zhang 2007, da Silva and

El-Tahawy 2008, Ebrahimi 2015). The effects of Λ_M/B and B/h_{av} can (indirectly) be inferred from subsequent considerations in this chapter.

6.4 BED DEFORMATION

6.4.1 Nature of deformed bed

It will be assumed in the following that the initial surface of the movable bed is flat (in the sense of footnote 4). An experimental "run" commences at time $t=0$. With the passage of time, the bed progressively deforms. Under steady-state conditions, the bed eventually reaches an equilibrium (or fully developed) state at time $t=T_b$, where net erosion and deposition reduce to zero and the bed ceases to deform further.

As is well-known, the bed development consists of the growth in the z-direction, of large-scale, laterally adjacent erosion pools and deposition bars (henceforth also termed pool-bar complexes), occurring once per half channel length. In the case of sine-generated streams, these have the length $L/2$. The location in flow plan of the pool-bar complexes remains invariant or very nearly so during the development process (Binns and da Silva 2011). As illustrated by numerous experimental and field observations (e.g. Hooke 1974, Bridge and Jarvis 1976, Harbor 1998, Whiting and Dietrich 1993b, Binns and da Silva 2009, 2015, Abad et al. 2013), bed forms such as ripples, dunes, and alternate and multiple bars, which are caused by reasons other than the meandering of the stream (and which thus can occur also in straight streams), can be superimposed on the pool-bar complexes, as an expression of the fact that often different phenomena co-exist in meandering streams. However, this (and the next) chapter is restricted to aspects that can be viewed as forming the essence of the meandering phenomenon. When referring to bed deformation or bed topography in subsequent parts of this book, it should thus be clear that we imply the pool-bar complexes described above, and which are directly induced (forced) by the curvature of the stream.

The equilibrium bed topography resulting from laboratory experiments carried out in sine-generated streams with different values of θ_0 has been reported by a number of authors (Hooke 1974, Losiyevskii, as reported by Makaveyvev 1975, Hasegawa 1983, Whiting and Dietrich 1993a,b, Termini 1996, da Silva and El-Tahawy 2008, da Silva et al. 2008, Binns and da Silva 2009, 2015). Examples of measured equilibrium bed topographies are shown in Figures 6.20a-c and 6.21. By considering together the results of these studies, it becomes clear that, similarly to the location of the [CD]'s, the location in flow plan of the pool-bar complexes is also not of a standard nature, but rather strongly dependent on sinuosity (or, equivalently, θ_0). For small values of θ_0, the pool-bar complexes are centred around the crossovers, with the most pronounced erosion and deposition occurring in their proximity (as in Figure 6.20a); for large values of θ_0, they are centred around the apexes, the most pronounced erosion and deposition occurring in the proximity of the latter (as in Figure 6.21). The position of the pool-bar complexes gradually shifts upstream as θ_0 increases from small to large (see Figures 6.20b,c).

The conditions just described are illustrated in the schematic Figure 6.22, where, like in Figure 6.15, the meandering streams are "straightened" for the sake of simplicity. They can also be inferred from Figure 6.23, where the values of λ_c determined from the

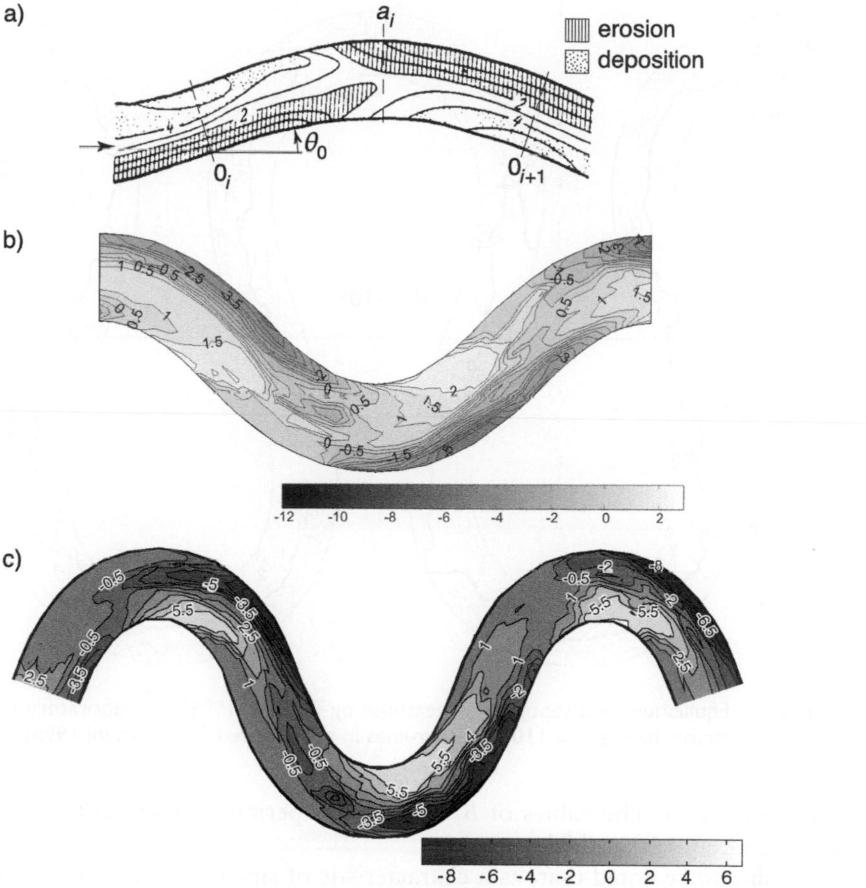

Figure 6.20 Examples of measured equilibrium bed topographies in laboratory sine-generated streams (contour-lines in cm). (a) $\theta_0 = 20°$; experiment by Losiyevskii (adapted from Makaveyvev 1975); (b) $\theta_0 = 45°$, Run R1 (from Binns and da Silva 2015, reprinted with permission from Elsevier); (c) $\theta_0 = 70°$, Run 2 (from da Silva and El-Tahawy 2008, reprinted with permission from IAHR).

measured bed topographies are plotted versus θ_0. Here λ_c is the distance (normalized by L) measured along the centreline from the upstream crossover of a meander loop (having positive R) to the upstream end (beginning) of the erosion-deposition zone partially or fully contained in that loop and exhibiting erosion at the left bank (for the definition of λ_c, see Figure 6.22, where such erosion-deposition zone is highlighted in bold). Figure 6.23 is an extended version of Figure 10 in da Silva and El-Tahawy (2008), produced by including also the recent experiments by Binns and da Silva (2015) in 45° and 95° sine-generated channels. In this figure, the experiments are sorted according to their values of Λ_M/B ($6 \leq \Lambda_M/B \leq 8$ and $10.3 \leq \Lambda_M/B \leq 10.6$); the solid and dashed lines are drawn so as to highlight the patterns of the data falling in the just indicated

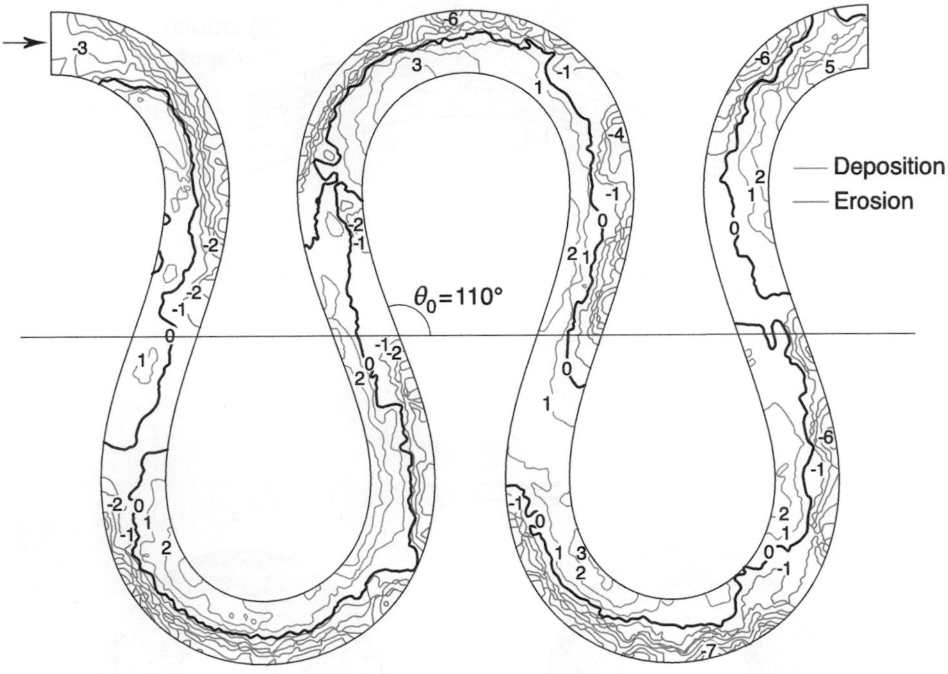

Figure 6.21 Equilibrium bed topography measured by Termini (1996) in a laboratory sine-generated stream having $\theta_0 = 110°$; contour-lines in cm (adapted from Termini 1996).

ranges of Λ_M/B. The values of B/h_{av} of the experiments represented in Figure 6.23 were between ≈ 10 and 30.

It should be noted that, as a characteristic of sine-generated alluvial streams, the location in flow plan of the erosion-deposition zones too, must be expected to depend not only on θ_0, but also on Λ_M/B, B/h_{av} and c_{av} (see Sub-section 6.2.2). The existing data suggest that such location is affected by Λ_M/B and B/h_{av}, albeit to a smaller extent than by θ_0 (Ebrahimi 2015). The dependency on Λ_M/B is evidenced by the sorting of the data corresponding to the different ranges of values of Λ_M/B in Figure 6.23. The form of dependency on B/h_{av} will be discussed in the next section. In comparison to the other independent variables, c_{av} appears to play only a minor role in determining the plan location of the erosion-deposition zones (da Silva and El-Tahawy 2008, Ebrahimi 2015).

Given that the findings above regarding the nature of the meandering deformed bed are derived from laboratory experiments in idealized (sine-generated) streams, it is valid to question whether or not the findings can be generalized to real meandering rivers. To address this matter, consider the following. As discussed later on in Sub-section 7.3.4, the regions where bank retreat occurs coincide with the regions where bed erosion takes place. But if so, and if the nature of the deformed bed changes with θ_0 as described above, then streams with small θ_0 must necessarily mainly migrate downstream, while those with large θ_0, mainly expand laterally. That this is so in rivers has perhaps been best illustrated by Kondratiev et al. (1982). These authors compiled

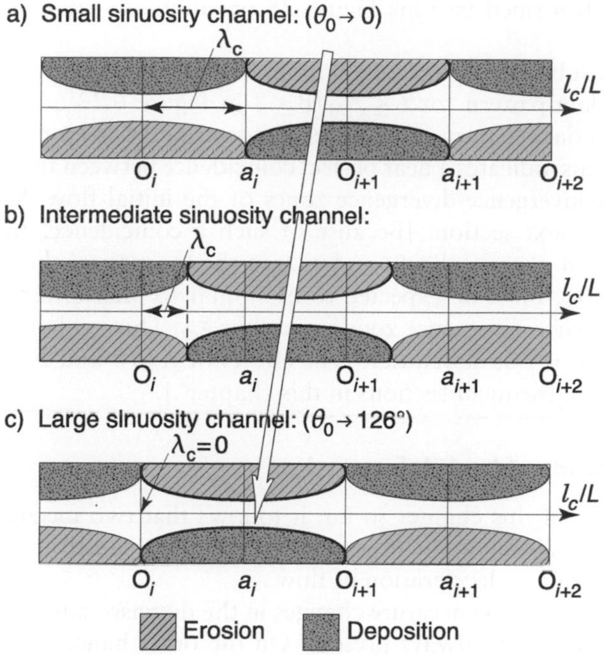

Figure 6.22 Schematic representation of the location of $L/2$-long erosion-deposition zones in streams of varying values of sinuosity. (a) Small θ_0; (b) intermediate θ_0; (c) large θ_0.

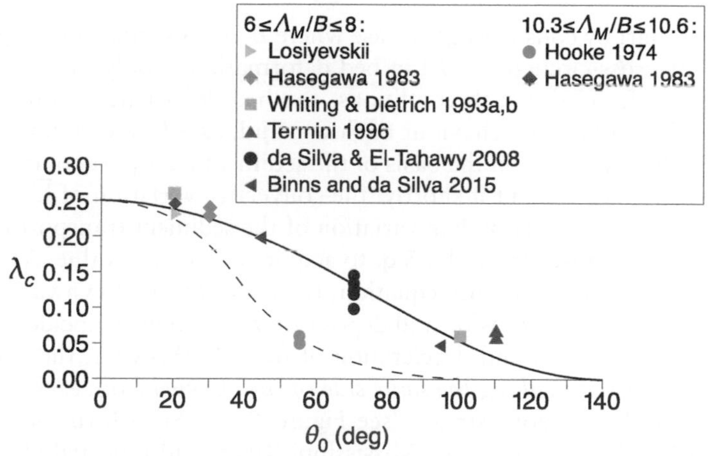

Figure 6.23 Plot of measured values of λ_c versus initial deflection angle (adapted from da Silva and El-Tahawy 2008).

and analyzed extensive series of river surveys carried out over long periods of time in Russian rivers, including the Dnieper, Oka, Irtish, etc., concluding that "at the early stages (small θ_0) it is the downstream migration that is mainly observable, at the latter stages (large θ_0) it is their expansion which dominates". Similar observations were made by Friedkin (1945) in his laboratory experiments in which the meandering streams were self-formed (see his Figure 25 or, alternatively, Figure 11 in da Silva 2006).

To end this sub-section, note the striking similarity between the solid line representing the λ_c data-pattern for $6 \leq \Lambda_M/B \leq 8$ in Figure 6.23 and the solid line representing the ξ_{c0} data-pattern in Figure 6.19 (determined from experiments in which $\Lambda_M/B = 2\pi$).[5] This indicates a near perfect coincidence between the erosion-deposition zones and the convergence-divergence zones of the initial flow. Such coincidence is explained in the next section. [Because of such a coincidence, the effect of Λ_M/B and B/h_{av} (and, of course, also θ_0 and c_{av}) on the location of the erosion-deposition zones, and thus λ_c, must be expected to be similar to their effect on the location of the flow convergence-divergence zones, and thus ξ_{c0}. This explains why at the end of Sub-section 6.3.4, it was stated that 'The effects of Λ_M/B and B/h_{av} can (indirectly) be inferred from subsequent sections in this chapter'.]

6.4.2 Mechanics of bed deformation

From the content of this chapter so far, it follows that two mechanisms are responsible for the bed deformation in meandering streams, namely cross-circulation and convective acceleration-deceleration of flow.

As long as the stream curvature changes in the downstream direction, the convective behaviour of flow is always present. On the other hand, the intensity of cross-circulation, as discussed earlier, progressively decreases with B/h_{av}, while depending on θ_0. This means that, for any θ_0, the extent to which cross-circulation (Γ) will play a role in the bed deformation depends on the value of B/h_{av}.

(i) Let us then start by considering the case where B/h_{av} is sufficiently large so that the role played by cross-circulation (Γ) in bed deformation is only of secondary importance or negligible. Under such conditions, the bed deformation is to be attributed primarily to the convective behaviour of flow – and, as follows from Chapter 1 (Section 1.7), can be explained on the basis of the sediment transport continuity equation $(1-p)\partial z_b/\partial t = -\nabla \mathbf{q}_s$. To put it shortly: the convective variation of $\overline{\mathbf{U}}$ in a flow zone inevitably causes the corresponding variation of the sediment transport rate vector $\mathbf{q}_s$ in that zone, i.e., it causes the scalar $\nabla \mathbf{q}_s$ to acquire a non-zero value. As follows from the sediment transport continuity equation, the zones of the downward and upward bed displacements (i.e. the erosion and deposition zones) must coincide with the zones of convective acceleration and deceleration of flow. In this case, the bed material is transported downstream *along the same side of the channel* from each pool primarily to the bar immediately downstream (see Figure 6.24). Such form of transport was observed by Matthes (1941) in the Mississippi River; and reported also by Leopold and Langbein (1966) for the case of other rivers and Friedkin (1945) for his laboratory experiments.

To explain the process in more detail, let $(z_b)_0$ be the elevation of the flat initial bed at a point P at $t = 0$ (beginning of experiment), and $(z_b)_T$, the elevation of the deformed (developed) bed at that point at $t = T_b$ (see Figure 6.25a). The (positive or negative)

[5]Hence, Eq. (6.47) can be viewed also as the equation giving λ_c when $6 \leq \Lambda_M/B \leq 8$.

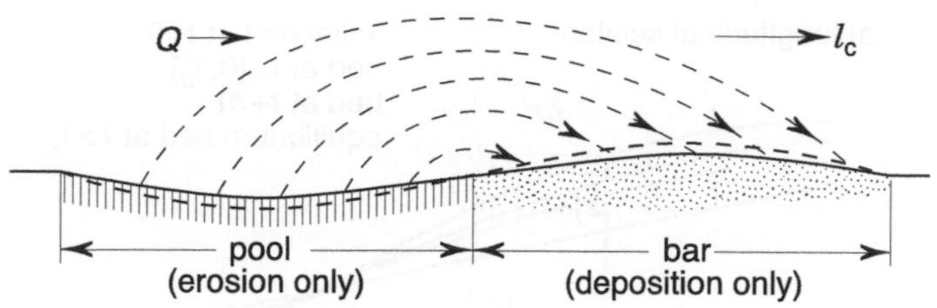

Figure 6.24 Downstream transport of bed material along the same side of the stream from a pool to a bar (longitudinal view along a coordinate line *l*).

final vertical displacement $z'_T = (z_b)_T - (z_b)_0$ of the bed surface at *P* is determined as the sum

$$z'_T = \lim_{N\to\infty} \sum_{i=1}^{N} (\delta z')_i = \lim_{N\to\infty} \sum_{i=1}^{N} W_i \delta t_i = \int_0^{T_b} W \cdot dt, \qquad (6.49)$$

where *W* is a monotonous function of *t*. Since $W \sim \nabla \mathbf{q}_s = (h\overline{\mathbf{U}})\nabla\psi_q$ (see Eqs. (1.73) and (1.79)), and ψ_q is an increasing function of $\tau_0 \sim \overline{U}^2$ (see Eq. (1.78)), the largest $|z'_T|$-values must occur in those flow-regions where the "convectivity" $|\partial\overline{U}/\partial s|$ is the largest. It should thus be clear that the cross-sectional characteristic $\overline{\omega}_c$ can be viewed as a "measure" of the degree of convergence-divergence of flow in "its" section.

The following should be emphatically pointed out here. The time-growth of z', i.e. of pools and bars, causes the flow to deviate from its original flat-bed pattern, for the bars obstruct and repel the flow, while the pools attract it ("topographic steering", Nelson and Smith 1989a,b). Hence, the flow velocities $\overline{U}$ over the bars (where *h*-values are small) decrease, while those over the pools (where *h* are large) increase with the passage of time. The position of the free surface does not vary appreciably while the flow "tries" to avoid bars in favour of pools. [Recall that the natural rivers having large values of B/h_{av} and θ_0, and a strongly deformed bed, have their largest $\overline{U}$ invariably at the outer bank around the apex (a_i) where the pool depth is most prominent, and *h* is the largest (and not at the crossovers (O_i) as in the case of a flat bed)].

It should be borne in mind, however, that the most intense vertical displacement of the bed surface, i.e. the largest values of $W \sim \delta z'$, occur at the early stages of the time interval $0 < t < T_b$ (Figures 6.25a,b). And this is why the location of bars and pools in flow plan is approximately predictable by the convective nature of flow past the flat initial bed (at $t = 0$). At the later stages, the bars and pools increase mainly in their prominence (amplitude), their configuration in plan view being hardly affected. This cannot be said with regard to the flow pattern which is strongly affected by the prominence of bars and pools. The development of the bed surface terminates when it acquires such a topography (at $t = T_b$) which yields $\nabla \mathbf{q}_s \equiv 0$ all over (see Section 1.7 (ii)). Figures 6.26 and 6.27, which correspond to large and intermediate θ_0, convey schematically the nature of the initial flow pattern (at or just after $t = 0$) and the final flow pattern (at $t = T_b$). The analogous flow patterns can, of course, be drawn also for flows having small θ_0.

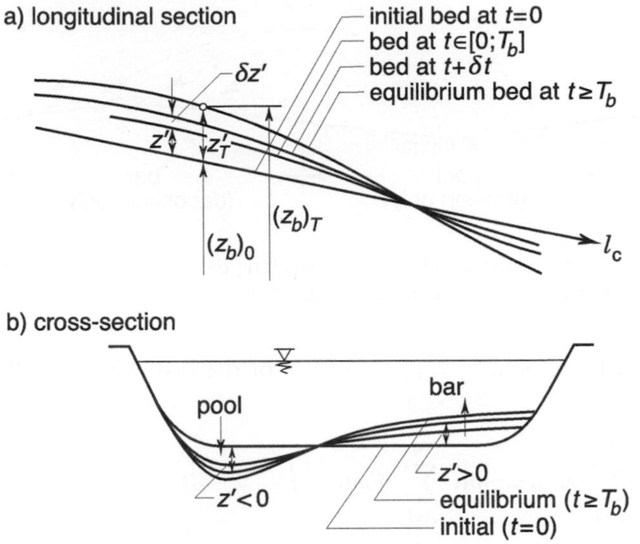

Figure 6.25 Schematic representation of meandering bed deformation over time. (a) Longitudinal section; (b) cross-sectional view.

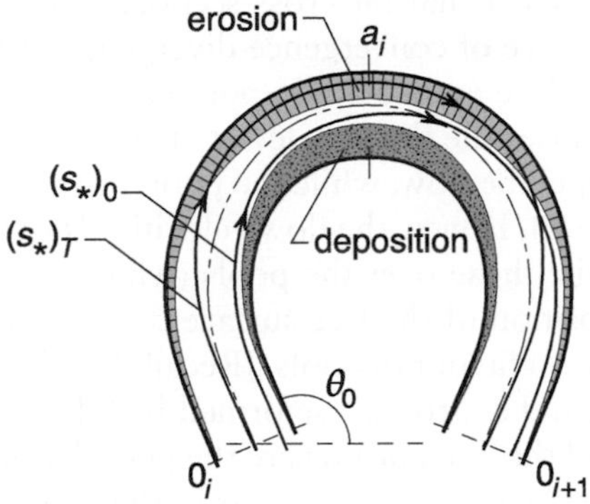

Figure 6.26 Vertically-averaged streamlines s_* at or just after $t=0$ ($(s_*)_0$), and at $t=T_b$ ($(s_*)_T$), in the case of a channel having large θ_0. Bed is flat at $t=0$.

(ii) From the explanations above, it follows that the approximate coincidence between the location of the erosion-deposition zones and the convergence-divergence zones of the initial flow can very adequately be described by the convective behaviour of flow arising from the streamwise variation in curvature.

Let us now consider again the cross-circulatory motion. In contrast to the convective behaviour of flow, this causes the transport of bed material across the channel from the outer to the inner bank. Since cross-circulation is the most intense where curvature is the largest, it must necessarily produce maximum erosion at the outer bank

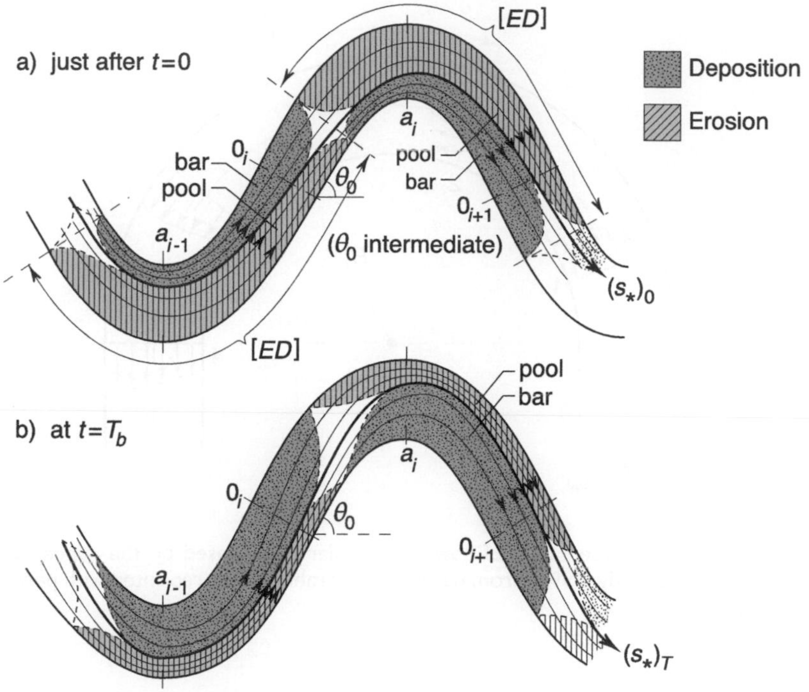

Figure 6.27 Patterns of the vertically-averaged flow in the case of a channel having intermediate value of θ_0: (a) at $t = 0$, or just after $t = 0$ when the bed deformation is still small; (b) at $t = T_b$, when the bed is strongly deformed.

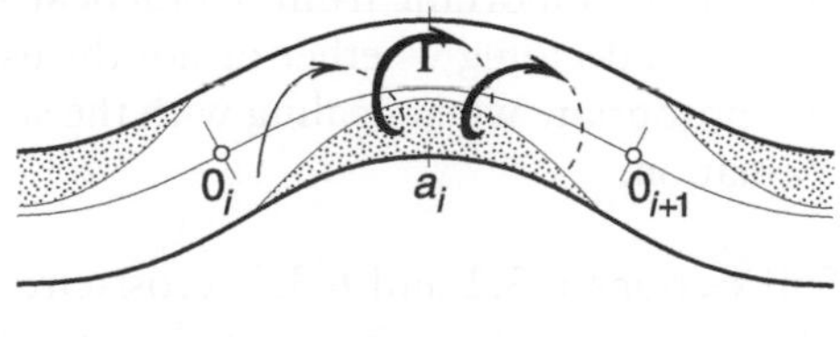

Figure 6.28 Deposition pattern due to cross-circulatory motion.

around the apex or at least in its proximity (see Figure 6.28). That is, cross-circulation (by itself) is associated with a "standard" erosion-deposition pattern. Clearly, the bed topography in streams having small θ_0 as described in Sub-section 6.4.1 with maximum erosion-deposition occurring near the crossovers, cannot be explained by such means. This is further supported by the fact that in several of the experiments described in Sub-section 6.4.1, it was observed that material eroded from a given loop was mainly transported downstream along the same side of the channel and eventually deposited so as to form the bar in the meander loop immediately downstream (e.g. Hooke 1974, Whiting and Dietrich 1993a,b, da Silva and El-Tahawy 2008, Binns and da Silva 2009). This brings us to the question: "What is the 'critical value' of B/h_{av}, $(B/h_{av})_{cr}$ say, beyond which the effect of cross-circulation becomes of secondary importance

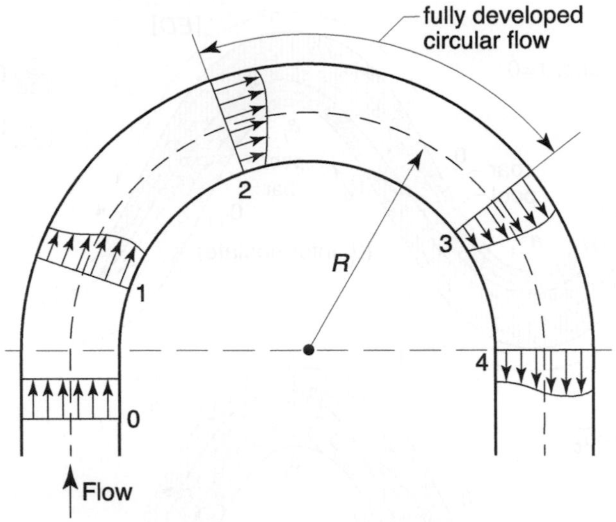

Figure 6.29 Schematic representation of flow in a circular bend, based on the conceptualization by Yeh and Kennedy (1993) (from da Silva and Ebrahimi 2017, reprinted with permission from ASCE).

where the bed topography is concerned?". In the following, this question is explored further.

From the content of this chapter so far, it should be clear that the vertically-averaged approach cannot reflect the influence of Γ (see Eq. (6.29)). Gaining insight into the above question is thus important from a practical standpoint, as this is a crucial aspect to consider when deciding whether or not the use of vertically-averaged hydrodynamic models is appropriate when dealing with the simulation and prediction of meandering bed deformation.

(iii) As follows from Sub-sections 6.3.2 and 6.3.3, cross-circulation transports high-speed momentum fluid towards the outer bank at the upper fluid layers, and low-momentum fluid towards the inner bank at the lower layers. This results in the downstream transport of momentum from the inner to the outer bank, or, to put it in other words, a redistribution of velocity in the downstream direction (de Vriend 1981, Yeh and Kennedy 1993, Lien et al. 1999, Blanckaert and de Vriend 2004). Such velocity redistribution is illustrated in Figure 6.29, showing a schematic representation of a flow in a curved circular bend ($1/R = const$). Observe that between sections 0 and 2, the flow continuously changes in the downstream direction. In particular, after an adjustment region at the channel entrance (between sections 0 and 1), the flow changes between sections 1 and 2 in such a way that the locus of maximum flow velocity moves from the inner bank at section 1 to the outer bank at section 2.[6] In this region, the flow

[6]From cross-section 2 onwards (up to cross-section 3), the flow is fully developed.

convectively decelerates along the inner bank and convectively accelerates along the outer bank – as a result of cross-circulation, and also longitudinal pressure gradients due to the free surface superelevation (Demuren and Rodi 1986). That is, convective behaviour of flow can also occur in the absence of stream curvature for the reasons just stated.

Consider now a curved stream (sine-generated, say) having a sufficiently large B/h_{av} so that $\Gamma \to 0$ and the convective flow patterns (assumed to be as described in Sub-section 6.3.4) can be attributed strictly to the streamwise variation in stream curvature; and consider also the convective flow patterns occurring in the same stream but under conditions where cross-circulation is not negligible. On the basis of the content of the previous paragraph, it is here hypothesized that, in comparison to the case where $\Gamma \to 0$, as the effect of Γ increases, the location of the [CD]'s will be shifted further upstream, and the more so the smaller B/h_{av}. Such hypothesis, of course, applics only to the cases of small and intermediate values of θ_0 – as for large θ_0, the [CD]'s are centred around the apexes, no matter what the value of B/h_{av} (i.e. no matter what the strength of Γ might be).

In addition to affecting the convective flow patterns, and as previously mentioned, cross-circulatory motion directly acts on the grains by moving them from the outer to the inner bank. This means that cross-circulation determines the bed topography (or plays a role in determining it) indirectly via its action on the convective flow pattern as well as directly by inducing the transport of bed material across the channel. Both of these effects result in deposition at the inner bank in the proximity of the apex – while for the case of small and intermediate values of θ_0, the erosion-deposition zones due to the streamwise variation in curvature are centred downstream from the apex (and the more so, the smaller is θ_0). But if so, one would expect the location of the bed erosion-deposition zones too, to be shifted further upstream in a case where the effect of cross-circulation is not negligible when compared to one where it is negligible – and the more so, the smaller B/h_{av}. Thus, we further hypothesize that a means to reveal the aforementioned critical value of B/h_{av} is to investigate differences in the location of the erosion-deposition zones for decreasing values of θ_0 (for $\theta_0 <\approx 70$–$80°$, say, as for large θ_0 the location of the flow [CD]'s is around the apex no matter what the value of B/h_{av}).

The above hypothesis was tested in da Silva et al. (2008), who accordingly determined the values of λ_c for two series of laboratory experiments in a 70° sine-generated channel ($B = 0.80m$, $\Lambda_M = 2\pi B$). This consisted of a total of eight experimental runs; the bed was flat at the beginning of the experiments; the runs were carried out until the equilibrium stage had been achieved. The values of B/h_{av} in the experiments varied from ≈ 10 to 70; $c_{av} \approx 11$. The resulting values of λ_c are plotted versus θ_0 in Figure 6.30 (in which the solid line highlights the data-pattern). As follows from this figure, the location in flow plan of the erosion-deposition zones did not change as long as $B/h_{av} >\approx 20$ – as reflected by the fact that $\lambda_c = const \approx 0.15$ when $B/h_{av} >\approx 20$. The value of λ_c was found to indeed gradually decrease with decreasing values of B/h_{av} for $B/h_{av} <\approx 20$. The experiments did not include tests with $B/h_{av} < 10$, but the data trend clearly suggests that for such values of B/h_{av} the location in flow plan of the erosion-deposition zones would be even further shifted upstream.

The aforementioned suggests that, in the experiments under consideration, the critical value of B/h_{av} beyond which the effect of cross-circulation became of secondary importance where the bed deformation is concerned was $(B/h_{av})_{cr} \approx 20$.

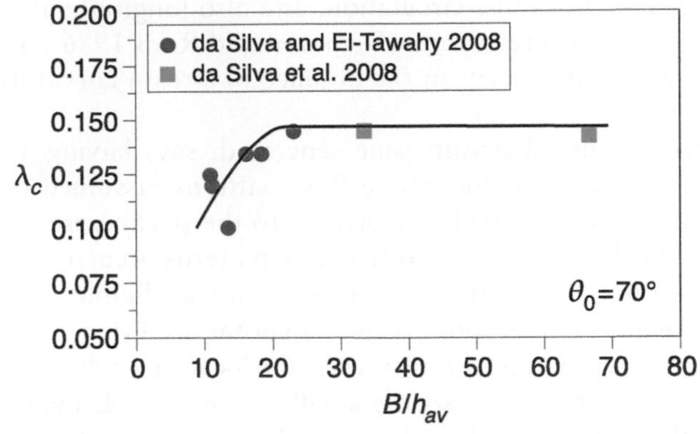

Figure 6.30 Plot of measured values of λ_c versus width-to-depth ratio resulting from experiments in a 70° sine-generated channel (adapted from da Silva et al. 2008).

On the basis of the content of Sub-section 6.2.2, it should be clear that $(B/h_{av})_{cr}$ too, must be expected to depend on θ_0, Λ_M/B and c_{av}, i.e. $(B/h_{av})_{cr} = \phi(\theta_0, \Lambda_M/B, c_{av})$. Thus the solid line in Figure 6.30 is to be viewed as the curve representing this function when $\theta_0 = 70°$, $\Lambda_M/B = 2\pi$ and $c_{av} = 11$. Further laboratory and/or numerical research is needed to fully reveal the function $(B/h_{av})_{cr} = \phi(\theta_0, \Lambda_M/B, c_{av})$.

6.5 RESISTANCE FACTOR OF MEANDERING STREAMS

(i) Consider the unit-prism $h \times 1 \times 1$ at a point $(l_c; r)$ of a meandering flow (Figure 6.31). The prism-bed is subjected to the action of the bed shear stress $\vec{\tau}_0$; the vertical prism-faces, to the action of the stresses σ_{ll}, σ_{rr} and $\tau_{rl} = \tau_{lr}$ caused by the flow curvature. Let $\vec{\chi}$ be the resultant of the stress forces $\overline{\sigma}_{ll}h$, $\overline{\sigma}_{rr}h$ and $\overline{\tau}_{rl}h = \overline{\tau}_{lr}h$ acting on the vertical prism-faces (where $\overline{\sigma}_{ll}$, $\overline{\sigma}_{rr}$, ..., are the vertically-averaged values of σ_{ll}, σ_{rr}, ...).[7] The total resistance force **T** acting on the fluid contained in the unit-prism is then

$$\mathbf{T} = \vec{\tau}_0 + \vec{\chi}. \tag{6.50}$$

The force **T**, which opposes the fluid motion along the vertically-averaged streamlines s, is acting along $-\mathbf{i}_s$. The bed shear stress $\vec{\tau}_0$, which opposes the fluid motion along the streamlines s_b at the bed, is acting along $-\mathbf{i}_b$. However, in the case of the wide channels under study the discrepancy between $\mathbf{i}_s$ and $\mathbf{i}_b$ is negligible. This gives the possibility to express Eq. (6.50) as

$$\mathbf{T} = \mathbf{i}_s T = \mathbf{i}_s \tau_0 + \vec{\chi} = \mathbf{i}_s(\tau_0 + \chi_s). \tag{6.51}$$

Here, χ_s is the scalar component of $\vec{\chi}$ along the streamline s (at the location $(l_c; r)$ of the prism).

[7]The expression of $\vec{\chi}$ in terms of $\overline{\sigma}_{ll}$, $\overline{\sigma}_{rr}$ and $\overline{\tau}_{lr} = \overline{\tau}_{rl}$ is given by Eqs. (7.14) and (7.15). Observe from these relations that the terms forming the components of $\vec{\chi}$, viz $\partial(\overline{\tau}_{rl}h)/\partial r$, $2\overline{\tau}_{rl}h/r$, ..., have the dimension of stress.

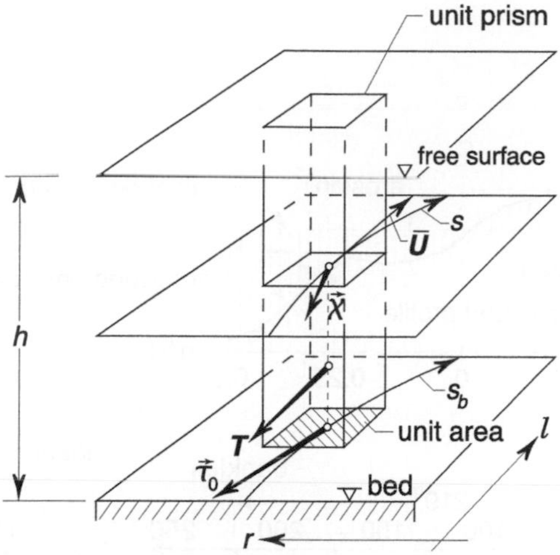

Figure 6.31 Forces acting on a prism of the height *h* and unit base area at a point of a meandering flow.

Let $\mathbf{\overline{U}} = \mathbf{i}_s\overline{U}$ be the vertically-averaged flow velocity vector at the same location. The dimensionless ratio

$$c_M = \frac{\overline{U}}{\sqrt{T/\rho}} \tag{6.52}$$

can be taken as the definition of the *local resistance factor* of a meandering flow (da Silva 1995, 1999). We go over now to consider how the channel average and cross-sectional average values of c_M can be evaluated on the basis of information supplied by experiment.

(ii) Since the "in" and "out" transport rates of a loop (at O_i and O_{i+1}) are treated, at any t, as identical, the average bed of a loop must, as has already been mentioned, always be the same as the initial flat bed at $t = 0$ (pp. 14-15, da Silva 1995).

Moreover, the field and laboratory measurements indicate (see e.g. Figures 6.32a,b) that the areas of the cross-sectional erosion and deposition are (nearly) identical (at any cross-section and at any t).[8] Hence, in the following, we will invariably assume that

$$h_m = h_{av} \qquad \text{and thus that} \qquad u_m = u_{av} \tag{6.53}$$

(for $u_m = Q/(Bh_m) = Q/(Bh_{av}) = u_{av}$).

[8] Consider the flow plan area confined between two sections ab and cd in the insert of Figure 6.32. Since $\widehat{ad} > \widehat{bc}$, the cross-sectional deposition area adjacent to the inner bank $\widehat{bc}$ must, in fact, be somewhat larger than the erosion area adjacent to the outer bank $\widehat{ad}$ (for the average bed level remains unchanging).

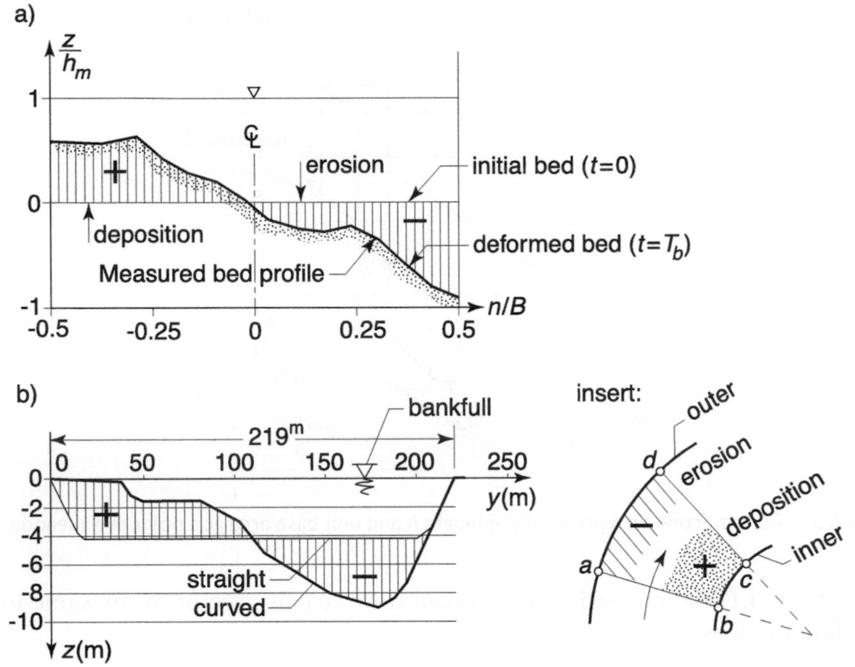

Figure 6.32 Examples of cross-sectional bed profiles in meandering streams. (a) Bed profile in a section of a sine-generated stream (adapted from Ikeda and Nishimura 1986); (b) typical straight and curved regions of the Syr-Darya River in lower reaches (adapted from Levi 1957). Insert: plan view of a meandering region abcd.

Consider now the bed slopes. The channel average value S_{av} of the longitudinal bed slope S (averaged through a large number of anti-symmetrically identical loops, or through a limited even number of loops) is equal to the centreline slope S_c:

$$S_{av} = S_c. \tag{6.54}$$

For the cross-sectional average value S_m of S one determines, by taking into account Eq. (6.5),

$$\begin{aligned} S_m &= \frac{1}{B}\int_{-B/2}^{B/2} S dn = \frac{S_c}{B}\int_{-B/2}^{B/2} \frac{R}{R+n} dn = S_c \frac{R}{B} \ln\left(\frac{1+B/2R}{1-B/2R}\right) \\ &= S_c\left[1+\frac{1}{12}\left(\frac{B}{R}\right)^2+\frac{1}{80}\left(\frac{B}{R}\right)^4+\ldots\right] \approx S_c. \end{aligned} \tag{6.55}$$

Here the last step is due to the fact that in the sine-generated channels the largest B/R, viz B/R_a, does not exceed ≈ 0.8 for any θ_0 (see Figure 6.3a); yet, using $B/R = 0.8$ in Eq. (6.55), we obtain $S_m = S_c[1 + 0.053 + 0.005 + \cdots] \approx 1.058 S_c$. Thus, with an accuracy sufficient for all practical purposes,

$$S_m \approx S_c = S_{av} \tag{6.56}$$

can be adopted. From Eqs. (6.53) and (6.56), it follows that

$$\frac{u_m}{\sqrt{gS_m h_m}} \approx \frac{u_{av}}{\sqrt{gS_{av} h_{av}}}, \tag{6.57}$$

where the left- and right-hand sides approximate the cross-sectional and channel average values of c_M, respectively. i.e.

$$(c_M)_m \approx (c_M)_{av} \,. \tag{6.58}$$

(iii) Much research has been carried out on hydraulic energy losses and the channel average resistance factor of meandering streams – a topic of a considerable interest to hydraulic engineers (see e.g. Mockmore 1944, Rozovskii 1957, Bagnold 1960, Leopold et al. 1960, Hayat 1965, Onishi et al. 1976, Chang 1983, da Silva and Binns 2009; an informative review of the available methods can be found in James 1994). From field and laboratory measurements, it follows (see the works just mentioned) that the value of the channel average resistance factor $(c_M)_{av}$ of a wide meandering stream does not differ appreciably, nor systematically, from the constant value c of its straight-channel counterpart.[9] Thus, one can write

$$(c_M)_{av} \approx c, \tag{6.59}$$

which, considering Eq. (6.58), can be augmented into

$$(c_M)_m \approx (c_M)_{av} \approx c. \tag{6.60}$$

The result (6.60) is, in fact, not surprising; for nature creates a meandering stream as to possess, under ideal conditions, the minimum variation of its curvature $(d(1/R)/dl \rightarrow min)$ – and the sine-generated plan shape can be even derived on this basis (Yalin 1992; see Sub-section 6.2.1 (i)). Moreover, the erosion and deposition zones ("pools" and "bars") created by a meandering stream do not have any abrupt faces (like the down-stream faces of ripples and dunes): these zones manifest themselves as "small-amplitude waves" superimposed on the flat average bed. In wide natural meandering streams, the largest transversal and longitudinal slopes of a deformed bed, viz $|\partial z'_T/\partial l_c|$ and $|\partial z'_T/\partial n|$, usually are smaller than, say, $\approx 1/20$ and $\approx 1/120$, respectively (e.g. in the curved region of the Syr-Darya River in Figure 6.32b, the transverse slope is $\approx 1/25$). But since the resistance of a wide meandering stream is thus almost totally due to ripples and dunes, the value of $(c_M)_{av}$ can be computed as c given by the straight-channel expression (3.12). This expression must be evaluated, of course, by using, in lieu of h, the channel average flow depth (h_{av}) of the meandering flow. In fact, some of the experimental points used to verify the c-curves representing this expression (Figures 3.2a-c) stem from straight-, some others from meandering-streams – no grouping of the points depending on channel curvature is detectable.

From Eqs. (6.59) and (6.60) it does not follow, and it is in fact false, that c_M remains constant in the $(l_c; r)$-plan (as has been assumed by some authors). The determination of the *local* c_M will be dealt with in the next chapter.

[9] It is assumed that the flow in the straight-channel counterpart is uniform, h and K_s being the same all over: $c = c_m = c_{av} = const.$

6.6 ADDITIONAL REMARKS

A systematic investigation of meandering streams, which has been started at the end of the nineteenth century, was motivated mainly by practical needs. Hence, it is not surprising that the first observations and measurements were mostly due to river engineers such as L. Fargue, M.P. Du Boys, N. de Leliavsky, H. Engels, C.C. Inglis, S. Leliavskii and others (see e.g. Fargue 1868, de Leliavsky 1894, 1905, Engels 1929, Du Boys 1933, Inglis 1947, Leliavskii 1959), and that they were carried out in the field.

Yet, much of the research aiming to reveal the mechanics of meandering streams over the past 60 years or so has been conducted in laboratories. Classically, in this kind of laboratory "meandering research" either single circular channel bends (of various central angles) or sequences of adjacent circular bends were used. In some cases, the channels were formed alternately by circular bends and straight segments. Examples of such works throughout the years include Rozovskii (1957), Ippen and Drinker (1962), Fox and Ball (1968), Yen and Yen (1971), Francis and Asfari (1971), Martvall and Nilsson (1972), Mosonyi and Goetz (1973), Varshney and Ramchandra (1975), Kikkawa et al. (1976), Choudhary and Narasimhan (1977), de Vriend and Koch (1978), Tamai et al. (1983), Odgaard (1984), Steffler (1984), Almquist and Holley (1985), among many others. However, in recent experimental research on meandering the streams increasingly tend to be idealized as sine-generated curves (Hooke 1974, Hasegawa 1983, Hasegawa and Yamaoka 1983, Ikeda and Nishimura 1986, Whiting and Dietrich 1993a,b,c, da Silva 1995, 1999, Termini 1996, 2009, da Silva et al. 2006, da Silva and El-Tahawy 2008, Binns and da Silva 2009, 2011, 2015, Termini and Piraino 2011, Xu and Bai 2013). It appears that, so far, only Hasegawa (1983), Abad and Garcia (2009a,b) and Abad et al. (2013) used Kinoshita curves as idealizations of meandering streams. Just like sine-generated curves, 'Kinoshita' curves are periodic and exhibit a downstream variation in stream curvature. In contrast to sine-generated curves they are, however, irregular in plan shape, in the sense that they are not symmetric in plan view with regard to the axis of bend (see Figure 6.2 for the definition of axis of bend).

(i) The $1/R$-diagram of a curved channel formed by circular segments exhibits along l_c a sequence of discontinuous "jumps" at the loop-end points (crossovers) O_i (Figure 6.33a), which causes the moving fluid elements to experience strong jolts at O_i. This renders the flow patterns to deviate substantially from those of sine-generated channels, whose $1/R$-diagrams, and all their l_c-derivatives, vary continuously along l_c (Figure 6.33b).

(ii) For a conventional laboratory flow to be observable and measurable, as well as to ensure that it is turbulent, h cannot be less than 2 to $4cm$, say. On the other hand, the laboratory area is often limited, and therefore $B \sim \Lambda_M$ cannot be too large either; hence, more often than not, the laboratory values of the aspect ratio B/h do not exceed 10 to 15, say. Yet, the B/h-value of an extensively meandering (low land) river often is comparable with 100 to 120, say. But this means that the aspect ratios of laboratory meandering streams usually are rather distorted. This high distortion of B/h inevitably leads to a disproportionate magnification of the relative cross-circulatory velocity $v_\Gamma/\overline{u} \sim \sqrt{h/B}$ (see Eq. (6.43)) and, consequently, to the exaggeration of the importance of cross-circulation (Γ) itself.

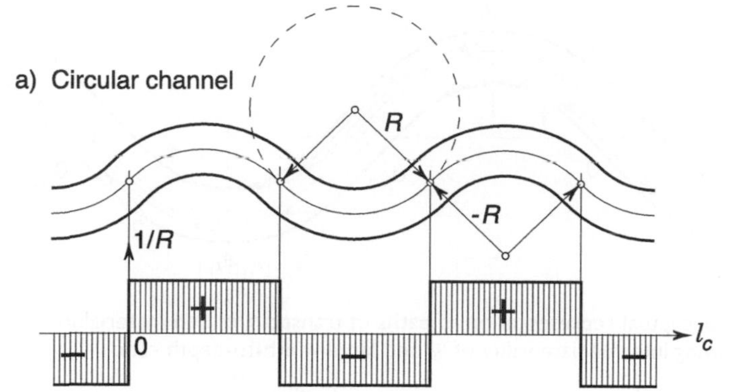

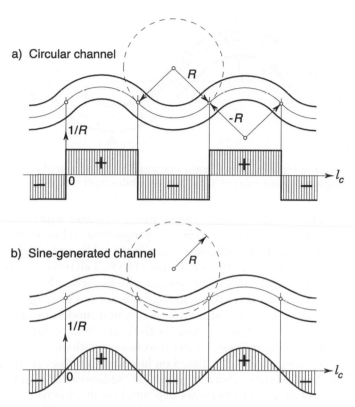

Figure 6.33 Diagrams of curvature along the channel centreline in: (a) a channel formed by circular segments; (b) a sine-generated channel.

(iii) In an idealized meandering stream (whose loops are assumed to be sufficiently far from the beginning and end of the flow) all the geometric and mechanical characteristics of flow have (anti-symmetrically) identical distributions in the cross-sectional areas A_i and A_{i+1} at the loop end-points O_i and O_{i+1}. The realization of this identity (which forms one of the most relevant boundary conditions in the solution of the differential equations of a meandering flow) is hardly possible in a laboratory experiment conducted with one or two circular loops having straight approach- and exit-channels. Yet, as mentioned above, many of the well known laboratory experiments aiming to reveal the mechanism of meandering flows were conducted exactly in this way.

In conclusion, it should be pointed out that perhaps the most pervasive and lasting misconception inherited from the literature stemming from the laboratory research on meandering, is the exaggerated importance of the cross-circulation (Γ). Indeed, even nowadays any so far unknown "radial manifestation" of a meandering stream or its boundaries, is routinely attributed by the laboratory researchers and theoreticians to the cross-circulation – no matter what the values of θ_0 and B/h might be. "The importance of cross-circulation in determining the geometry of river beds in meanders

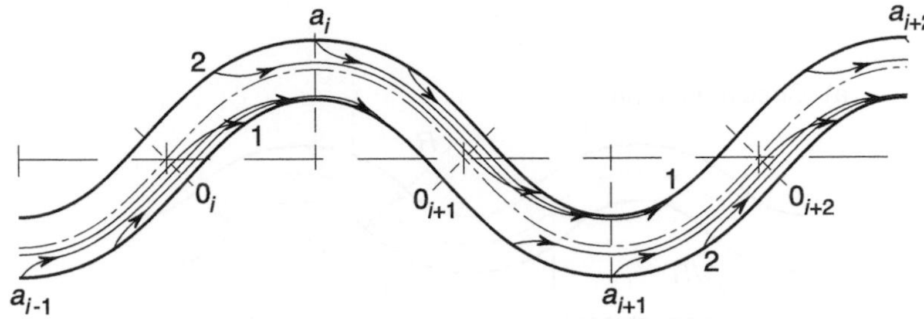

Figure 6.34 Conceptual representation of paths of transport of bed material in a meandering stream having intermediate value of θ_0 and large width-to-depth ratio.

has been over-emphasized for many years, and it will take some time to bring the significance of such flow patterns into proper perspective" (Hooke 1980).

The above mentioned "Γ-enthusiasm" is not shared by as eminent field research-engineers as Matthes (1941, 1948), Velikanov (1955), Leliavskii (1959), Makaveyev (1975) and Kondratiev et al. (1982). These research engineers, which were dealing with large natural rivers (having large B/h and θ_0), have noticed long ago that "transverse circulation only takes place in distorted experimental models, and in such natural channels whose width is small as compared with the depth ..." (Matthes 1941; see also Matthes 1948). Observe that this statement is consistent with the schematic form (6.43) derived for v_Γ in Sub-section 6.3.3. Based on his extensive observations, especially in the Mississippi River, Matthes also mentions that "sediment entering the stream from the scour of the concave banks becomes deposited on the downstream convex bank on the *same side* of the river, i.e. as shown schematically in Figure 6.34, and only a small portion of the eroded material crosses the channel" (see also Kondratiev et al. 1959). The fact that in the case of large rivers the influence of cross-circulation is only secondary was also noticed by Makaveyvev (1975), who points out that "the influence of circulating flows on the formation of bends consists mainly in that they (merely) *facilitate* the growth of convex banks". In this context, Hooke (1974) points out that "while the existence of point bar (that is, the deposition zone at the inner bank in the apex region) cannot be attributed to the secondary currents (i.e. to the cross-circulation), the detailed geometry of the bar is at least partly controlled by them". It should thus be clear that in the case of large natural streams (having large B/h and θ_0) the bed deformation is mainly due to the convective behaviour of the flow and not so much due to Γ. Following Engelund (1974), Nelson and Smith (1989a,b), Struiksma et al. (1985), Struiksma and Crosato (1989), Shimizu and Itakura (1989), Shimizu (1991), Jia and Wang (1999), the bed deformation is nowadays invariably computed by taking into account the convective behaviour of (vertically-averaged) flow.

PROBLEMS

6.1 The ratio of the sinuosities of two sine-generated meandering channels (channels 1 and 2) is $\sigma_1/\sigma_2 = 2.202$; the ratio of their deflection angles is $(\theta_0)_1/(\theta_0)_2 = 2.0$. What are the values of $(\theta_0)_1$ and $(\theta_0)_2$?

Note: The following polynomial approximation can be used to compute $J_0(\theta_0)$:

$$J_0(\theta_0) \approx 1 - 2.2499997(\theta_0/3)^2 + 1.2656208(\theta_0/3)^4 - 0.3163866(\theta_0/3)^6 + 0.0444479(\theta_0/3)^8 - 0.0039444(\theta_0/3)^{10} + 0.0002100(\theta_0/3)^{12}.$$

6.2 For which values of θ_0 do sine-generated meandering streams have B/R_a (i.e. B/R at the apex-section) equal to 0.37?

6.3 Write a general computer program to draw the centreline of a sine-generated meandering stream from O_i to O_{i+1} (i.e. between two consecutive crossovers). The deflection angle θ_0 and the flow width B are given; take $\Lambda_M = 2\pi B$.
Hint: Determine the cartesian coordinates of several points along the centreline.

6.4 Use the computer program of Problem 6.3 to draw the centreline of a sine-generated meandering stream having $\theta_0 = 110°$. The flow width is $B = 0.40m$.

6.5 A river is "endeavouring" to achieve its regime state by meandering. The valley slope is $S_0 = 1/500$ and the regime slope is $S_R = 1/1350$. The centreline of the river closely follows a sine-generated curve.
a) At a certain stage Θ of its regime development, the slope of the river is $S = 1/670$. What is the value of the deflection angle θ_0 at that stage?
b) What is the maximum θ_0 this river can possess?

REFERENCES

Abad, J.D., Frias, C.E., Buscaglia, G.C. and Gárcia, M.H. (2013). Modulation of the flow structure by progressive bedforms in the Kinoshita meandering channel. *Earth Surface Processes and Landforms*, 38(13), 1612-1622.

Abad, J.D. and Gárcia, M.H. (2009a). Experiments in a high-amplitude Kinoshita meandering channel: 1. Implications of bend orientation on mean and turbulent flow structure. *Water Resources Research*, 45, W02401, doi:10.1029/2008WR007016.

Abad, J.D. and Gárcia, M.H. (2009b). Experiments in a high-amplitude Kinoshita meandering channel: 2. Implications of bend orientation on bed morphodynamics. *Water Resources Research*, 45, W02402, doi:10.1029/2008WR007017.

Almquist, C.W. and Holley, E.R. (1985). *Transverse mixing in meandering laboratory channels with rectangular and naturally varying cross-sections*. Report CRWR 205, The University of Texas at Austin.

Bagnold, R.A. (1960). Some aspects of the shape of river meanders. *U.S. Geological Survey Professional Papers*, 282-E, 135-144.

Binns, A.D. and da Silva, A.M.F. (2015). Meandering bed development time: formulation and related experimental testing. *Advances in Water Resources*, 81, 152-160.

Binns, A.D. and da Silva, A.M.F. (2011). Rate of growth and other features of the temporal development of pool-bar complexes in meandering streams. *Journal of Hydraulic Engineering*, 137(12), 1565-1575.

Binns, A.D. and da Silva, A.M.F. (2009). On the quantification of the bed development time of alluvial meandering streams. *Journal of Hydraulic Engineering*, 135(5), 350-360.

Blanckaert, K. (2010). Topographic steering, flow recirculation, velocity redistribution, and bed topography in sharp meander bends. *Water Resources Research*, 46, W09506, doi: 10.1029/2009WR008303.

Blanckaert, K. and de Vriend, H.J. (2004). Secondary flow in sharp open-channel bends. *Journal of Fluid Mechanics*, Vol. 498, 353-380.

Bridge, J.S. and Jarvis, J. (1976). Flow and sedimentary processes in the meandering river South Esk, Glen Clova, Scotland. *Earth Surface Processes and Landforms*, 1(4), 303-336.

Chang, H.H. (1988). *Fluvial processes in river engineering*. John Wiley and Sons, Inc.

Chang, H.H. (1983). Energy expenditure in curved open channels. *Journal of Hydraulic Engineering*, 109(7), 1012-1022.

Choudhary, U.K. and Narasimhan, S. (1977). Flow in 180° open channel rigid boundary bends. *Journal of the Hydraulics Division*, ASCE, 103(6), 651-657.

da Silva, A.M.F. and Ebrahimi, M. (2017). Meandering morphodynamics: insights from laboratory and numerical experiments. *Journal of Hydraulic Engineering*, 143(9), doi: 10.1061/(ASCE)HY.1943-7900.0001324.

da Silva, A.M.F. (2015). Recent advances from research on meandering and directions for future work. Chapter 14, in *Rivers – Physical, Fluvial and Environmental Processes*, edited by P. Rowińsky and A. Radecki-Pawlik, GeoPlanet: Earth and Planetary Book Series, Springer International Publishing, Switzerland, 373-401.

da Silva, A.M.F. and Binns, A.D. (2009). On the resistance to flow of alluvial meandering streams. *Proceedings of the 33rd IAHR Congress*, Vancouver, BC, Canada, 1388-1395.

da Silva, A.M.F. and El-Tahawy, T. (2008). On the location in flow plan of erosion-deposition zones in sine-generated meandering streams. *Journal of Hydraulic Research*, 46, Extra Issue 1, 49-60.

da Silva, A.M.F., Holzwarth, S., Pasche, E. and El-Tahawy, T. (2008). Bed topography of alluvial meandering streams under varying width-to-depth-ratios. *Proceedings of River Flow 2008, 4th International Conference on Fluvial Hydraulics*, Cesme-Izmir, Turkey, edited by M.S. Altinakar, M.A. Kokpinar, I. Aydin, S. Cokgor and S. Kirkgoz, Ankara: Kubaba Congress Department and Travel Services, 1363-1372.

da Silva, A.M.F. (2006). On why and how do rivers meander. *Journal of Hydraulic Research*, 44(5), 579-590.

da Silva, A.M.F., El-Tahawy, T. and Tape, W.D. (2006). Variation of flow pattern with sinuosity in sine-generated meandering channels. *Journal of Hydraulic Engineering*, 132(10), 1003-1014.

da Silva, A.M.F. (1999). Friction factor of meandering flows. *Journal of Hydraulic Engineering*, 125(7), 779-783.

da Silva, A.M.F. (1995). *Turbulent flow in sine-generated meandering channels*. Ph.D. Thesis, Department of Civil Engineering, Queen's University, Kingston, Canada.

da Silva, A.M.F. (1991). *Alternate bars and related alluvial processes*. M.Sc. Thesis, Department of Civil Engineering, Queen's University, Kingston, Canada.

Davidsen, T.S. (2007). *Numerical studies of flow in curved channels*. Ph.D. Thesis, Department of Mathematics, University of Bergen, Norway.

de Leliavsky, N. (1905). Results obtained by dredging on river-shoals. *10th International Congress on Internal Navigation*, Milan, Italy.
de Leliavsky, N. (1894). Currents in streams and the formation of stream beds. *6th International Congress on Internal Navigation*, The Hague, The Netherlands.
Demuren, A.O. and Rodi, W. (1986). Calculation of flow and pollutant dispersion in meandering channels. *Journal of Fluid Mechanics*, Vol. 172, 63-92.
de Vriend, H.H. (1981). Velocity redistribution in curved rectangular channels. *Journal of Fluid Mechanics*, Vol. 107, 423-439.
de Vriend, H.J. and Koch, F.G. (1978). *Flow of water in a curved open channel.* Delft Hydraulics Laboratory/Delft University of Technology, TOW-Report R657-VII/M141S-III.
de Vriend, H.J. (1977). A mathematical model of steady flow in curved shallow channels. *Journal of Hydraulic Research*, 15(1), 37-54.
Du Boys, M.P. (1933). *The Rhone and rivers of shifting bed. (1879).* Translated by H.G. Doke, Memphis, Tennessee, U.S. Engineer Office.
Ebrahimi, M. (2015). *Flow patterns, bank erosion and planimetric changes in meandering streams: an experimental study*. Ph.D. Thesis, Department of Civil Engineering, Queen's University, Kingston, Canada.
Engels, H. (1929). Movement of sedimentary materials in river bends. In *Hydraulic Laboratory Practice*, edited by J.R. Freeman, ASME, New York.
Engelund, F. (1974). Flow and bed topography in channel bends. *Journal of the Hydraulics Division*, ASCE, 100(11), 1631-1648.
Falcon-Ascanio, M. and Kennedy, J.F. (1983). Flow in alluvial-river curves. *Journal of Fluid Mechanics*, Vol. 133, 1-16.
Fargue, L. (1868). Étude sur la correlation entre la configuration du lit et la profondeur d'eau dans les rivières à fond mobile. *Annales des Ponts et Chaussées*, 38, 34-92.
Fox, J.A. and Ball, D.J. (1968). The analysis of secondary flow in bends in open channels. *Proceedings of the Institution of Civil Engineers*, Vol. 39, 467-475.
Francis, J.R.D. and Asfari, A.F. (1971). Velocity distributions in wide, curved open-channel flows. *Journal of Hydraulic Research*, 9(1), 73-90.
Friedkin, J.F. (1945). *A laboratory study of the meandering of alluvial rivers.* U.S. Waterways Experiment Station, Vicksburg, Mississippi.
Frothingham, K.M. and Rhoads, B.L. (2003). Three-dimensional flow structure and channel change in an asymmetrical compound meander loop, Embarras River, Illinois. *Earth Surface Processes and Landforms*, 28(6), 625-644.
Grishanin, K.V. (1979). *Dynamics of alluvial streams.* Gidrometeoizdat, Leningrad.
Hasegawa, K. (1983). A study on flows and bed topographies in meandering channels. (In Japanese) *Proceedings of the Japanese Society of Civil Engineers*, 338, 105-114.
Hasegawa, K. and Yamaoka, I. (1983). Phase shifts of pools and their depths in meander bends. In *River Meandering, Proceedings of Rivers '83*, edited by C.M. Elliott, ASCE, 885-895.
Harbor, D.J. (1998). Dynamics of bedforms in the Lower Mississippi River. *Journal of Sedimentary Research, Section A: Sedimentary Petrology and Processes*, 67(5), 750-762.
Hayat, S. (1965). *The variation of loss coefficient with Froude number in an open-channel bend.* M.Sc. Thesis, University of Iowa.

Hooke, R.L. (1980). Shear-stress distribution in stable channel bends. Discussion, *Journal of the Hydraulics Division*, ASCE, 106(7), 1271-1272.
Hooke, R.L. (1974). *Distribution of sediment transport and shear stress in a meander bend*. Report 30, Uppsala Univ. Naturgeografiska Inst., 58.
Ikeda, S. and Nishimura, T. (1986). Flow and bed profile in meandering sand-silt rivers. *Journal of Hydraulic Engineering*, 112(7), 562-579.
Inglis, C.C. (1947). Meanders and their bearing on river training. *Proceedings of the Institution of Civil Engineers, Maritime and Waterways Engineering Division*, Paper No. 7, 3-24.
Ippen, A.T. and Drinker, P.A. (1962). Boundary shear stress in curved trapezoidal channels. *Journal of the Hydraulics Division*, ASCE, 88(5), 143-179.
James, C.S. (1994). Evaluation of methods for predicting bend loss in meandering channels. *Journal of Hydraulic Engineering*, 120(2), 245-253.
Jia, Y. and Wang, S.S.Y. (1999). Numerical model for channel flow and morphological change studies. *Journal of Hydraulic Engineering*, 125(9), 924-933.
Kalkwijk, J.P.Th. and de Vriend, H.J. (1980). Computation of the flow in shallow river bends. *Journal of Hydraulic Research*, 18(4), 327-342.
Kashyap, S., Constantinescu, G., Rennie, C.D., Post, G. and Townsend, R. (2012). Influence of channel aspect ratio and curvature on flow, secondary circulation and bed shear stress in a bend. *Journal of Hydraulic Engineering*, 138(12), 1045-1059.
Kikkawa, H., Ikeda, S. and Kitagawa, A. (1976). Flow and bed topography in curved open channels. *Journal of the Hydraulics Division*, ASCE, 102(9), 1327-1342.
Kondratiev, N., Popov, I. and Snishchenko, B. (1982). *Foundations of hydromorphological theory of fluvial processes*. (In Russian) Gidrometeoizdat, Leningrad.
Kondratiev, N.E., Lyapin, A.N., Popov, I.V., Pinkovskii, S.I., Fedorov, N.N. and Yakunin, I.N. (1959). *River flow and river channel formation*. Gidrometeoizdat, Leningrad. Translated from Russian by the Israel Program for Scientific Translations, Jerusalem, 1962.
Langbein, W.B. and Leopold, L.B. (1966). River meanders – theory of minimum variance. *U.S. Geological Survey Professional Papers*, 422-H, 1-15.
Leliavskii, S. (1959). *An introduction to fluvial hydraulics*. Constable and Company.
Leopold, L.B. and Langbein, W.B. (1966). River meanders. *Scientific American*, 214(6), 60-70.
Leopold, L.B., Bagnold, R.A., Wolman, M.G. and Brush, L.M.Jr. (1960). Flow resistance in sinuous or irregular channels. *U.S. Geological Survey Professional Papers*, 282-D, 111-135.
Levi, I.I. (1957). *Dynamics of alluvial streams*. State Energy Publishing, Leningrad.
Lien, H.C., Hsieh, T. and Yang, J.C. (1999). Bend-flow simulation using 2D depth-averaged model. *Journal of Hydraulic Engineering*, 125(10), 1097-1108.
Makaveyvev, N.I. (1975). *River bed and erosion in its basin*. Press of the Academy of Sciences of the USSR, Moscow.
Martvall, S. and Nilsson, G. (1972). *Experimental studies of meandering. The transport and deposition of material in curved channels*. UNGI Report 20, University of Uppsala, Sweden.
Matthes, G.H. (1948). Mississippi River cutoffs. *Transactions of the American Society of Civil Engineers*, 113, 16-39.

Matthes, G.H. (1941). Basic aspects of stream meanders. *Transactions of the American Geophysical Union*, 632-636.
Mockmore, C.A. (1944). Flow around bends in stable channels. *Transactions of the American Society of Civil Engineers*, 109, 593-628.
Mosonyi, E. and Goetz, W. (1973). Secondary currents in subsequent model bends. *Proceedings of the International Symposium on River Mechanics*, IAHR, Bangkok, Thailand, 1, 191-201.
Movshovitz-Hadar, N. and Shmukler, A. (2006). River meandering and a mathematical model of this phenomenon. *Physica Plus*, Online magazine of the Israel Physical Society (IPS), physicaplus.org.il , Issue No. 7.
Nelson, J.M. and Smith, J.D. (1989a). Evolution and stability of erodible channel beds. In *River Meandering: Water Resources Monograph*, 12, edited by S. Ikeda and G. Parker, American Geophysical Union, 321-377.
Nelson, J.M. and Smith, J.D. (1989b). *Flow in meandering channels with natural topography*. In *River Meandering: Water Resources Monograph*, 12, edited by S. Ikeda and G. Parker, American Geophysical Union, 69-102.
Odgaard, A.J. (1984). Flow and bed topography in alluvial channel bend. *Journal of Hydraulic Engineering*, 110(4), 521-536.
Onishi, Y., Jain, S.C. and Kennedy, J.F. (1976). Effects of meandering in alluvial streams. *Journal of the Hydraulics Division*, ASCE, 102(7), 899-917.
Rozovskii, J.L. (1957). *Flow of water in bends of open channels*. The Academy of Sciences of the Ukrainian SSR, Translated from Russian by the Israel Program for Scientific Translations, Jerusalem, 1962.
Sedov, L.I. (1960). *Similarity and dimensional methods in hydraulics*. Academic Press Inc., New York.
Shimizu, Y. (1991). *A study on prediction of flows and bed deformation in alluvial streams*. (In Japanese) Report, Civil Engineering Research Institute, Hokkaido Development Bureau, Sapporo, Japan.
Shimizu, Y. and Itakura, T. (1989). Calculation of bed variation in alluvial channels. *Journal of Hydraulic Engineering*, 115(3), 367-384.
Smith, J.D. and McLean, S.R. (1984). A model for flow in meandering streams. *Water Resources Research*, 20(9), 1301-1315.
Steffler, P.M. (1984). *Turbulent flow in a curved rectangular channel*. Ph.D. Thesis, University of Alberta, Alberta, Canada.
Struiksma, N. and Crosato, A. (1989). Analysis of a 2-D bed topography model for rivers. In *River Meandering: Water Resources Monograph*, 12, edited by S. Ikeda and G. Parker, American Geophysical Union, 153-180.
Struiksma, N., Olesen, K.W., Flokstra, C. and de Vriend, H.J. (1985). Bed deformation in curved alluvial channels. *Journal of Hydraulic Research*, 23(1), 57-79.
Tamai, N., Ikeuchi, K. and Yamazaki, A. (1983). Experimental analysis of the open channel flow in continuous bends. (In Japanese) *Proceedings of the Japanese Society of Civil Engineers*, 331, 83-94.
Termini, D. (2009). Experimental observations of flow and bed processes in large-amplitude meandering flume. *Journal of Hydraulic Engineering*, 135(7), 575-587.
Termini, D. and Piraino, M. (2011). Experimental analysis of cross-sectional flow motion in a large amplitude meandering bend. *Earth Surface Processes and Landforms*, 36(2), 244-256.

Termini, D. (1996). *Evoluzione di un canale meandriforme a fondo inizialmente piano: studio teorico-sperimentale del fondo e le caratteristiche cinematiche iniziali della corrente*. Ph.D. Thesis, Department of Hydraulic Engineering and Environmental Applications, University of Palermo, Italy.

Varshney, D.V. and Ramchandra, J.G. (1975). Shear distribution in bends in rectangular channels. *Journal of the Hydraulics Division*, ASCE, 101(8), 1053-1066.

Velikanov, M.A. (1955). *Dynamics of alluvial streams. Vol. II. Sediment and bed flow*. State Publishing House for Theoretical and Technical Literature, Moscow.

Von Schelling, H. (1951). Most frequent particle path in a plane. *Transactions of the American Geophysical Union*, 32(2), 222-226.

Whiting, P.J. and Dietrich, W.E. (1993a). Experimental studies of bed topography and flow patterns in large-amplitude meanders. 1. Observations. *Water Resources Research*, 29(11), 3605-3614.

Whiting, P.J. and Dietrich, W.E. (1993b). Experimental studies of bed topography and flow patterns in large-amplitude meanders. 2. Mechanisms. *Water Resources Research*, 29(11), 3615-3622.

Whiting, P.J. and Dietrich, W.E. (1993c). Experimental constraints on bar migration through bends: implications for meander wavelength. *Water Resources Research*, 29(4), 1091-1102.

Xu, D. and Bai, Y (2013). Experimental study on the bed topography evolution in alluvial meandering rivers with various sinuosnesses. *Journal of Hydro-environment Research*, 7(2), 92-102.

Yalin, M.S. (1992). *River mechanics*. Pergamon Press, Oxford.

Yalin, M.S. (1970). *Theory of hydraulic models*. MacMillan and Co., Ltd., London.

Yeh, K.-C. and Kennedy, J.F. (1993). Moment model of non-uniform channel bend flow. *Journal of Hydraulic Engineering*, 119(7), 776-795.

Yen, B.C. (1972). Spiral motion of developed flow in wide curved open channels. Chapter 22, in *Sedimentation (Einstein)*, edited by H.W. Shen, P.O. Box 606, Fort Collins, Colorado, 22:1-33.

Yen, C.L. and Yen, B.C. (1971). Water surface configuration in channel bends. *Journal of the Hydraulics Division*, ASCE, 97(2), 303-321.

Zhang, Y. (2007). *On the computation of flow and bed deformation in alluvial meandering streams*. Ph.D. Thesis, Department of Civil Engineering, Queen's University, Kingston, Canada.

Chapter 7

Meandering-related computations

The rapid development of computer hardware and software in recent decades has enabled the solution of the complex meandering flow governing equations (e.g. Kalkwijk and de Vriend 1980, Demuren and Rodi 1986, Smith and McLean 1984, Vasquez et al. 2008); and the coupling of advanced hydrodynamic models to sediment transport and morphodynamic models to determine meandering bed deformation (e.g. Struiksma 1985, Struiksma et al. 1985, Shimizu et al. 1987, Shimizu and Itakura 1989, Struiksma and Crosato 1989, Nelson and Smith 1989a,b, Shimizu 1991, Lien et al. 1999, Jia and Wang 1999, Wu et al. 2000, Kassem and Chaudry 2002, Duc et al. 2004, Vasquez et al. 2011; see also Rousseau et al. 2016) as well as meandering planimetric evolution (e.g. Nagata et al. 2000, Darby et al. 2002, Olsen 2003; Chen and Duan 2006, Rüther and Olsen 2007, Crosato 2008, Zolezzi et al. 2012, Nasermoaddeli 2012, Motta et al. 2012, 2014, Eke 2013).

Nowadays, meandering flows are commonly resolved with the aid of fully 3-D RANS (Reynolds Averaged Navier-Stokes) hydrodynamic models. A significant recent development is the application of LES (Large Eddy Simulation) and DES (Detached Eddy Simulation) to meandering flows, yielding descriptions of the flow and turbulence that so far have not been attainable by other means (see e.g. van Balen et al. 2010, Stoesser et al. 2010, Kang and Sotiropoulos 2011, Constantinescu et al. 2013, Dallali 2016). Yet, in practice, vertically-averaged hydrodynamic models continue to be routinely used as the foundation of computer models for the simulation and prediction of meandering morphological evolution. Considering this, in this chapter we consider special aspects of the computation of the vertically-averaged flow and bed deformation in meandering streams. Finally, a physically-based, yet simple method for the computation of the planimetric evolution of meandering streams is presented.

7.1 VERTICALLY-AVERAGED FLOW IN MEANDERING CHANNELS

7.1.1 Equations of motion and continuity; evaluation of the resistance factor c_M

As follows from Eq. (6.29), the vertically-averaged approach *cannot* reflect the influence of the three-dimensional cross-circulatory motion (Γ). The computation of the vertically-averaged meandering flow thus is usually achieved by parameterizing the effect of Γ (see e.g. Blanckaert and de Vriend 2004, Blanckaert 2009, 2010, Ottevanger et al. 2012). Such an approach assumes that the width-to-depth ratio is "sufficiently small" so that the effect of Γ on the flow cannot be disregarded (see Sub-section 6.4.2). In contrast, in the following we consider the case of wide channels, where the effect of Γ is only of secondary importance or negligible.

(i) The (longitudinal and radial) equations of motion and the equation of continuity of a vertically-averaged steady state meandering flow can be expressed in the channel-fitted coordinates l_c and n as follows (see e.g. Smith and McLean 1984, Nelson and Smith 1989a,b)

$$\overbrace{\frac{R}{R+n}\frac{\partial(\overline{u}^2 h)}{\partial l_c}+\frac{\partial(\overline{u}\,\overline{v}h)}{\partial n}+2\frac{(\overline{u}\,\overline{v}h)}{R+n}}^{\text{I}} = \overbrace{-gh\frac{R}{R+n}\left(\frac{\partial z_b}{\partial l_c}+\frac{\partial h}{\partial l_c}\right)}^{\text{II}} \overbrace{-\frac{\overline{u}^2}{c_M^2}}^{\text{III}} \tag{7.1}$$

$$\frac{R}{R+n}\frac{\partial(\overline{u}\,\overline{v}h)}{\partial l_c}+\frac{\partial(\overline{v}^2 h)}{\partial n}-\frac{\overline{u}^2 h}{R+n} = -gh\left(\frac{\partial z_b}{\partial n}+\frac{\partial h}{\partial n}\right) - \frac{\overline{u}\,\overline{v}}{c_M^2} \tag{7.2}$$

$$\frac{R}{R+n}\frac{\partial(\overline{u}h)}{\partial l_c}+\frac{\partial(\overline{v}h)}{\partial n}+\frac{\overline{v}h}{R+n} = 0. \tag{7.3}$$

Here the terms I, II, III (in the equations of motion) are the l- and r-components of the (reversed) inertia force **I**, the tractive force **F**, and the total resistance force $\mathbf{T}=\rho\overline{\mathbf{U}}^2/c_M^2$ (as defined in Section 6.5 (i)) – all acting on the unit fluid prism $h\times 1\times 1$ at $(l_c;r)$. Since the elevation of the free surface is given by $z_f=z_b+h$, the expressions in round brackets on the right-hand sides of Eqs. (7.1) and (7.2) imply the l_c- and n-derivatives of z_f, respectively.

Using $\overline{v}=\overline{\omega}\cdot\overline{u}$, one determines, after some algebraic operations (see da Silva 1995), the following equivalent of the system of equations above – where $\overline{\omega}$ appears instead of $\overline{v}$:

$$\overline{u}^2\frac{\partial\overline{\omega}}{\partial n} = g\frac{R}{R+n}\left(\frac{\partial z_b}{\partial l_c}+\frac{\partial h}{\partial l_c}\right)+\frac{\overline{u}^2}{hc_M^2} \tag{7.4}$$

$$\overline{u}^2\frac{R}{R+n}\frac{\partial\overline{\omega}}{\partial l_c}-\frac{\overline{u}^2}{R+n} = -g\left(\frac{\partial z_b}{\partial n}+\frac{\partial h}{\partial n}\right)-\frac{\overline{u}^2}{hc_M^2}\overline{\omega} \tag{7.5}$$

$$\frac{R}{R+n}\frac{\partial(\overline{u}h)}{\partial l_c}+\frac{\partial(\overline{\omega}\,\overline{u}h)}{\partial n}+\frac{\overline{\omega}\,\overline{u}h}{R+n}=0. \tag{7.6}$$

In the present analysis it is the latter equivalent (Eqs. (7.4)-(7.6)) which will be mostly used.

The bank-friction is not taken into account by the equations of motion above, and therefore the functions $\overline{u}$, h and $\overline{\omega}$ (or $\overline{v}$) supplied by these equations cannot be regarded as valid throughout the flow width B: in the neighbourhood of the banks, they must be appropriately "corrected" (some methods for this purpose can be found e.g. in da Silva 1995).

In the current literature the resistance factor c_M is invariably treated as a "known" function, and the three "unknown" functions $h=f_h(l_c,n)$, $\overline{u}=f_u(l_c,n)$ and $\overline{\omega}=f_\omega(l_c,n)$

(or $\overline{v}=f_v(l_c,n)$) are solved, numerically, from the three equations (7.4)-(7.6) (or (7.1)-(7.3)): the solution domain is the plan area of one meander loop O_iO_{i+1}. The functions h, $\overline{u}$ and $\overline{\omega}$ (or $\overline{v}$) must satisfy the following boundary conditions:

1 No flow across the banks:

$$f_\omega(l_c,\pm B/2)=0 \quad (\text{or } f_v(l_c,\pm B/2)=0); \tag{7.7}$$

2 Cross-sectional average values of $\overline{u}$ and h are, at any l_c, equal to their channel average values (see Eq. (6.53)):

$$\frac{1}{B}\int_{-B/2}^{+B/2}\overline{u}dn=u_{av} \quad \text{and} \quad \frac{1}{B}\int_{-B/2}^{+B/2}hdn=h_{av}; \tag{7.8}$$

3 The $\overline{u}$-distribution diagrams at the "in" and "out" sections O_i and O_{i+1} are anti-symmetrically identical:

$$f_u(0,n)=f_u(L/2,-n); \quad f_h(0,n)=f_h(L/2,-n); \tag{7.9}$$

$$f_\omega(0,n)=-f_\omega(L/2,-n).$$

The sine-generated meandering channel at hand is specified by some constant values of θ_0, Λ_M, B, S_c and $(c_M)_{av}\approx c$: the flow rate Q is given. Thus the values of u_{av} $(=u_m)$ and h_{av} $(=h_m)$, which are determined as

$$u_{av}=c\sqrt{gS_ch_{av}} \quad \text{and} \quad Q=Bh_{av}u_{av}, \tag{7.10}$$

are also known. For any specific plan geometry of the channel, R is to be considered as given (e.g. if the channel is sine-generated, it is given by Eq. (6.8)). Although, as mentioned above, in the current literature the local c_M is supposed to be known, this knowledge is not as straightforward as it is often assumed, and the following is an attempt to clarify this point.

(ii) The relation (6.52) defining the local resistance factor c_M can be expressed, with the aid of Eq. (6.51), as

$$\frac{1}{c_M^2}=\frac{T}{\rho\overline{U}^2}=\frac{[\tau_0+\chi_s]}{\rho\overline{U}^2}=\frac{\tau_0}{\rho\overline{U}^2}\left(1+\frac{\chi_s}{\tau_0}\right)=\frac{1}{c_{M0}^2}\left(1+\frac{\chi_s}{\tau_0}\right), \tag{7.11}$$

i.e. as

$$\frac{T}{\tau_0}=\left(\frac{c_M}{c_{M0}}\right)^{-2}=1+\alpha \quad (\text{with } \alpha=\chi_s/\tau_0). \tag{7.12}$$

Here $c_{M0}^{-2}=\tau_0/\rho\overline{U}^2$ is that part of c_M^{-2} which is due to the bed resistance only (Eq. (3.1)), and its value can be computed (as $c_{M0}=c$) e.g. with the aid of Eq. (3.12), by using the local h, or of any other c-determination method (see Sub-section 3.1.3). Clearly, c_{M0} varies with location (for τ_0 and $\overline{U}$ do so). The term $\alpha=\chi_s/\tau_0$ reflects the influence of the stress forces $\overline{\sigma}_{ll}h$, $\overline{\sigma}_{rr}h$ and $\overline{\tau}_{rl}h$ caused by the channel curvature (and therefore it also varies from one location to another).

If $1/R \equiv 0$ (straight channel) and the flow is uniform, then the stress forces due to channel curvature are zero. Consequently, $\chi_s \equiv 0$, $\alpha \equiv 0$, and c_M reduces into c_{M0} all over. If, on the other hand, $1/R \neq 0$ (curved channel), then the curvature stress forces are finite (see Eqs. (7.16) and (7.17)), and so is α. In this case c_M is determined by c_{M0} as well as by the non-zero α. Yet, Eqs. (7.1)-(7.3) are often solved for meandering flows simply by identifying c_M with c_{M0}, and ignoring α, i.e. by adopting[1]

$$c_M = c_{M0}, \qquad \text{i.e.} \quad \alpha = 0. \tag{7.13}$$

(Smith and McLean 1984, Nelson and Smith 1989a,b, Struiksma 1985, Struiksma et al. 1985, Struiksma and Crosato 1989, etc.). The solutions thus obtained are not satisfactory even for the case of a flat bed. This is illustrated e.g. by the $\overline{u}$-diagrams in Figure 7.1. They were computed for a flat bed from Eqs. (7.1)-(7.3) on the basis of Eq. (7.13) (Smith and McLean 1984). Note that the largest values of $\overline{u}$ are always at the inner bank around the apex a_i, that is, irrespective of what the value of θ_0 might be. That is, the utilization of Eq. (7.13) renders Eqs. (7.1)-(7.3) (or Eqs. (7.4)-(7.6)) to yield only the ingoing flows (as defined in Sub-section 6.3.4 (ii)). Clearly, such solutions are obviously not realistic, for the ingoing flows are present only if θ_0 is "small": if θ_0 is "large", the meandering flows are outgoing (Figures 6.16a,b). The authors of the references mentioned could obtain an outgoing flow only after an artificial "bump" (simulating the point-bar) was introduced on the bed at the inner bank at a_i.

(iii) At the first glance, one might think that the ratio $\alpha = \chi_s/\tau_0 = \chi_s(c_{M0}^2/\rho\overline{U}^2)$ can be revealed simply by computing $\chi = \sqrt{\chi_l^2 + \chi_r^2}$ with the aid of the relations

$$-\chi_l = \frac{\partial(\overline{\sigma}_{ll}h)}{\partial l} + \frac{\partial(\overline{\tau}_{rl}h)}{\partial r} + 2\frac{\overline{\tau}_{rl}h}{r} \tag{7.14}$$

and

$$-\chi_r = \frac{\partial(\overline{\tau}_{lr}h)}{\partial l} + \frac{\partial(\overline{\sigma}_{rr}h)}{\partial r} + \frac{(\overline{\sigma}_{rr} - \overline{\sigma}_{ll})h}{r}. \tag{7.15}$$

The stresses involved in these relations are determined by

$$\overline{\sigma}_{ll} = 2\rho\overline{\nu}_t\left(\frac{\partial\overline{u}}{\partial l} + \frac{\overline{\upsilon}}{r}\right); \qquad \overline{\sigma}_{rr} = 2\rho\overline{\nu}_t\frac{\partial\overline{\upsilon}}{\partial r} \tag{7.16}$$

and

$$\overline{\tau}_{rl} = \rho\overline{\nu}_t\left(\frac{\partial\overline{\upsilon}}{\partial l} + \frac{\partial\overline{u}}{\partial r} - \frac{\overline{u}}{r}\right), \tag{7.17}$$

where $\overline{\nu}_t$ is the vertically-averaged turbulent kinematic viscosity (at $l; r$).[2]

[1] Apparently this kind of approach was inspired by the fact that the value of the channel average resistance factor of a meandering stream is nearly the same as that of a straight stream (i.e. by $(c_M)_{av} \approx c$ (see Section 6.5 (iii))).

[2] The relations (7.14)-(7.17) are given (as they stand in the textbooks of fluid mechanics) in terms of $l; r$: use Eq. (6.4) to express them in terms of $l_c; n$.

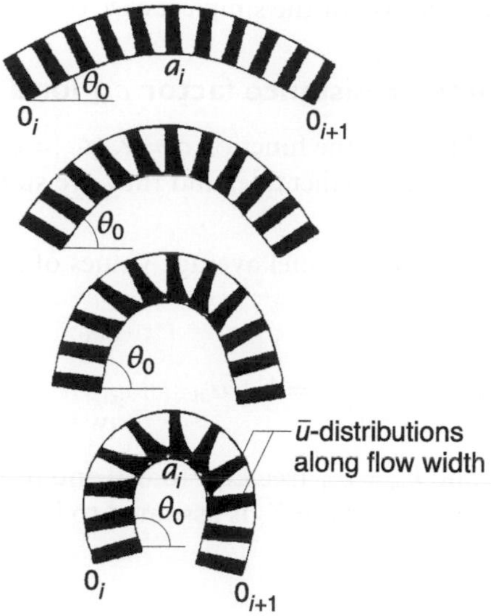

Figure 7.1 Plan views of cross-sectional distributions of vertically-averaged longitudinal flow velocity (as computed by Smith and McLean 1984), in channels having $\theta_0 = 30°, 55°, 80°$ and $105°$ (from Smith and McLean 1984, reprinted with permission from John Wiley and Sons).

In fact, however, the determination of χ_s and thus c_M by this method is not simple at all. Indeed, no information is available, at the present, as to how the turbulent kinematic viscosity ν_t (diffusion coefficient) is distributed within the body of a meandering turbulent flow. The usual tendency to express ν_t of *any* turbulent open-channel flow by the parabolic form $\nu_t = \kappa v_* z(1 - z/h)$ – which, in fact, is valid *only* for a straight two-dimensional open-channel flow – cannot be followed here. This is because the form mentioned, which has been derived for the vertical turbulence ($\sim h$), is meaningless with regard to the horizontal turbulence ($\sim B$), i.e. with regard to the (predominant) horizontal momentum exchange taking place in a wide meandering flow.

The determination of velocity and shear stress fields with the aid of the currently popular $k - \epsilon$ method is also not always satisfactory. Indeed, in some cases the results supplied by this method appear to diverge considerably from the corresponding patterns of experimental points (see e.g. Rodi 1980, p. 54). Moreover, the $k - \epsilon$ method involves five constants which are not of a universal nature, and it is thus not readily applicable to all flows. This method will be rendered more realistic if "... some of its *constants* are replaced by *functions* of suitable flow parameters" (Rodi 1980, p. 29), that is, by certain functions of θ_0, Λ_M/B, B/h_{av} and K_s/h_{av} (in the case of a meandering flow). [The influence of B/h_{av} on the internal structure of turbulence, even in straight open-channels, has already been detected (see Okoye 1970). And the fact that the results supplied by the $k - \epsilon$ method are not always satisfactory for meandering flows may be inferred e.g. from Demuren (1993).]

It is, of course, possible to "construct" a function $\overline{\nu}_t$ that would accord with the available data, and use it to compute $[\overline{\sigma}_{ll}; \overline{\sigma}_{rr}; \overline{\tau}_{rl}]$, and subsequently $[\chi_l\,; \chi_r]$, and finally

c_M. But then, why not to construct an appropriate expression for α in the first place? One of the possibilities to do so, for the simplest case of a flat bed, is explained below.

7.1.2 Expression of the resistance factor c_M for a flat bed

(i) Although the detailed form of the function $\alpha = \chi_s/\tau_0$ $(=\phi_\alpha(l_c, r))$ is so far unknown, some of its general aspects are predictable, and they are stated below.

1 For the cross-sectional and channel average values of c_M we have, on the basis of Eq. (7.12),

$$\frac{(c_{M0})_m^2}{(c_M)_m^2} = 1 + \alpha_m; \qquad \frac{(c_{M0})_{av}^2}{(c_M)_{av}^2} = 1 + \alpha_{av}. \tag{7.18}$$

But $(c_M)_{av} \approx (c_M)_m$ and $h_m = h_{av}$ (see Eqs. (6.60) and (6.53)) suggest that $(c_{M0})_{av} \approx (c_{M0})_m$ (for $c_{M0} = \phi(h/K_s)$, where K_s is assumed to be constant), and thus that

$$\alpha_{av} \approx \alpha_m \approx 0. \tag{7.19}$$

The relation (7.19) implies that the cross-sectional distribution of α consists of positive and negative parts whose areas can be taken as equal.

2 In da Silva (1999) it has been demonstrated that the system of equations (7.1)-(7.3) can yield an outgoing flow only if $(c_M/c_{M0})^{-2}$, i.e. α, progressively decreases along r (that is, along $\eta \in [-0.5; 0.5]$). Considering this, and taking into account that the cross-sectional average of $(c_M/c_{M0})^{-2}$ is unity (or of α is zero), one infers that the cross-sectional diagram of $(c_M/c_{M0})^{-2} = 1 + \alpha$ must be of the form implied by Figure 7.2.

3 The intensity of the decrement of α along η, viz $|\partial\alpha/\partial\eta|$, must be expected to increase with $1/R$. Hence the largest $|\partial\alpha/\partial\eta|$ should be at the apex-section, where the channel curvature $1/R$ is the largest; at the crossovers, where $1/R = 0$, $|\partial\alpha/\partial\eta|$ should be zero.

4 As can be inferred from Sub-section 6.3.4, the transition from ingoing to outgoing flows is strongly correlated with θ_0. Hence α must be a monotonously increasing function of θ_0.

The following relation, which satisfies all of the four requirements above, has been proposed in da Silva (1995):

$$\alpha \approx \alpha_\chi(\underbrace{[(R/r)^2 - 1]}_{\text{I}} - \underbrace{[4(R/B)^2 - 1]^{-1}}_{\text{II}}). \tag{7.20}$$

The graph of this function is shown schematically in Figure 7.2. The term I of Eq. (7.20) provides the cross-sectional decrement of α with $r = R + n$ (i.e. with η); the term II lowers the curve I so as to render the (+) and (−) areas equal. Note that the terms I and II are known functions of ξ_c, η and θ_0.

(ii) Clearly, the method described above can be used to compute the vertically-averaged flow (i.e. the $\overline{u}$-, $\overline{\omega}$- and h-fields) only if the "unknown function" α_χ is revealed first.

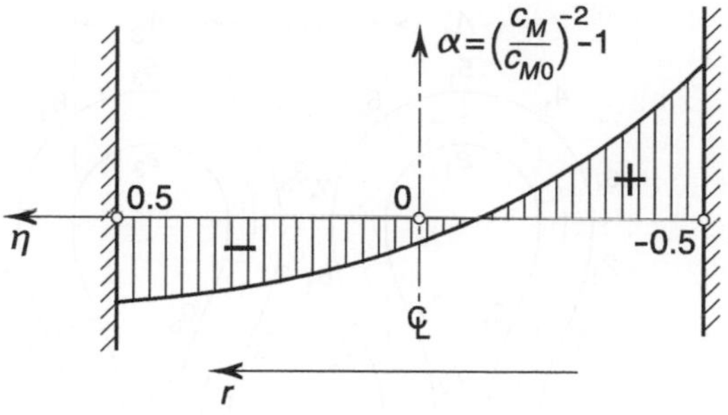

Figure 7.2 Schematic representation of the cross-sectional distribution of α.

It is likely that α_χ varies with θ_0, Λ_M/B and B/h_{av} only, and therefore for a given run (i.e. for some specified θ_0, Λ_M/B and B/h_{av}) it must be expected to have a certain constant value. This means that, for each given run, there must be *one* value of α_χ that yields as solution of Eqs. (7.4)-(7.6) the realistic $\overline{u}$-, $\overline{\omega}$- and h-fields – i.e. that produces the computed $\overline{u}$-, $\overline{\omega}$- and h-fields in agreement with their measured counterparts. To put it differently, one can say that if for an experimental run the values of h, $\overline{u}$ and $\overline{\omega}$ are measured, and the point-patterns implying their ξ_c- and/or η-distributions are produced, then the appropriate adjustment of α_χ alone should be sufficient to render the computed (from Eqs. (7.4)-(7.6)) curves of h, $\overline{u}$ and $\overline{\omega}$ to pass through their measured point-patterns. This method was adopted by da Silva (1995) and El-Tahawy (2004) in their first attempts to reveal the function α_χ, as summarized below.

(iii) The experiments used by the aforementioned authors to evaluate α_χ were already briefly described in Sub-section 6.3.4 (ii). These consist of five runs carried out in five distinct sine-generated channels having $\theta_0 = 30°$, $50°$, $70°$, $90°$ and $110°$. In all channels, the flow cross-section was rectangular; the flow width was $B = 0.40m$; the meander wavelength was $\Lambda_M = 2\pi B = 2.51m$. The effective length of each channel consisted of three consecutive meander loops. The general layout of the experimental set-ups is illustrated in Figures 7.3a,b, showing as examples schematics of the 30° and 110° channels. The channel walls were made of plexiglass; the channel bed was formed by a cohesionless granular material having $D_{50} = 2.2mm$. The bed surface was scraped so as to have along the channel centreline the desired slope S_c; the bed slope in radial direction was zero; $h_{av} \approx 0.03m$. The grains of the uppermost layers of the bed were immobilized by spraying the bed surface with a diluted varnish. [For further details on the experimental conditions, see da Silva et al. 2006.]

The values of $\overline{u}$, $\overline{\omega}_c$ and h were measured in all five channels. These measurements were carried out in each of the three meander loops Λ_1, Λ_2 and Λ_3 (see Figure 7.3), and in each loop at several r-points of eight equally spaced l_c-sections. The resulting $\overline{u}$-fields in the 30°, 70° and 110° channels have already been shown in Figure 6.14. (The complete set of measurements can be found in da Silva 1995 and El-Tahawy 2004).

For the present purposes, the $\overline{u}$-, $\overline{\omega}$- and h-fields were computed from the system of Eqs. (7.4)-(7.6), by adopting a finite difference discretization scheme, with a "marching forward" procedure. This can be summarized (for $\overline{u}$) as follows. Adopt for

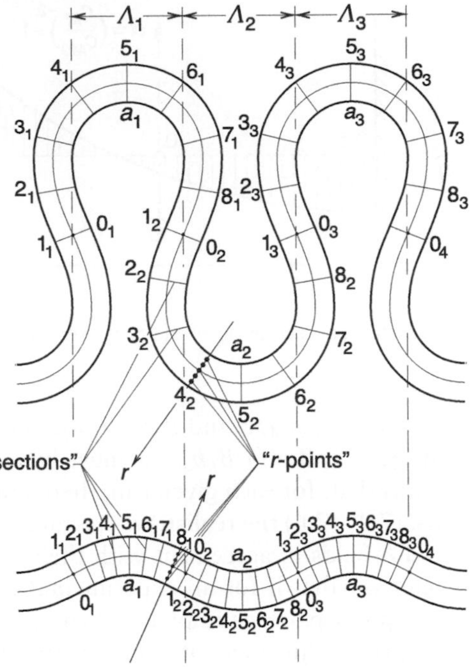

Figure 7.3 Schematics of the sine-generated laboratory channels having 110° (top figure) and 30° (bottom figure) used for the measurements leading to the evaluation of the coefficient α_χ.

the "in" section O_i a distribution $\overline{u}_1 = f_{u_1}(0, n)$ (e.g. uniform distribution $\overline{u}_1 = const$). Moving forward along l_c, arrive at O_{i+1} with a distribution $\overline{u}_2 = f_{u_2}(L/2, n)$. Invert this distribution (replace n by $-n$) and use it as input at O_i. Arrive again at O_{i+1} with $\overline{u}_3 = f_{u_3}(L/2, n)$, ..., etc. The computation stops when $\overline{u}_j = \overline{u}_{j+1}$ is achieved. Eqs. (7.4)-(7.6) were solved by taking into account the following flat bed relations:

$$\frac{\partial z_b}{\partial n} = 0 \quad \text{and} \quad \frac{\partial z_b}{\partial l} = S = \frac{R}{R+n} S_c. \tag{7.21}$$

In the calculations, the resistance factor c_M was evaluated in accordance with Eq. (7.12), with α given by Eq. (7.20). The values of α_χ (in Eq. (7.20)) found to yield the best agreement between measured and computed flow fields were as follows: $\alpha_\chi = 0.05$, 0.6, 1.4, 1.7 and 1.7 for $\theta_0 = 30°$, 50°, 70°, 90° and 110°, respectively (see El-Tahawy 2004). These values of α_χ are plotted versus θ_0 in Figure 7.4.

The agreement between computed flow characteristics and their measured counterparts can be inferred from the $\overline{u}$-fields in Figure 7.5 (compare with Figure 6.14), as well as Figures 7.6, 7.7 and 7.8. The solid curves in Figures 7.6 and 7.8 are the computed cross-sectional distributions of $\overline{u}/u_{av}$ and h/h_{av}, respectively; in Figure 7.7, they are the computed downstream distribution of $\overline{\omega}_c$. For the sake of briefness, in Figures 7.6 and 7.8 only the results obtained for the crossover- and apex-sections are shown; the measured cross-sectional velocity distributions in loop Λ_2 are inverted so as to produce a single point-pattern when plotted together with those measured in

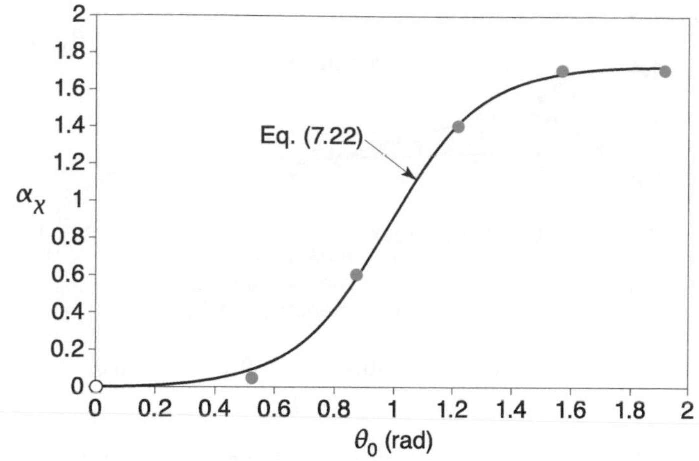

Figure 7.4 Plot of experimentally determined values of α_χ (green circles) versus initial deflection angle and graph of Eq. (7.22).

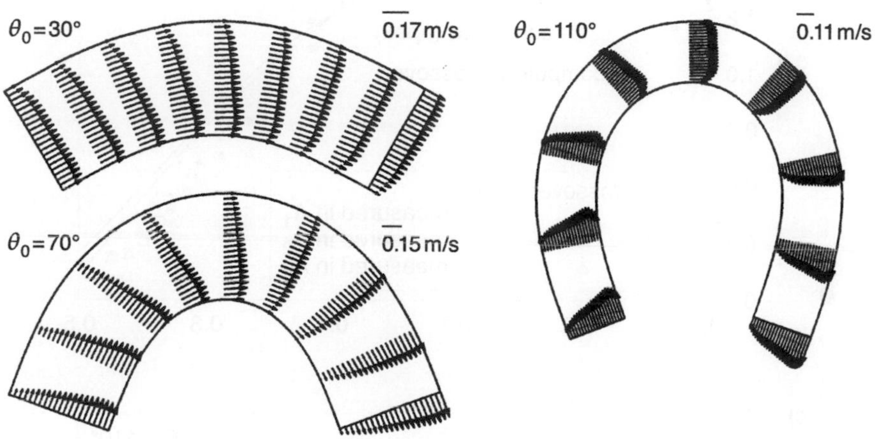

Figure 7.5 Computed $\overline{u}$-fields in three different sine-generated channels having $\theta_0 = 30°, 70°$ and $110°$.

loops Λ_1 and Λ_3. Note also that the angle $\overline{\omega}_c$ at any centreline point of the loop Λ_2 has opposite sign of the angles $\overline{\omega}_c$ at the corresponding points of the loops Λ_1 and Λ_3. To facilitate the comparison, the signs of all measured $\overline{\omega}_c$-values were unified, by inverting the signs of $\overline{\omega}_c$ in Λ_1 and Λ_3.

(iv) The point pattern in Figure 7.4 follows a S-like curve, which is well approximated by the following equation

$$\alpha_\chi = 1.72 - \frac{1.72}{1 + e^{(\theta_0 - 1)/0.16}}. \tag{7.22}$$

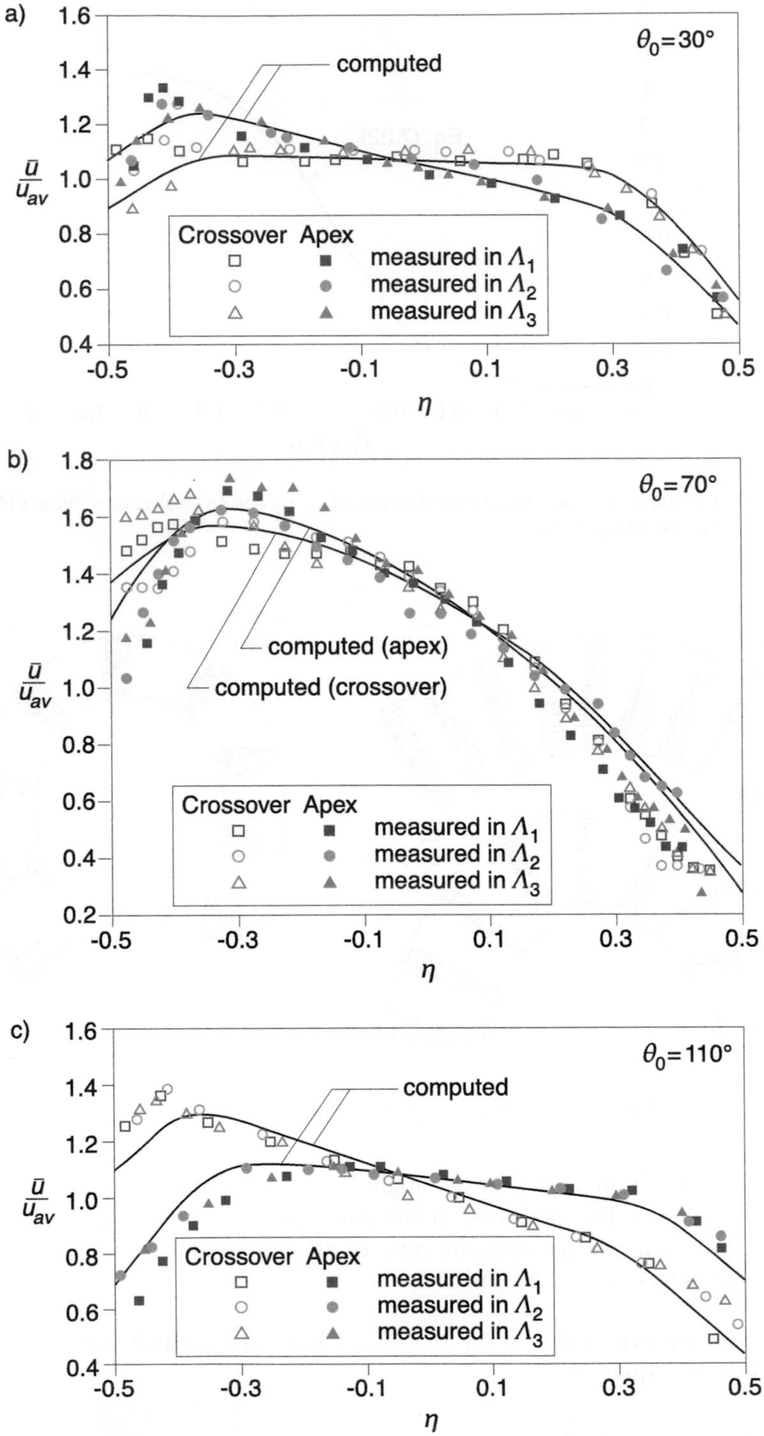

Figure 7.6 Plots of measured and computed $\bar{u}/u_{av}$-values at the crossover- and apex-sections of three distinct sine-generated channels having: (a) $\theta_0 = 30°$; (b) $\theta_0 = 70°$; (c) $\theta_0 = 110°$. (From da Silva et al. 2006, reprinted with permission from ASCE).

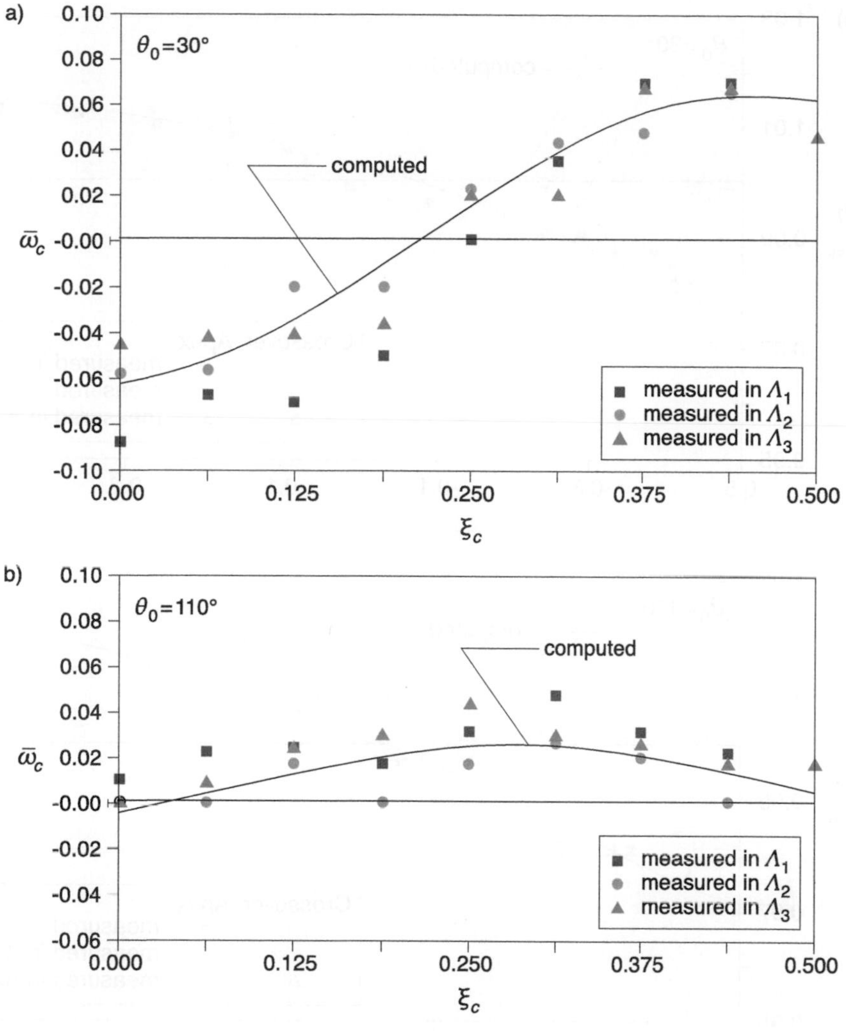

Figure 7.7 Plots of measured and computed $\overline{\omega}_c$-values along ξ_c and corresponding to two distinct sine-generated channels having: (a) $\theta_0 = 30°$; (b) $\theta_0 = 110°$.

Here θ_0 is in radians. The solid line in Figure 7.4 is the graph of this equation. It should be clear that such line is the graphical representation of the function of three variables $\alpha_\chi = \phi(\theta_0, \Lambda_M/B, B/h_{av})$ for the special case $\Lambda_M = 2\pi B$ and $B/h_{av} \approx 13$.

7.1.3 Variational approach to the determination of a meandering flow

(i) In the preceding sub-section c_M was treated, in accordance with convention, as a *known* function, while h, $\overline{u}$ and $\overline{\omega}$ (or $\overline{v}$)[3] as *unknown* functions. The system of

[3]Henceforward we will use only $\overline{\omega}$, and we will refer to Eqs. (7.4)-(7.6) containing it.

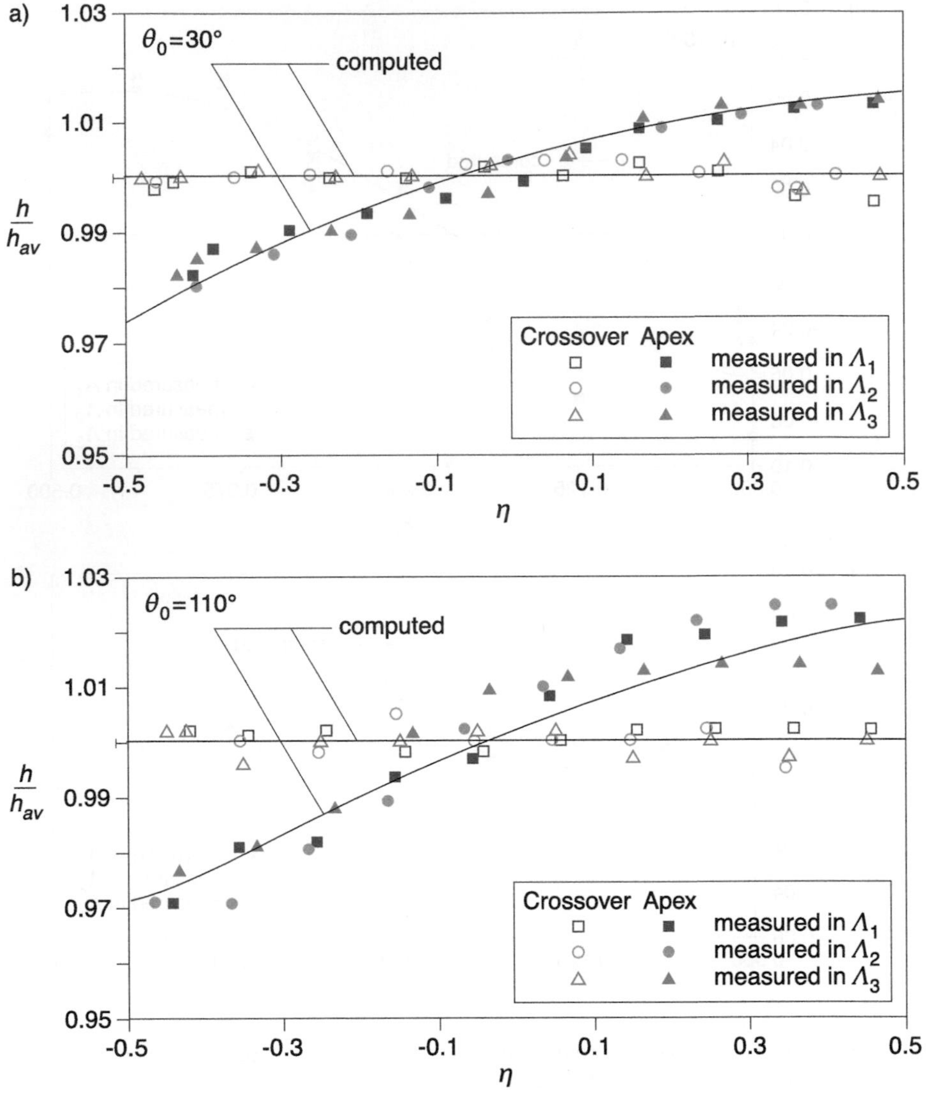

Figure 7.8 Plots of measured and computed h/h_{av}-values at the crossover- and apex-sections of three distinct sine-generated channels having: (a) $\theta_0 = 30°$; (b) $\theta_0 = 110°$.

three equations (7.4)-(7.6) determining these three unknown functions can be shown symbolically as

$$\phi_1(h, \overline{u}, \overline{\omega}) = 0 \qquad \text{(Eq. of Motion (7.4) along } l_c\text{)} \tag{7.23}$$

$$\phi_2(h, \overline{u}, \overline{\omega}) = 0 \qquad \text{(Eq. of Motion (7.5) along } n\text{)} \tag{7.24}$$

$$\phi_3(h, \overline{u}, \overline{\omega}) = 0 \qquad \text{(Continuity Eq. (7.6)).} \tag{7.25}$$

In fact, however, as should be clear from the content of preceding sub-sections, c_M is not a known quantity – even in the case of a flat bed. Rather, it is just another unknown function that must be computed, just like h, $\overline{u}$ and $\overline{\omega}$. But this means that we have actually not *three* but *four* unknown functions, viz

$$h, \overline{u}, \overline{\omega}, \text{ and } c_M. \tag{7.26}$$

Hence four equations are needed to reveal a vertically averaged sine-generated meandering flow past the bed of any given geometry (i.e. of any given function $z' = f_{z'}(l_c, n)$). What can the additional fourth equation be? In an attempt to answer this question we recall that no matter what θ_0 might be, i.e. irrespective of whether the meandering flow is ingoing, outgoing or intermediate, the elevation z_f of the free surface always increases when the outer bank is approached. The rate of this increment is the largest at the flow sections around the apex a, and it is approximately zero at the crossovers O:

$$\mathcal{J} = \frac{\partial z_f}{\partial r} \geq 0 \quad (\text{for any } \theta_0). \tag{7.27}$$

If the bed is flat, then $\partial z_f/\partial r = \partial h/\partial r$.

The increment of z_f with r is an inevitable consequence of the channel curvature, and the only query on the score is *how* does such an increment take place. It would be reasonable to postulate that the free surface is shaped by nature in such a way as to render its variations (in every direction) to occur in the "smoothest" possible manner; for only in this case the pressure within the fluid will vary from one location to another in the least perceptible form and fluid will flow most "comfortably". In mathematical terms this would mean that the free surface should be such, that the integral of its $\mathcal{J}$-values, or rather of its $\mathcal{J}^2$-values, over the plan-area Ω of a meander loop, should be the smallest possible:[4]

$$\int_{\Omega} \mathcal{J}^2 d\Omega \to \min. \tag{7.28}$$

Note that the involvement of the longitudinal free surface slope $J = \partial z_f/\partial l$, in addition to $\mathcal{J}$, is redundant. For $\mathcal{J}$ and J are interrelated ($\partial J/\partial r = \partial \mathcal{J}/\partial l = \partial^2 z_f/\partial l \partial r$), and if one of them is determined, the other is determined automatically.

(ii) The differential $d\Omega$ of the plan area Ω (see Figure 7.9) is given by

$$d\Omega = dl \cdot dn = \left(1 + \frac{n}{R}\right) dl_c \cdot dn = \left(1 + \eta\frac{B}{R}\right) LB d\xi_c d\eta, \tag{7.29}$$

where B/R is a known function of ξ_c (see Eq. (6.13)), viz

$$\frac{B}{R} = [\theta_0 J_0(\theta_0)] \sin(2\pi\xi_c) \quad (= f(\xi_c)). \tag{7.30}$$

[4]The reason for the preference of f^2 to f in this kind of minimization problems, is explained in the textbooks of calculus of variations.

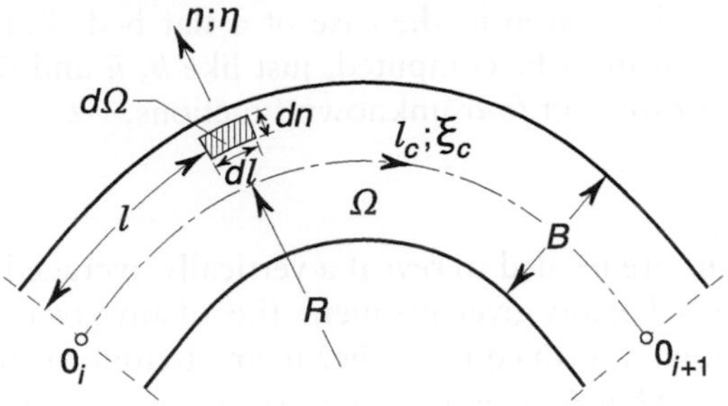

Figure 7.9 Plan view of meandering stream including infinitesimal area $d\Omega$.

Using the notations

$$\frac{\partial A}{\partial \eta} = A_\eta, \qquad H = \frac{z_f}{B} \quad \left(= \frac{z_b}{B} + \frac{h}{B} \right), \tag{7.31}$$

we determine

$$\mathcal{J} = \frac{\partial z_f}{\partial r} = \frac{\partial z_f}{\partial n} = \frac{\partial H}{\partial \eta} = H_\eta, \tag{7.32}$$

while the consideration of Eqs. (7.29), (7.30) and (7.32) makes it possible to express Eq. (7.28) as

$$\int_\Omega H_\eta^2 d\Omega = LB \int_0^{1/2} \left(\int_{-1/2}^{+1/2} H_\eta^2 (1 + f(\xi_c) \cdot \eta) d\eta \right) d\xi_c \to \min. \tag{7.33}$$

Since ξ_c appears only in the known function $f(\xi_c)$, the minimization above implies

$$\int_{-1/2}^{+1/2} \underbrace{H_\eta^2 (1 + f(\xi_c) \cdot \eta)}_{F} d\eta \to \min. \tag{7.34}$$

In this expression, which is assumed to be valid for all ξ_c-sections, the integrand F is to be treated as a function of H_η and η only. Substituting this F into the Euler equation

$$\frac{\partial F}{\partial H} - \frac{\partial}{\partial \eta} \left(\frac{\partial F}{\partial H_\eta} \right) = 0 \quad \text{i.e.} \quad \frac{\partial}{\partial \eta} \left(\frac{\partial F}{\partial H_\eta} \right) = 0, \tag{7.35}$$

one obtains

$$\frac{\partial}{\partial \eta} (2H_\eta (1 + f(\xi_c) \cdot \eta)) = 0, \tag{7.36}$$

i.e.

$$\frac{\partial^2 H}{\partial \eta^2} + \left(\frac{f(\xi_c)}{1 + f(\xi_c) \cdot \eta} \right) \frac{\partial H}{\partial \eta} = 0 \tag{7.37}$$

or

$$\frac{\partial^2 z_f}{\partial n^2} + \frac{1}{R+n}\frac{\partial z_f}{\partial n} = 0, \tag{7.38}$$

which is the additional (fourth) equation sought.

(iii) Integrating Eq. (7.36), we determine

$$\frac{\partial H}{\partial \eta}(1 + f(\xi_c)\cdot\eta) = C_1, \tag{7.39}$$

i.e.

$$\frac{\partial H}{\partial \eta} = \mathcal{J} = \frac{C_1}{1 + f(\xi_c)\cdot\eta} \tag{7.40}$$

where C_1 reflects the variation of $\mathcal{J} = \partial H/\partial\eta$ along the centreline l_c (where $\eta = 0$): $C_1 = \mathcal{J}(\xi_c; 0)$. Integrating now Eq. (7.40), we obtain

$$H = \frac{C_1}{f(\xi_c)} \ln(1 + f(\xi_c)\cdot\eta) + C_2 \tag{7.41}$$

where C_2 reflects the variation of H along l_c: $C_2 = H(\xi_c; 0)$.

The cross-sectional average value $\mathcal{J}_m$ of $\mathcal{J}$ can be expressed in the (well known) form

$$\mathcal{J}_m = \alpha_* \frac{u_m^2}{gR} \quad \text{i.e.} \quad \mathcal{J}_m = \alpha_* \frac{u_m^2}{gB}\cdot f(\xi_c), \tag{7.42}$$

where α_* is close to unity. On the other hand, $\mathcal{J}_m$ can be determined by averaging the expression (7.40) of $\mathcal{J}$ along the flow width $-1/2 \leq \eta \leq 1/2$:

$$\mathcal{J}_m = C_1 \int_{-1/2}^{+1/2} \frac{d\eta}{1 + f(\xi_c)\cdot\eta} = \frac{C_1}{f(\xi_c)} \ln\left(\frac{1 + f(\xi_c)}{1 - f(\xi_c)}\right) \tag{7.43}$$

where $f(\xi_c)$ is always smaller than unity, and the logarithmic multiplier in Eq. (7.43) is thus always positive.

Equating the two expressions of $\mathcal{J}_m$ above, we determine

$$C_1 = \mathcal{J}(\xi_c; 0) = \alpha_* Fr \cdot \frac{h_m}{B} \cdot \frac{f^2(\xi_c)}{\ln(a/b)}, \tag{7.44}$$

where

$$Fr = \frac{u_m^2}{g h_m}; \quad a = 1 + f(\xi_c); \quad b = 1 - f(\xi_c). \tag{7.45}$$

The substitution of Eq. (7.44) in Eq. (7.41) gives

$$H = \alpha_* Fr \frac{h_m}{B} \frac{f(\xi_c)}{\ln(a/b)} \ln(1 + f(\xi_c)\cdot\eta) + C_2. \tag{7.46}$$

Since

$$HB = z_f = z_b + h = z_b + h_m(h/h_m), \tag{7.47}$$

the expression (7.46) of H can be converted into the expression of h/h_m, viz

$$\frac{h}{h_m}=\alpha_* Fr\frac{f(\xi_c)}{\ln(a/b)}F(\xi_c,\eta)+C_2' \tag{7.48}$$

where

$$F(\xi_c,\eta)=\ln(1+f(\xi_c)\cdot\eta);\qquad C_2'=\frac{B}{h_m}C_2-\frac{z_b}{h_m}. \tag{7.49}$$

The term C_2' reflects the variation of h/h_m along the centreline l_c.

Averaging the expression (7.48) of h/h_m over the flow width, we obtain

$$1=\alpha_* Fr\frac{f(\xi_c)}{\ln(a/b)}\int_{-1/2}^{+1/2}F(\xi_c,\eta)d\eta+\frac{B}{h_m}C_2-\frac{1}{h_m}\int_{-1/2}^{+1/2}z_b d\eta. \tag{7.50}$$

But $z_b=(z_b)_m+z'$ where $(z_b)_m$ varies (in a known manner) with ξ_c, and z' with ξ_c and η. Hence

$$\frac{1}{h_m}\int_{-1/2}^{+1/2}z_b d\eta=\frac{(z_b)_m}{h_m}+\frac{1}{h_m}\int_{-1/2}^{+1/2}z' d\eta. \tag{7.51}$$

Substituting Eq. (7.51) in Eq. (7.50) and subtracting the resulting equation from Eq. (7.48), we obtain

$$\frac{h}{h_m}-1=\alpha_* Fr\frac{f(\xi_c)}{\ln(a/b)}\left[F(\xi_c,\eta)-\int_{-1/2}^{+1/2}F(\xi_c,\eta)d\eta\right]+K \tag{7.52}$$

where

$$K=-\frac{z_b}{h_m}+\frac{(z_b)_m}{h_m}+\frac{1}{h_m}\int_{-1/2}^{+1/2}z' d\eta \tag{7.53}$$

i.e.

$$K=-\frac{1}{h_m}\left(\int_{-1/2}^{+1/2}z' d\eta-z'\right). \tag{7.54}$$

Note that if, as mentioned in Section 6.5 (ii), the cross-sectional deposition and erosion areas (formed by (+) and (−) z') are equal, then the integral in Eq. (7.54) is zero and this equation reduces into

$$K=-\frac{z'}{h_m}. \tag{7.55}$$

Note also that if the bed is flat (or z'/h are "small"), then $K=0$ (or $K\to 0$).

Figure 7.10 shows how the h/h_m-values computed from Eq. (7.52) compare with the laboratory data. The solid lines are the computed distributions of h/h_m across the channel width for the apex- and crossover-sections. The computations were carried out for the flow in the 110° channel described in Sub-section 7.1.2 (i): $K=0$ (flat bed)

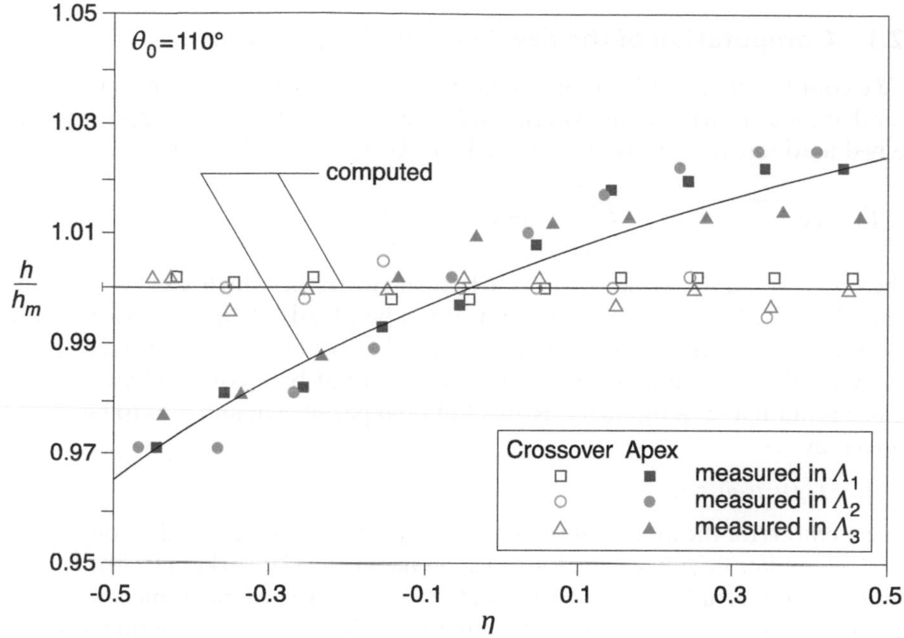

Figure 7.10 Measured and computed (on the basis of Eq. (7.52)) values of h/h_m in a 110° sine-generated channel.

and $\alpha_* = 1.2$ were used. The data in Figure 7.10 are the same as those in Figure 7.8b. The agreement of the computed lines in Figure 7.10 with the data is favourable.

(iv) The variational approach presented in this sub-section indicates that the functions $h = f_h(l_c, n)$ and $z' = f_{z'}(l_c, n)$ of a meandering flow are interrelated: they must satisfy a certain condition, which can be denoted symbolically as $\phi_4(h, z') = 0$, and which is in fact the variational equation (7.38) or (7.52). If the geometry of the bed surface is given, i.e. if $z' = f_{z'}(l_c, n)$ is a known function, then $\phi_4(h, z') = 0$ reduces into a condition which must be satisfied by the unknown function $h = f_h(l_c, n)$ only: it reduces into $\phi_4(h) = 0$.

It follows that in the case of a specified geometry of the bed surface the four unknown functions h, $\overline{u}$, $\overline{\omega}$ and c_M of a vertically-averaged sine-generated meandering flow can be revealed by solving numerically the following system of four equations:

$$\phi_1(h, \overline{u}, \overline{\omega}, c_M) = 0 \quad \text{(Eq. Motion (7.4) along } l_c\text{)} \tag{7.56}$$

$$\phi_2(h, \overline{u}, \overline{\omega}, c_M) = 0 \quad \text{(Eq. Motion (7.5) along } n\text{)} \tag{7.57}$$

$$\phi_3(h, \overline{u}, \overline{\omega}) = 0 \quad \text{(Continuity Eq. (7.6))} \tag{7.58}$$

$$\phi_4(h) = 0 \quad \text{(Variational Eq. (7.38) or (7.52));} \tag{7.59}$$

the boundary conditions are still those stated in Sub-section 7.1.1 (i), for no derivatives of c_M are present.

7.2 BED DEFORMATION BY A MEANDERING FLOW

7.2.1 Computation of the developed bed topography

(i) We confine our considerations to the case of bed-load only. From Eq. (3.49), it is clear that for a given granular material and fluid (and thus for a given $(\tau_0)_{cr} = \gamma_s D\Psi(\Xi)$), the bed-load rate $q_s = q_{sb}$ is determined by h, $\overline{U}$, τ_0 and λ_c. But since

$$\overline{U} = \overline{u}\sqrt{1+\overline{\omega}^2}, \qquad \tau_0 = \frac{\rho \overline{U}^2}{c_{M0}^2} \quad \text{and} \quad \lambda_c = \frac{c_{M0}}{c_f} \tag{7.60}$$

(where $c_f = \phi_c(Z; \Psi(\Xi))$, and c_{M0} is a known function, which can be computed as c from Eq. (3.12)), for a given experiment the bed-load rate $q_s = q_{sb}$ varies from one location of flow to another depending on the unknown functions h, $\overline{u}$ and $\overline{\omega}$. However, strictly speaking, Bagnold's formula is valid for a flat bed (covered by bed forms). If the bed is not flat ($z' \neq 0$), then z' is an additional parameter, and q_s is to be considered as given by

$$q_s = f_{q_s}(h, \overline{u}, \overline{\omega}, z'). \tag{7.61}$$

Some methods for the modification of the transport formulae depending on z' (or rather on $\partial z'/\partial l_c$ and/or $\partial z'/\partial n$) can be found e.g. in Ikeda (1982), Parker (1984), Nelson and Smith (1989a), Struiksma and Crosato (1989). However, as mentioned in Section 6.5 (iii), the erosion-deposition zones forming in wide natural meandering streams are, as a rule, rather flat. Consequently, the error should not be appreciable if, for such streams, q_s is computed from the regular (non-modified) transport formulae – as long as these formulae are evaluated by the affected-by-z' values of h, $\overline{u}$ and $\overline{\omega}$.

(ii) Consider the meandering flow past the developed bed surface at $t \geq T_b$. This surface, implying the unknown beforehand function $z'_T = f_{z'_T}(l_c, n)$, is superimposed on (the known) flat initial bed surface $(z_b)_0$. The flow past the surface z'_T is, of course, also unknown. Hence, we have five unknown functions, viz

$$h, \overline{u}, \overline{\omega}, c_M, z'_T, \tag{7.62}$$

and therefore we need a fifth equation in addition to the four Eqs. (7.56)-(7.59).

From the content of Sub-section 6.4.2 (i), it should be clear that z' becomes z'_T when $W \sim \nabla \mathbf{q}_s$ reduces to zero (at $t = T_b$). Thus the fifth equation is $\nabla \mathbf{q}_s = 0$, which, in view of Eq. (7.61), implies the interrelation

$$\nabla \mathbf{q}_s = \phi_5(h, \overline{u}, \overline{\omega}, z'_T). \tag{7.63}$$

Since the unknown function z'_T must now be present in the variational condition ϕ_4, as well as in the equations of motion ϕ_1 and ϕ_2, the symbolic system of equations becomes

$$\phi_1(h, \overline{u}, \overline{\omega}, c_M, z'_T) = 0 \quad \text{(Eq. Motion (7.4) along } l_c) \tag{7.64}$$

$$\phi_2(h, \overline{u}, \overline{\omega}, c_M, z'_T) = 0 \quad \text{(Eq. Motion (7.5) along } n) \tag{7.65}$$

$$\phi_3(h, \overline{u}, \overline{\omega}) = 0 \quad \text{(Continuity Eq. (7.6))} \tag{7.66}$$

$$\phi_4(h, z'_T) = 0 \quad \text{(Variational Eq. (7.52))} \tag{7.67}$$

$$\phi_5(h, \overline{u}, \overline{\omega}, z'_T) = 0 \quad \text{(Sediment Transport Continuity Eq. (7.63))} \tag{7.68}$$

(The presence of z'_T in ϕ_1 and ϕ_2 is due to $z_f = (z_b)_T + h = (z_b)_0 + z'_T + h$; see Sub-section 7.1.1 (i)).[5]

(iii) As previously mentioned, the computation of bed topography of a meandering stream has already been carried out by several authors. However, in all such works the developed bed topography (corresponding to $t \geq T_b$ and thus $\nabla \mathbf{q}_s = 0$) has not been computed directly. Rather, T_b was approached by means of a sequence of time steps δt_i, each of which yielding a corresponding increment $(\delta z')_i = W_i\, \delta t_i$ (where both W_i and $(\delta z')_i$ progressively approach zero with the increment of i). The developed z'_T at a location of flow plan was then obtained as $\sum (\delta z')_i$.

The δt_i-method described above, though more cumbersome as far as the determination of the developed conditions (at $t \geq T_b$) is concerned, has the advantage of being able to reveal the conditions corresponding to any $t \in [0; T_b]$. This is illustrated in Figures 7.11a,b showing the bed topography computed by the δt_i-method for small and large θ_0 (viz $\theta_0 = 30°$ and $110°$), respectively. The computations correspond to the "early stages" of bed development – only one, the first time step (δt_1), was used. q_s was computed from the non-modified Bagnold's formula; c_M, from Eqs. (7.12) and (7.20), with $\alpha_\chi = 0.08$ and 1.71 when $\theta_0 = 30°$ and $110°$, respectively, as implied by Eq. (7.22). The computations were carried out for the same values of the characteristic parameters specifying the flows in the $\theta_0 = 30°$ and $110°$ channels described in Sub-section 7.1.2 (iii), except for the value of γ_s, and thus of $(\eta_*)_{av}$. In order to ensure that the bed material was able to move, it was assumed that $(\eta_*)_{av} = 1.5$, i.e. that γ_s was 9.4 times lighter than that of sand.

Note that although the erosion-deposition zones in Figures 7.11a,b were computed so as to correspond to the "early stages", their locations in the flow plan are, as far as one can judge, the same as those of a developed bed (compare with Figures 6.20a and 6.21). This further justifies the statements made in Sub-section 6.4.2 (i).

7.2.2 Computation of bed deformation over time

Recent contributions

In the following, we briefly describe two relatively recent numerical models for the computation of the meandering bed deformation, namely the models by Wu et al. (2000) and Nagata et al. (2000). These are presented merely as noteworthy examples of the conventional "δt_i-method" – and the interested readers are here directed also to models by other authors, several of which are listed at the beginning of this chapter. The differences between the models by different authors are primarily due to the adoption of different numerical schemes to solve the governing differential equations of fluid motion, turbulence models, and sediment transport formulations.

1 The method by Wu et al. (2000) enables the *three-dimensional* computation of flow and bed deformation (rigid banks). This 3D-numerical model rests on the

[5] z'_T is present in ϕ_5 if the transport formula is modified (e.g. as indicated in Nelson and Smith 1989a, Parker 1984, etc.); z'_T is not present in ϕ_5 if no modification is undertaken.

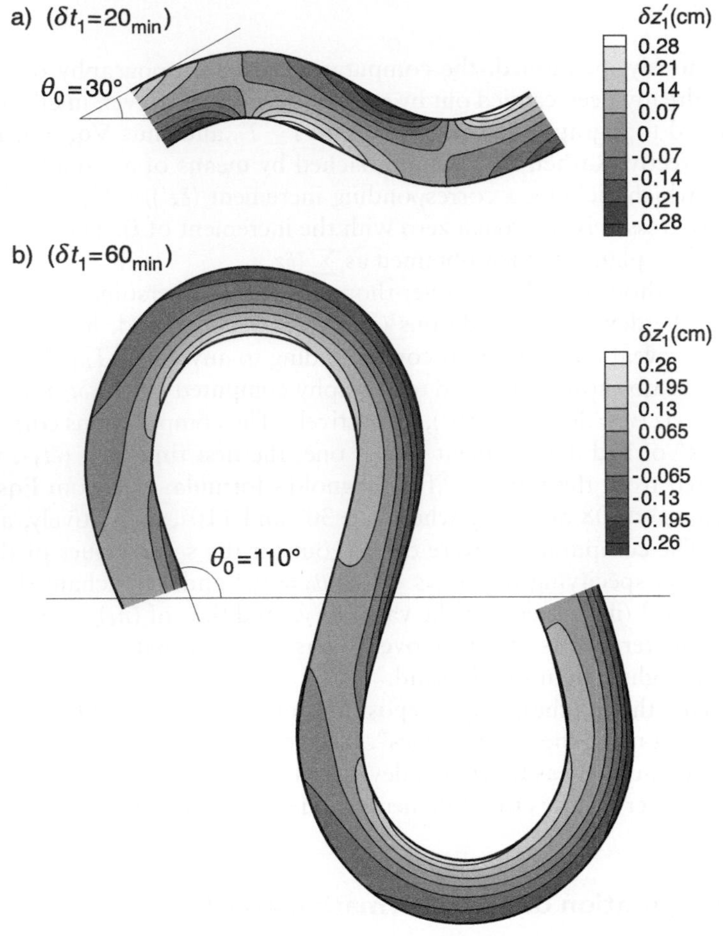

Figure 7.11 Computed bed topography at the early stages of bed development in two sine-generated channels having: (a) $\theta_0 = 30°$; (b) $\theta_0 = 110°$. (From da Silva et al. 2006, reprinted with permission from ASCE).

conventional three equations of fluid motion and continuity, and on the sediment transport continuity equation, which is essentially the same as Eq. (1.73). The flow resistance factor c_M is not used. Instead, the stresses τ_{ij} are considered and they are evaluated with the aid of the $k - \epsilon$ model (see Sub-section 7.1.1 (ii), (iii)), by adopting for the required five coefficients the constant values $c_\mu = 0.09$, $c_{\epsilon 1} = 1.44$, $c_{\epsilon 2} = 1.92$, $\sigma_k = 1.0$, $\sigma_\epsilon = 1.3$, irrespective of what the plan and cross-sectional geometry of the open-channel flow might be. In spite of this, the computed characteristics appear to compare favourably with those of the laboratory flows tested. The effective bed roughness was evaluated with the aid of the van Rijn (1984) expression (Eq. (3.29)).

2 The metod by Nagata et al (2000) consists of a 2D-model of bed *and also* bank deformation. It is assumed that the material (p_s) removed from the bank surface

is deposited almost totally on the bed surface (p_d). The value of p_s (per unit area of the flow surface and per unit time) is determined according to the formula of Nakagawa et al. (1986). The removed material is deposited (mainly) on the bed surface, so that p_d (also per unit area and unit time) decreases exponentially with the distance to the source. The flow is determined with the aid of the conventional equations of motion and continuity. The vertically-averaged turbulent kinetic energy (k) and the eddy viscosity (ν_t) are identified (for any flow geometry) with $2.07v_*^2$ and $\alpha h v_*$, respectively. The computed results compare favourably with the laboratory experiments.

Pertinent information on the time scales of bed development

In order to calculate the meandering bed deformation with the aid of the "δt_i"-method, it is necessary to select the "size" of the time-steps δt_i. For this purpose, it would be very useful to *a priori* have information on the time-scales of the bed development, and in particular the bed development time (or development duration) T_b and the rate of growth of the bed deformation.

To the best knowledge of the authors, the only attempt so far to develop a predictive equation for the bed development time is by Binns and da Silva (2009, 2015). In their derivations, it was assumed that sediment transport is by bed-load only. On the basis of dimensional and physical considerations (see Binns and da Silva 2009), the just mentioned authors arrived at the following equation for T_b:

$$T_b = \frac{B^2}{(q_{sb})_{av}} \cdot \phi_T(\theta_0, \Lambda_M/B), \tag{7.69}$$

in which $(q_{sb})_{av}$ is the bed-load rate of the channel-averaged flow. The function $\phi_T(\theta_0, \Lambda_M/B)$ was experimentally determined for the case $\Lambda_M/B = 2\pi$, which yielded

$$\phi_T(\theta_0, \Lambda_M/B = 2\pi) = \left[0.275(\theta_0 - 1.18)^2 + 0.05\right] + 2.1e^{-5.60\theta_0} + \left(e^{0.00025\theta_0^{9.20}} - 1\right). \tag{7.70}$$

The rate of growth of the bed deformation was analyzed in Binns and da Silva (2011), by plotting the ratio $(D_S)_t/(D_S)_{T_b}$ versus the normalized time t/T_b. Here $(D_S)_t$ is a measure of the bed deformation at any time t, and $(D_S)_{T_b}$, a similar measure of the bed deformation at time T_b (see Binns and da Silva 2011 for the precise meaning of both $(D_S)_t$ and $(D_S)_{T_b}$). The resulting plot is shown in Figure 7.12. This is derived from a series of experimental runs carried out in a 70° sine-generated channel. As implied by Figure 7.12, the bed development was invariably very rapid at the early stages of any run, with $\approx 70\%$ of the bed development being accomplished by $T_b/4$ and $\approx 90\%$ by $T_b/2$. The experiments revealed that only minor adjustments took place in the last half of the bed development time.

The growth rate of the pools and bars in the different runs, when scaled to their individual bed development times, was found to be identical. This is reflected in the fact that in the normalized plot of $(D_S)_t/(D_S)_{T_b}$ versus t/T_b in Figure 7.12, the data-points of the different runs collapse into one single point-pattern. This point-pattern is reasonably well represented by the following equation

$$\frac{(D_S)_t}{(D_S)_{T_b}} = 1 - \left[e^{-2.5(t/T_b)^{0.86}}\right]^{1.6}. \tag{7.71}$$

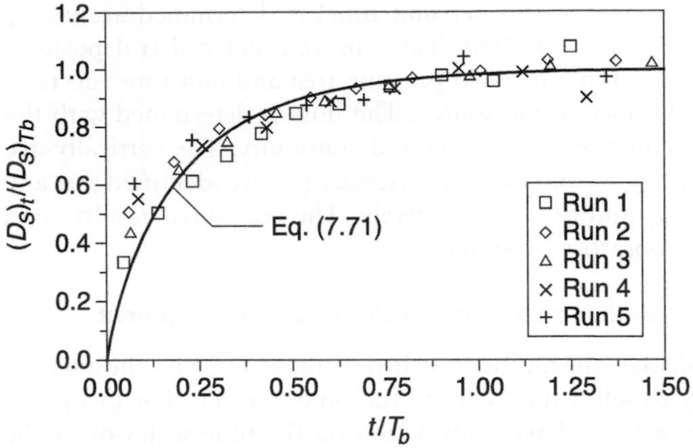

Figure 7.12 Plot of $(D_S)_t/(D_S)_{T_b}$ versus t/T_b (from Binns and da Silva 2011, reprinted with permission from ASCE).

The solid line in Figure 7.12 is the graph of this equation. It should be noted that the growth rate of $(D_S)_t/(D_S)_{T_b}$ is a direct measure of the growth rate of the pool-bar complexes. Thus, Eq. (7.71), describing the growth rate of $(D_S)_t/(D_S)_{T_b}$, is to be viewed as a (percentual) relation describing the growth rate of the observed pool-bar complexes.

It should be emphatically pointed out that Eq. (7.71) is based on a rather limited series of experiments, involving only one value of θ_0 and a rather narrow range of flow conditions. In the absence of further tests, when applied to conditions different from those of the just mentioned experiments, Eq. (7.71) should be simply viewed as a rough means of estimating the growth rate of the pool-bar complexes.

7.3 MIGRATION AND EXPANSION OF MEANDER LOOPS

7.3.1 General

In the preceding section we were dealing with the bed deformation, assuming that the plan geometry of the flow, and thus the banks, are virtually rigid (θ_0 virtually constant). In this section we will be dealing with the displacement of banks (their downstream migration and lateral expansion). The bank erosion associated with such displacement is illustrated in Figure 7.13.

We will aim mainly at *large* natural streams, where the B/h_{av}-ratio is often a three-digit number (Jansen et al. 1979, Yalin 1992), and any effect of the cross-circulation Γ is negligible (see Eq. (6.43)). Consequently, the displacement of banks will be attributed solely to the convective behaviour of flow and the regime trend. Such an approach would be consistent with the (already mentioned) fact that "sediment entering the stream from the scour of the concave banks becomes deposited on the downstream convex bank on the *same side* of the river, and only a small portion of the eroded material crosses the channel" (Matthes 1941, 1948, Friedkin 1945, Velikanov 1955, Levi 1957, Kondratiev et al. 1982) – see Figures 6.24 and 6.34.

Figure 7.13 Erosion of outer bank in the Hardebek-Brokelander Au River, in northern Germany (photo courtesy of Dr.-Ing. M. Hassan Nasermoaddeli, Landesbetrieb Straßen, Brücken und Gewässer (LSBG), Hamburg, Germany).

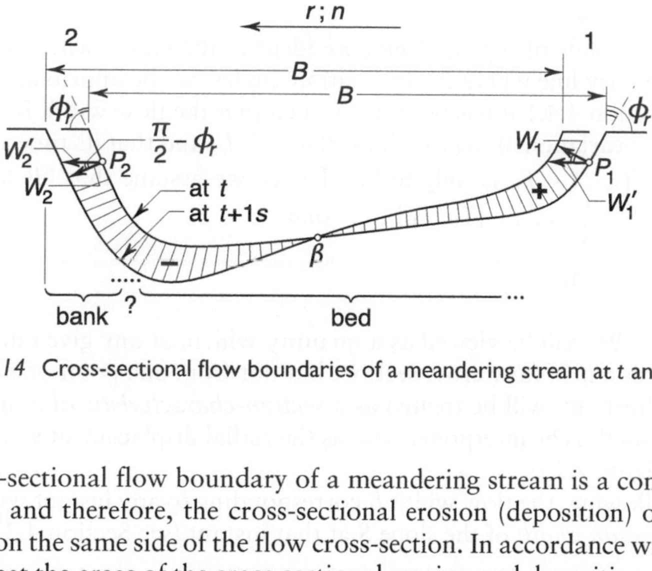

Figure 7.14 Cross-sectional flow boundaries of a meandering stream at *t* and $t + 1$s.

The cross-sectional flow boundary of a meandering stream is a continuous curve (Figure 7.14), and therefore, the cross-sectional erosion (deposition) of the bed and banks occurs on the same side of the flow cross-section. In accordance with Section 6.5 (ii), we will treat the areas of the cross-sectional erosion and deposition zones as equal (at any l_c and $t \sim \Theta$). This gives the possibility to identify $(q_s)_m$ with $(q_s)_{av}$ $(= const)$.

It will be assumed throughout this section that the decrement of the slope S is by meandering only (and not e.g. by meandering *and* degradation-aggradation).

7.3.2 Normal and radial bank displacement velocities

Consider the cross-sectional diagram of the *normal* boundary displacement velocities W sketched in Figure 7.14 (the positive and negative areas of the W-diagram are equal). The *radial* (horizontal) bank displacement velocities W_1' and W_2', at the bank

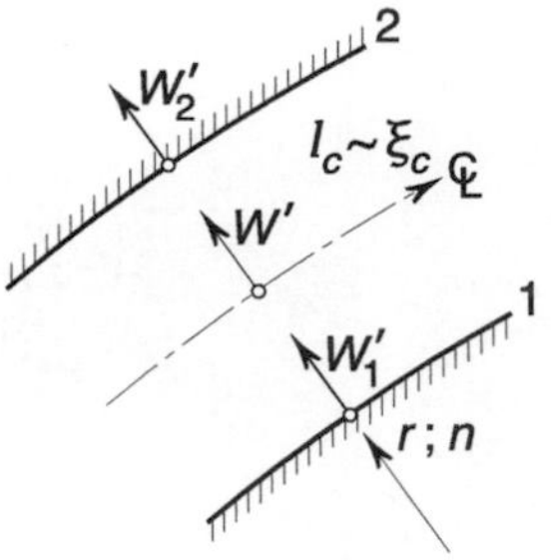

Figure 7.15 Plan view of stretch of a meandering stream, showing radial displacement velocities of banks and stream centreline.

points P_1 and P_2, are determined by the normal displacement velocities W_1 and W_2, at those points, as

$$W_1' = W_1/\sin\phi_r \quad \text{and} \quad W_2' = W_2/\sin\phi_r \tag{7.72}$$

where ϕ_r is the angle of repose. Here, we identify the banks with those end-parts of the flow boundary line where the inclination angles can be approximated by ϕ_r.

In Sub-section 4.4.1 it has been mentioned that the flow width B does not exhibit any systematic variation along the channel length l_c; and that its time-variation (increment) during $\hat{T}_0 < t < T_R$ is only feeble. Hence we assume that $\partial B/\partial t = W_2' - W_1'$ is negligible (at any l_c and $t \in [\hat{T}_0; T_R]$), i.e. that

$$W_1' = W_2' \quad (= W'). \tag{7.73}$$

It follows that W' will be viewed as a quantity which, at any given instant, may vary with the location l_c of flow cross-section, but not from one point of a cross-section to another. In short, W' will be treated as a *section-characteristic* of a meandering flow. Clearly, W' can thus be interpreted also as the radial displacement velocity of the flow centreline (Figure 7.15).

In the following, the flow width B corresponding to any instant will be considered as the *equilibrium width* of the slope S at that instant (see Section 4.3 (i)).

7.3.3 Evaluation of radial bank displacement velocity

In this and the next sub-section we will consider the bank displacements as they are caused only by the convective action of the time-averaged meandering flow: the additional influence of the regime development will be considered in Sub-section 7.3.5.

(i) If sediment transport is present, then the function ϕ_A in the expression (6.28) must acquire an additional dimensionless variable, $\Pi_{q_s} = (q_s)_{av}/\mathcal{V}\mathcal{L}$ say. And if A is a section-characteristic, then η must be excluded. Consider the section-characteristics, W' and

$\overline{\omega}_c$. Identifying each of them with A, and using L and u_m in lieu of $\mathcal{L}$ and $\mathcal{V}$, we obtain on the basis of Eq. (6.28)[6]

$$\Pi_{W'} = \frac{W'}{u_m} = \overline{\overline{\phi}}_{W'}(\theta_0, B/h_{av}, c_{av}, \Pi_{q_s}, \xi_c) \tag{7.74}$$

and

$$\Pi_{\overline{\omega}_c} = \overline{\omega}_c = \overline{\overline{\phi}}_{\overline{\omega}_c}(\theta_0, B/h_{av}, c_{av}, \Pi_{q_s}, \xi_c), \tag{7.75}$$

where

$$\Pi_{q_s} = \frac{(q_s)_{av}}{u_m L}. \tag{7.76}$$

The elimination of ξ_c between Eqs. (7.74) and (7.75) yields the relation

$$\Pi_{W'} = \frac{W'}{u_m} = \overline{\phi}_{W'}(\theta_0, B/h_{av}, c_{av}, \Pi_{q_s}, \overline{\omega}_c), \tag{7.77}$$

where the flow-section is specified by the (varying along ξ_c) value of $\overline{\omega}_c$.

If $(q_s)_{av} = 0$, then $W' = 0$ as well (banks cannot deform in the absence of transport). But this means that Eq. (7.77) must be of the form

$$\Pi_{W'} = \frac{W'}{u_m} = \Pi_{q_s}^{n_q} \phi_{W'}(\theta_0, B/h_{av}, c_{av}, \overline{\omega}_c) \qquad \text{(where } n_q \geq 1\text{)}. \tag{7.78}$$

The normal bank displacement velocities W_1 and W_2 (in Figure 7.15), to which W' $(= W'_1 = W'_2)$ is related by a constant proportion (Eq. (7.72)), are merely the "end-values" of one finite and continuous W-diagram in Figure 7.14. But since (at least because of dimensional reasons) all W-values of this diagram must be proportional to each other, we have for W at any cross-sectional location η $(= n/B)$

$$W_1 = W_2 = \phi(\eta)\, W. \tag{7.79}$$

Considering Eq. (1.73), viz

$$W = -\frac{1}{1-p} \nabla \mathbf{q}_s \sim \frac{(q_s)_{av}}{L}, \tag{7.80}$$

and taking into account Eqs. (7.77), (7.72), (7.79) and (7.76), we determine

$$\Pi_{W'} \sim W' \sim (W_1 = W_2) \sim W \sim \frac{(q_s)_{av}}{L} \sim \Pi_{q_s}. \tag{7.81}$$

This relation indicates that the exponent n_q in Eq. (7.78) is unity, and that this equation must thus be of the form

$$\frac{W'}{u_m} = \frac{(q_s)_{av}}{u_m L} \phi_{W'}(\theta_0, B/h_{av}, c_{av}, \overline{\omega}_c), \tag{7.82}$$

[6]Here, merely for the sake of simplicity, it is assumed that the value of Λ_M/B is specified (e.g. $\Lambda_M/B = 6, 8, \ldots$). Hence Λ_M/B is not included in Eqs. (7.74) and (7.75).

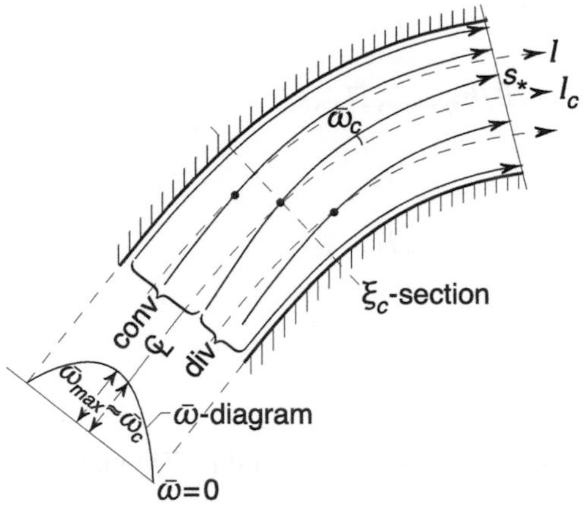

Figure 7.16 Plan view of a meandering stream showing laterally adjacent convergence and divergence flow zones.

which yields

$$W' = \frac{(q_s)_{av}}{L}\phi_{W'}(\theta_0, B/h_{av}, c_{av}, \overline{\omega}_c). \tag{7.83}$$

(ii) Consider Figure 7.16, which shows that at a section ξ_c of a meandering flow we have two zones, viz converging and diverging zones (*conv* and *div* in Figure 7.16). The largest $\overline{\omega}$, viz $\overline{\omega}_{max}$, is in the neighbourhood of the centreline; $\overline{\omega}=0$ is at the banks. Hence $\overline{\omega}_c$ does not differ much from $\overline{\omega}_{max}$ and it can thus be taken as a "measure" of the intensity of convergence-divergence of flow at a section. But this means that $\overline{\omega}_c$ can also be taken as a measure of the magnitude of $\nabla \mathbf{q}_s$ (which is an increasing function of the convergence-divergence of the vertically-averaged flow streamlines s), and consequently of the bed displacement velocity W (see Eq. (7.80)).

As follows from the content of Chapter 6, the above mentioned is consistent with the deformation patterns of natural meandering streams. Indeed, consider e.g. Figures 6.16a and 6.20a, which correspond to "small" values of θ_0. Figure 6.16a indicates that the smallest $\overline{\omega}_c$ (viz $(\overline{\omega}_c)_{min}$) is at the crossover-sections O_i; and Figure 6.20a shows that the most intense bed deformation occurs also around the crossover-sections O_i. The analogous is valid for the case of "large" θ_0, depicted in Figures 6.16b and 6.21b. According to Fig. 6.16b, the largest $\overline{\omega}_c$ (viz $(\overline{\omega}_c)_{max}$) is at the apex a_i, while Fig. 6.21b conveys that the most intense erosion-depositions are also around a_i. The same can be said with regard to the locations of zero-values of $\overline{\omega}_c$ and negligible bed deformations.

(iii) But if $\overline{\omega}_c$ can be taken as a "measure" of the bed displacement velocity W, then it can also be taken as a "measure" of the radial bank displacement velocities $W'_1 = W'_2$ ($= W'$) (for $W' \sim W$). And experiment shows that in the case of both, small and large values of θ_0, the largest and zero radial bank displacement velocities (W') occur indeed in those flow regions where the bed deformations (W) and thus $\overline{\omega}_c$ are largest and zero. This fact is depicted schematically in Figures 7.17a,b.

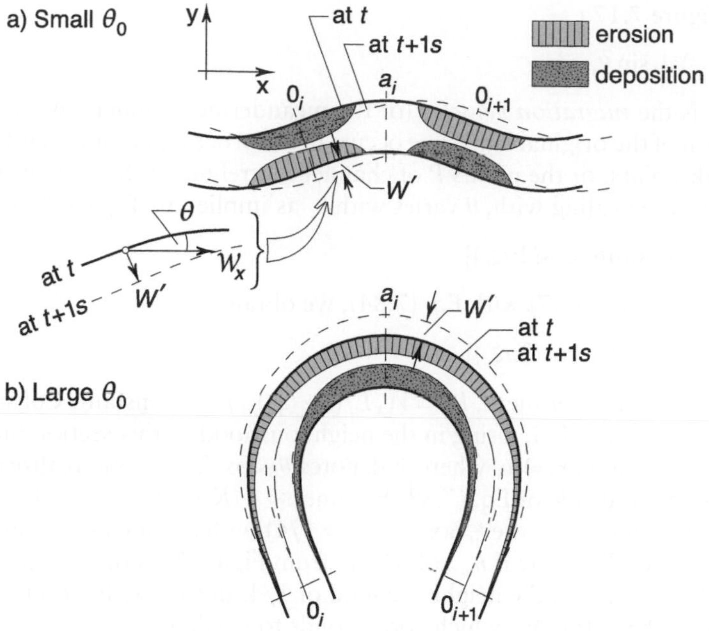

Figure 7.17 Bed erosion-deposition zones and bank displacement patterns in streams having: (a) small θ_0; (b) large θ_0.

The conditions described above will be satisfied if W' is treated as an increasing function of $\overline{\omega}_c$, which becomes zero when $\overline{\omega}_c$ is zero; i.e. if the relation (7.83) is of the form

$$W' = \alpha_W \frac{(q_s)_{av}}{L} \overline{\omega}_c^{n_\omega}, \tag{7.84}$$

where

$$n_\omega \geq 1 \quad \text{and} \quad \alpha_W = \phi_{W'}(\theta_0, B/h_{av}, c_{av}). \tag{7.85}$$

The value of n_ω will be revealed in the next sub-section; the so far unknown function α_W will be discussed in Sub-section 7.3.5.

7.3.4 Migration and expansion components of the radial bank displacement velocity

An attempt will be made now to establish how W' varies along ξ_c in the limiting cases of "small" and "large" values of θ_0.

(i) *If θ_0 is small*, then, under the idealized conditions depicted in Figure 7.17a, the bank-line "at $t+1s$" can be viewed as a shifted-along x bank-line "at t" (for the channel expansion is known to be negligible for small θ_0). Hence, as should be clear from the insert in Figure 7.17a,

$$W' = -\mathcal{W}_x \sin\theta, \tag{7.86}$$

where $\mathcal{W}_x$ is the *migration velocity* (of the meandering channel). We assume that no deformation of the original bank-line occurs in the process and thus that $\mathcal{W}_x$ is the same for all bank-points, or the points P of channel centreline. In the case of sine-generated channels we are dealing with, θ varies with ξ_c as implied by Eq. (6.7), and therefore

$$W' = -\mathcal{W}_x \sin[\theta_0 \cos(2\pi\xi_c)]. \tag{7.87}$$

Identifying Eq. (7.87) with Eq. (7.84), we obtain

$$\overline{\omega}_c^{n_\omega} = -K_1 \sin[\theta_0 \cos(2\pi\xi_c)] \tag{7.88}$$

where, for a given experiment, $K_1 = \mathcal{W}_x L/(\alpha_W (q_s)_{av})$ is a constant. Consider the apex-section a_i, where $\xi_c = 1/4$. Using in the neighbourhood of this section the well known relation $\lim_{x\to 0}(\sin x/x) = 1$, where x denotes $\theta_0 \cos(2\pi\xi_c)$, one realizes that, at $\xi_c = 1/4$, the right-hand side of Eq. (7.88) becomes $\approx -(K_1\theta_0)\cos(2\pi\xi_c)$. The graph of this cosine function intersects the ξ_c-axis (at $\xi_c = 1/4$) with a finite angle – and so must do $\overline{\omega}_c^{n_\omega}$, which is possible only if $n_\omega = 1$. (Note from Figure 7.7a that the point-pattern $\overline{\omega}_c$ intersects the ξ_c-axis, in the neighbourhood of 1/4, indeed with a finite angle).

Hence we have for W', which corresponds to small θ_0,

$$W' = \alpha_W \frac{(q_s)_{av}}{L}\overline{\omega}_c = -\mathcal{W}_x \sin[\theta_0 \cos(2\pi\xi_c)]. \tag{7.89}$$

(ii) *If θ_0 is large*, then the channel-migration is known to be negligible ($\mathcal{W}_x \to 0$), and one can assume that the ideal channel (Figure 7.17b) deforms only because of the expansion of its loops around the fixed inflection points O_i (Kondratiev et al. 1982, Yalin 1992). In this case the radial displacement velocity W' of a point P of the channel centreline is simply the *channel-expansion velocity* at that point.

In order to reveal the variation of W' with ξ_c we recall that the sine-generated function (6.7) is applicable to all $t \in [0; T_R]$, and thus to all θ_0. Accordingly, we can write for the channel centreline at t and $t^* = t + \delta t$

$$\theta = \theta_0 \cos(2\pi\xi_c) \quad \text{and} \quad \theta^* = \theta_0^* \cos(2\pi\xi_c^*), \tag{7.90}$$

respectively (Figure 7.18). The point P of the channel centreline at t travels during δt the distance $PP^* = W'\delta t$. Since δt and thus $\theta_0^* - \theta_0$ $(=\delta\theta)$ are treated as "small", PP^* can be assumed to be perpendicular to both, C_L and C_L^*. But this means that $l_c/L = \xi_c$ hardly varies during δt (i.e. that $\xi_c = \xi_c^*$), and one can write, on the basis of Eq. (7.90)

$$\frac{d\theta}{dt} = \frac{d\theta_0}{dt}\cos(2\pi\xi_c). \tag{7.91}$$

Let us now divide the (finite) distance l_c $(=\xi_c L)$ into a large number N of adjacent (small) intervals $(\delta l_c)_i$. The difference between the radial distances travelled by the end-points of each $(\delta l_c)_i$ during δt is

$$\frac{d\theta_i}{dt}\delta t(\delta l_c)_i \quad (=\lambda_i), \tag{7.92}$$

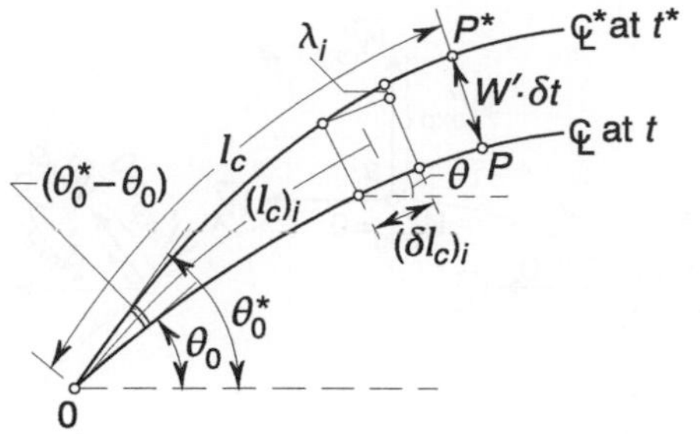

Figure 7.18 Stream centrelines at t and t*.

the distance $PP^* = W'\delta t$ being but the sum of all λ_i. Thus

$$W'\delta t = \lim_{N\to\infty}\sum_{i=1}^{N}\lambda_i = \lim_{N\to\infty}\sum_{i=1}^{N}\frac{d\theta_i}{dt}\delta t(\delta l_c)_i = \int_0^{l_c}\frac{d\theta}{dt}\delta t dl_c, \tag{7.93}$$

and since the integration extends along l_c (and not along t)

$$W' = \int_0^{l_c}\frac{d\theta}{dt}dl_c. \tag{7.94}$$

Substituting in Eq. (7.94) the value of $d\theta/dt$ given by Eq. (7.91), and taking into account that $dl_c = Ld\xi_c$, we determine

$$W' = \frac{d\theta_0}{dt}L\int_0^{\xi_c}\cos(2\pi\xi_c)d\xi_c = \frac{d\theta_0}{dt}\frac{L}{2\pi}\sin(2\pi\xi_c). \tag{7.95}$$

Consequently, for the expansion velocity W'_a at the apex a_i, where $\xi_c = 1/4$, we have

$$W'_a = \frac{d\theta_0}{dt}\frac{L}{2\pi}, \tag{7.96}$$

which makes it possible to express Eq. (7.95) as

$$W' = W'_a\sin(2\pi\xi_c). \tag{7.97}$$

This relation indicates that the knowledge of only W'_a is sufficient to know the expansion velocity at any section ξ_c of a meandering channel. The expansion velocity diagram is thus a sine-curve: its maximum is at the apex, its zero value at the crossovers.

Identifying Eq. (7.97) with Eq. (7.84) (as has been done in the preceding paragraph), we obtain

$$\overline{\omega}_c^{n_\omega} = K_2\sin(2\pi\xi_c), \tag{7.98}$$

where $K_2 = W'_a L/(\alpha_W(q_s)_{av})$ does not vary along ξ_c. Not surprisingly, this case too, indicates that n_ω must be equal to unity. Indeed, at the crossovers O_i, where $\xi_c = 0$, we have

$$\lim_{\xi_c\to 0}(\sin(2\pi\xi_c)) = 2\pi\xi_c \quad (=\overline{\omega}_c^{n_\omega}/K_2), \tag{7.99}$$

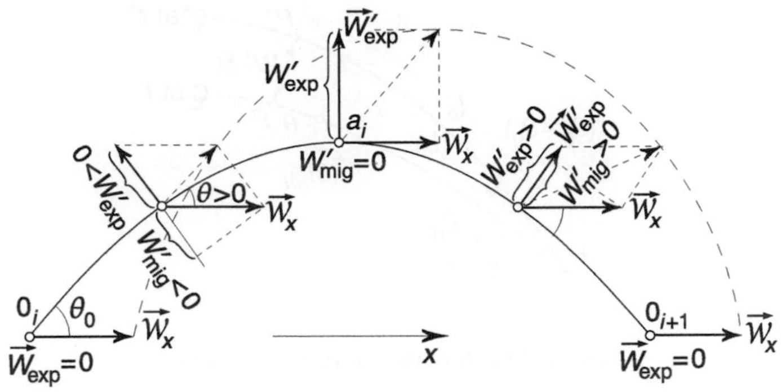

Figure 7.19 Local displacement vectors of meander loop downstream migration and lateral expansion.

which conveys that the graph of $\overline{\omega}_c^{n_\omega}$ intersects the ξ_c-axis at $\xi_c = 0$ with a finite angle, and thus that $n_\omega = 1$. Hence

$$W' = \alpha_W \frac{(q_s)_{av}}{L} \overline{\omega}_c = W'_a \sin(2\pi\xi_c). \tag{7.100}$$

Note from Figure 7.7b that in this case too, the point-pattern $\overline{\omega}_c$ intersects the ξ_c-axis with a finite angle.

(iii) *General case.* If θ_0 is neither "large" nor "small", then the radial displacement velocity W' at a point P of the channel centreline is the algebraic sum

$$W' = W'_{mig} + W'_{exp}, \tag{7.101}$$

where W'_{mig} and W'_{exp} are the migration- and expansion-components of W'. Clearly, W'_{mig} at a point P is but the scalar component of the vector $\mathbf{i}_x \mathcal{W}_x$ in the r-direction, viz $\mathbf{i}_r\left(\mathbf{i}_x \mathcal{W}_x\right)$. Indeed

$$W'_{mig} = \mathbf{i}_r\left(\mathbf{i}_x \mathcal{W}_x\right) = \cos\left(\frac{\pi}{2} + \theta\right) \mathcal{W}_x = -\mathcal{W}_x \sin\theta \tag{7.102}$$

which is the same as Eq. (7.86).

Hence, with regard to the loop $O_i a_i O_{i+1}$ (Figure 7.19), $W'_{mig} < 0$ along $O_i a_i$ (where $\theta > 0$), and $W'_{mig} > 0$ along $a_i O_{i+1}$ (where $\theta < 0$); W'_{exp} is, of course, positive throughout.

Substituting for W'_{mig}, W'_{exp}, and W' their expressions (7.87), (7.97) and (7.84), and considering that $n_\omega = 1$, we obtain for Eq. (7.101)

$$W' = \alpha_W \overline{\omega}_c \frac{(q_s)_{av}}{L} = -\mathcal{W}_x \sin[\theta_0 \cos(2\pi\xi_c)] + W'_a \sin(2\pi\xi_c). \tag{7.103}$$

Note from Eq. (7.103) (and Figure 7.19) that at the crossovers O_i, O_{i+1} (where $\xi_c = 0; 1/2$), we have $W'_{exp} = 0$ and thus

$$\mathcal{W}_x = -(\overline{\omega}_c)_O \frac{(q_s)_{av}}{L} \frac{\alpha_W}{\sin\theta_0}, \tag{7.104}$$

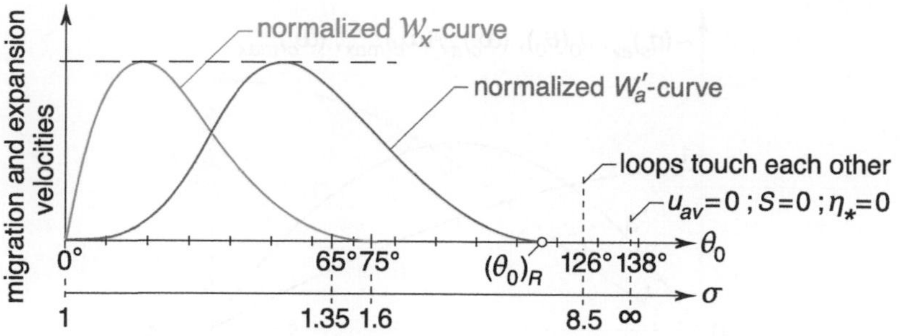

Figure 7.20 Schematic plot of bank migration and expansion velocities versus initial deflection angle.

while at the apex a_i (where $\xi_c = 1/4$), we have $W'_{mig} = 0$ and, consequently,

$$W'_a = (\overline{\omega}_c)_a \frac{(q_s)_{av}}{L} \alpha_W = (\overline{\omega}_c)_a \frac{(q_s)_{av}}{R_a} \frac{\alpha_W}{2\pi\theta_0}, \tag{7.105}$$

where the last step is by the use of Eq. (6.9). On the other hand, Eqs. (6.9) and (6.10) give

$$\frac{\Lambda_M}{R_a} = 2\pi\theta_0 \frac{1}{\sigma} = 2\pi\theta_0 J_0(\theta_0), \tag{7.106}$$

which makes it possible to express W'_a in terms of θ_0 (and Λ_M) as

$$W'_a = (\overline{\omega}_c)_a \frac{(q_s)_{av}}{\Lambda_M} J_0(\theta_0) \alpha_W. \tag{7.107}$$

(iv) From field observations and the (rather scant) measurements (Kondratiev et al. 1982), it follows that the curves representing the variations of $\mathcal{W}_x$ and W'_a with θ_0, i.e. the $\mathcal{W}_x$- and W'_a-curves, are of the form sketched in Figure 7.20; "at the early stages (small θ_0), it is the downstream migration ($\mathcal{W}_x$) of loops which is mainly observable, at the latter stages (large θ_0), it is their expansion which dominates" (Kondratiev et al. 1982, p. 108). Here we will be dealing only with the (regime-related) expansion velocity W'_a.

In Sub-section 6.3.4 (ii) it has been mentioned that the deviation angle $(\overline{\omega}_c)_a$ is always positive (Figure 6.18). And since the remaining characteristics on the right-hand side of Eq. (7.107) are always positive as well, the expansion velocity W'_a exists starting from the very beginning, i.e. from $\theta_0 = 0$. This is only to be expected, for the mere fact that the channel migration is detectable just after $\theta_0 = 0$, means that the meander loops (which are due to W'_a (> 0)) are already present. It should also be noted that the regime channel formation by meandering can be accomplished only if the regime value of θ_0, viz $(\theta_0)_R$, is smaller than 126°, where the adjacent meander loops begin to touch each other (Figure 6.3b). If $(\theta_0)_R > 126°$, then the formation perpetuates indefinitely. Clearly, $(\theta_0)_R$ is always smaller than 138°, which corresponds to $S = 0$ and thus $u_{av} = 0$ (Figure 7.20).

In order to reveal how the function (7.107) compares with the W'_a-curve in Figure 7.20, we consider the θ_0-variation of the multipliers forming the right-hand side of Eq. (7.107). Here, Λ_M does not depend on θ_0 while the possible θ_0-variation

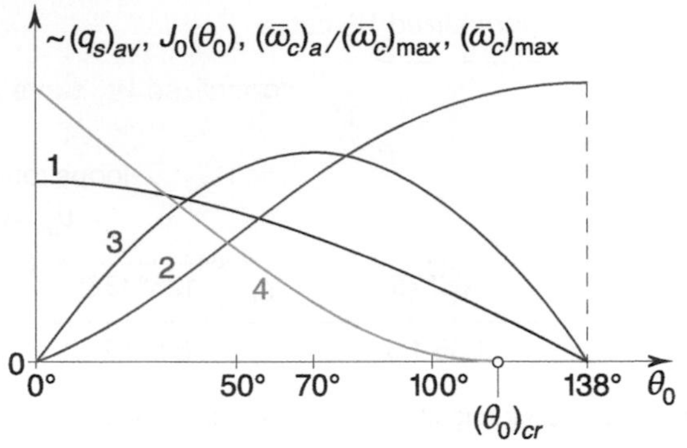

Figure 7.21 Schematic plot indicating the form of variation of pertinent quantities with the initial deflection angle.

of α_W (which maintains its positive sign and has always a finite value) can be ignored for the present purpose. $J_0(\theta_0)$ $(=1/\sigma)$ varies as indicated in Fig. 6.3a, i.e. as shown schematically in Figure 7.21 by the curve 1. Consider now $(\overline{\omega}_c)_a$. As follows from the considerations in Sub-section 6.3.4, the ratio $(\overline{\omega}_c)_a/(\overline{\omega}_c)_{max}$ monotonously increases from zero to unity when θ_0 increases from zero to 138°; curve 2 in Figure 7.21. Yet, with the increment of θ_0, the maximum deflection angle $(\overline{\omega}_c)_{max}$, at any ξ_c-section, first increases from zero onwards until it acquires its maximum value, and then decreases as to become zero again at $\theta_0 = 138°$ – for the relative channel curvature B/R (at any ξ_c) tends to vanish when $\sigma \to 1$ and $\sigma \to \infty$. The largest $(\overline{\omega}_c)_{max}$ must be expected to occur for that θ_0 where B/R_a $(=\theta_0 J_0(\theta_0))$ is the largest, i.e. for $\theta_0 \approx 70°$ (see Figure 6.19); curve 3 in Figure 7.21. Consider finally $(q_s)_{av}$. Its value must continually decrease when η_* $(\sim S)$ decreases (curve 4), as to vanish for that $\theta_0 = (\theta_0)_{cr}$ which corresponds to $(\eta_*)_{cr} = 1$. One can easily see that the multiplication of the curves 1, 2, 3, and 4 must yield a curve that is of the same type as the W'_a-curve in Figure 7.20.

7.3.5 Loop-expansion velocity and regime development

(i) If the regime-development is *gravel-like*, then this development must terminate, and the channel expansion must stop, when η_* reduces to $(\eta_*)_{cr} = 1$. And in this case, Eq. (7.107) – which contains $(q_s)_{av}$ – gives indeed $W'_a = 0$ when $\eta_* = 1$ and thus $(q_s)_{av} = 0$.

If, however, the regime-development is *sand-like*, then $(q_s)_{av}$ is finite $(\gg 0)$ even at the regime state ("live-bed" regime channel). Yet, in this case, Eq. (7.107) cannot yield the required $W'_a = 0$; for none of the multipliers on its right-hand side becomes zero when θ_0 acquires its regime value $(\theta_0)_R$. Hence Eq. (7.107), which has been derived solely on the basis of flow-kinematics and sediment transport, cannot be regarded as complete as it stands: it must be augmented so as to take into account also the conditions presented by the regime-development.

(ii) From the aforementioned it should be clear that Eq. (7.107) must be brought into a form that would yield $W'_a = 0$ when $\theta_0 = (\theta_0)_R$ (which corresponds to $(\eta_*)_R$), even if the regime formation is sand-like, and thus $(\eta_{*R}) \gg 1$. In this context, it is important to realize that although for a specified experiment, the regime slope S_R has a definite (computable) value, no definite value can be ascribed to the slope S_0 of the initial channel, which is to be treated as arbitrary. But this means that the regime state of a meandering stream (corresponding to a specified Q and a given granular material and fluid) *cannot* be associated with a certain plan-geometry, i.e. with a certain regime sinuosity σ_R, or a regime deflection angle $(\theta_0)_R$. Indeed, the relation

$$\frac{S_0}{S_R} = \frac{L_R}{\Lambda_M} = \sigma_R = \frac{1}{J_0((\theta_0)_R)}, \tag{7.108}$$

indicates that every arbitrarily selected S_0 has "its own" σ_R and $(\theta_0)_R$. If S_R is computed and S_0 is selected, then $(\theta_0)_R$ follows from Eq. (7.108). In particular, if S_0 is selected to be the same as S_R, then the straight initial channel remains as it is ($\sigma_R = 1$; $(\theta_0)_R = 0$). To put it differently, the initial (valley) slope S_0 manifests itself as an *additional parameter*, as far as the determination of the "regime-geometry" of a meandering stream in plan view is concerned.

The fulfillment of the requirement $W'_a = 0$ when $S = S_R$, can be achieved by introducing an additional multiplier-function, β_W say. That is, by augmenting Eq. (7.107) into

$$W'_a = (\overline{\omega}_c)_a \frac{(q_s)_{av}}{\Lambda_M} J_0(\theta_0)\alpha_W \beta_W, \tag{7.109}$$

where

$$\left.\begin{aligned} &\beta_W \equiv 1 && \text{(if gravel-like)} \\ &\beta_W = \phi_\beta((\theta_0)_R, \theta_0) && \text{(if sand-like)} \end{aligned}\right\} \tag{7.110}$$

From the considerations above it should be clear that for a given experiment (for a specified Q and materials), the loop expansion velocity W'_a varies as a function of two variables (θ_0 and S_0) as implied by the curve-family sketched in Figure 7.22 – every S_0 having "its own" $(\theta_0)_R$, which signifies the end of the meander (regime) development. One would expect that the W'_a-curves C_1, C_2, ..., C_i corresponding to various S_{0i} start to grow in the same manner (i.e. that they merge into each other when $\theta_0 = 0$ is approached). This means that the so far unknown function $\phi_\beta((\theta_0)_R, \theta_0)$ is likely to satisfy the conditions

$$\phi_\beta((\theta_0)_R, 0) = 1; \qquad \frac{\partial \phi_\beta((\theta_0)_R, 0)}{\partial \theta_0} = 0, \tag{7.111}$$

as well as, of course, $\phi_\beta((\theta_0)_R, (\theta_0)_R) = 0$. The expressions (7.107) and (7.109) can be used for the computation of $\mathcal{W}_x$ and W'_a only if the multiplier functions α_W and β_W are revealed first. A special laboratory research directed towards the determination of these functions (of B/h_{av}, c_{av}, S_0, and perhaps θ_0) would certainly be worthwhile.

7.3.6 Computational procedure and example of application

In this sub-section, and following da Silva and Bahar (2003), we present a computational procedure to determine the planimetric evolution of a meandering stream using the equations introduced in the previous sections. It is assumed that the valley slope S_v and the (constant) flow rate Q are given, and that the granular material is specified

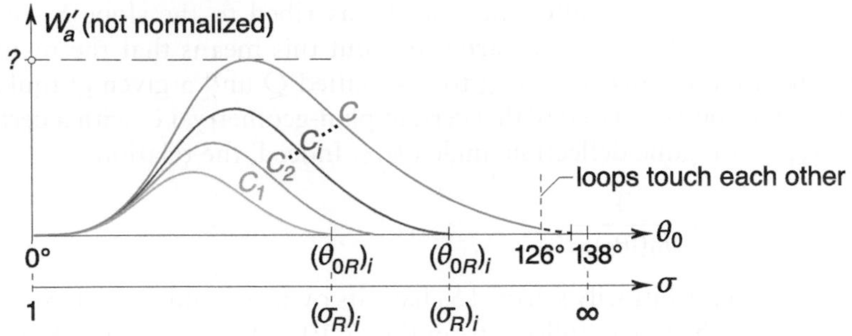

Figure 7.22 Schematic representation of the loop expansion velocity curve-family.

(by D and γ_s/γ). The (constant) flow width, which is identified with the regime width B_R, is determined using the computer program BHS-STABLE (see Sub-section 4.5.1). At every different stage of the meander development, associated with a given value of the channel centreline slope S_c, the flow depth h_{av} is computed from the resistance equation, viz $Q = Bh_{av}c\sqrt{gS_ch_{av}}$. The resistance factor c in this equation is evaluated by taking into account the presence of bed forms (dunes and/or ripples), with the aid of Eq. (3.12). The bed form length Λ_i and steepness δ_i ($i=d$ if dunes; $i=r$ if ripples) are computed from the equations in Section 2.3. Owing to the presence of $c=f_c(h_{av}, S_c)$ in the resistance equation, a numerical procedure must be adopted to determine the flow depth h_{av} from it: the bisection method is adopted for this purpose. Given that at present no information is available on the nature of β_W (in Eq. (7.109)), Eq. (7.107) is used in the calculations instead of Eq. (7.109).

Knowing thus Q, D, γ_s/γ, B ($=B_R$), and knowing also the position of the channel centreline at time t_i (and thus the value of θ_0 at t_i), the following steps are adopted to determine and draw the channel centreline at time $t_{i+1} = t_i + \Delta t$:

1 Compute $S_c = S_v/\sigma = S_vJ_0(\theta_0)$.
2 Knowing S_c, compute h_{av} from the resistance equation.
3 Knowing thus S_c and h_{av}, compute $(q_s)_{av}$. Sediment transport is assumed to be by bed-load only; Bagnold's equation is used to calculate $(q_s)_{av}$.
4 Compute $(\overline{\omega}_c)_O$ and $(\overline{\omega}_c)_a$ from Eqs. (6.46)-(6.48) (note that $l_c/L=0$ at the crossover O_i and $l_c/L=1/4$ at the apex a_i).
5 Knowing thus $(q_s)_{av}$, $(\overline{\omega}_c)_O$ and $(\overline{\omega}_c)_a$, compute $\mathcal{W}_x$ from Eq. (7.104) and W' from $W' = W'_a \sin(2\pi l_c/L)$, where W'_a is given by Eq. (7.107).
6 For the time interval Δt, determine the displacements Δx and Δy due to both migration and expansion of the points $P(x_{t,i}; y_{t,i})$ on the channel centreline, and compute the new coordinates of P: $x_{t,i+1} = x_{t,i} + \Delta x$ and $y_{t,i+1} = y_{t,i} + \Delta y$. The displacement Δx and Δy can be computed (for $l_c/L \in [0; 0.5]$) from $\Delta x = [\mathcal{W}_x + W' \cos(90° + \theta)] \cdot \Delta t$ and $\Delta y = (W' \cos\theta) \cdot \Delta t$
7 Using the coordinates of the (displaced) points P, draw the channel centreline at time t_{i+1}.

To exemplify the above method, this is next applied to a (hypothetical) stream having the following characteristics: $Q = 15000 m^3/s$, $D = 0.5 mm$, $\gamma_s/\gamma = 1.65$,

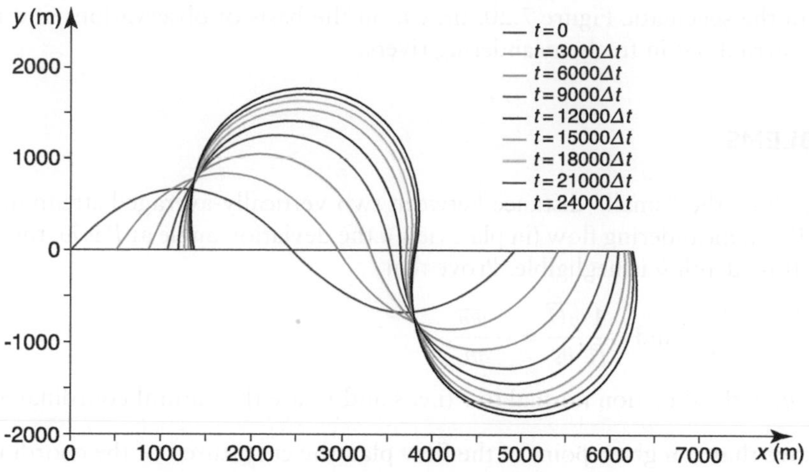

Figure 7.23 Computed plan evolution over time of channel centreline.

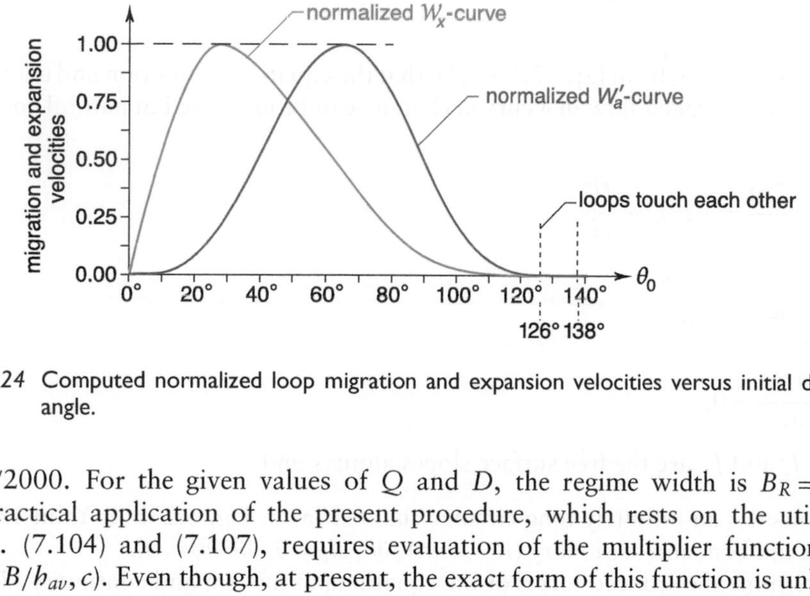

Figure 7.24 Computed normalized loop migration and expansion velocities versus initial deflection angle.

$S_v = 1/2000$. For the given values of Q and D, the regime width is $B_R = 780m$. The practical application of the present procedure, which rests on the utilization of Eqs. (7.104) and (7.107), requires evaluation of the multiplier function $\alpha_W = \phi_W(\theta_0, B/h_{av}, c)$. Even though, at present, the exact form of this function is unknown, preliminary results indicate that for given B/h_{av} and c, α_W is an increasing function of θ_0 when $\theta_0 <\approx 30°$ ($\alpha_W = 0$ for $\theta_0 = 0$), and that α_W remains nearly constant ($=(\alpha_W)_{max}$) when $\theta_0 \geq 30°$. Accordingly, α_W is here expressed by the following equation: $\alpha_W = (\alpha_W)_{max} \sin(3\theta_0)$, in which θ_0 is in degrees, if $\theta_0 \leq\approx 30°$; and $\alpha_W = (\alpha_W)_{max}$ if $\theta_0 >\approx 30°$. The value $(\alpha_W)_{max} = 80$ is adopted as this value yields a (seemingly) reasonable time-scale for the problem under consideration. The calculations were carried out using a time-step $\Delta = 5 \times 10^4 s \approx 0.6$ days.

Figure 7.23 shows the computed plan evolution of the channel centreline. In this simulation, at $t = 0$, $\theta_0 = 45°$; the simulation ends at $t = 24000\Delta t \approx 40$ years ($\theta_0 = 98.5°$). The graphs of the computed W_x and W'_a are shown in Figure 7.24. Observe that the (normalized) $\mathcal{W}_x$- and W_a'-curves in this figure are in complete agreement with those in the schematic Figure 7.20, drawn on the basis of observations and measurements carried out in freely meandering rivers.

PROBLEMS

7.1 Let b be the "small" distance between two vertically-averaged streamlines s at a point P of a meandering flow (in plan view): the deviation angle at P is $\overline{\omega}$, the variation of the flow depth h is negligible. Prove that

$$\frac{1}{b}\frac{\partial b}{\partial s}=\frac{\partial \overline{\omega}}{\partial n_s} \quad \text{and} \quad \frac{1}{\overline{U}}\frac{\partial \overline{U}}{\partial s}=-\frac{\partial \overline{\omega}}{\partial n_s},$$

where n_s is the direction normal to s (i.e. s and n_s are the natural coordinates).

7.2 Prove that at a given point of the flow plan the curvature r of the coordinate line l and the curvature r_s of the (vertically-averaged) streamlines s are interrelated as follows

$$\frac{1}{r_s}=-\frac{\partial \overline{\omega}}{\partial s}+\frac{1}{r}\cos\overline{\omega}.$$

7.3 Prove (starting from Eqs. (7.1)-(7.3)) that the equations of motion and continuity of a vertically-averaged flow in a curved channel can be expressed in natural coordinates $(s; r_s)$ as

$$\frac{\partial(\overline{U}^2 h)}{\partial s}=-ghJ_s-\frac{\overline{U}^2}{c_M^2} \tag{7.112}$$

$$\frac{\overline{U}^2}{r_s}=-ghJ_{r_s}. \tag{7.113}$$

$$\frac{\partial(\overline{U}h)}{\partial s}=0, \tag{7.114}$$

where J_s and J_{r_s} are the free surface slopes along s and r_s.

7.4 Consider an infinitely long circular channel (as in Fig. 6.23b in Yalin 1992): the flow is uniform and stationary ($R=const$, $\partial h/\partial l_c=0$, $\partial\overline{u}/\partial l_c=0$).
a) Determine the reduced forms of the equations of motion and continuity (7.4)-(7.6) which correspond to this special case.
b) Determine the expressions of $\overline{u}$ and h as functions of n/R.

7.5 Prove on the basis of the equations of motion and continuity that if c_M is constant, or if it varies only as a function of h/K_s, then, in the case of a flat bed, these equations can yield only an ingoing flow – no matter how large the value of θ_0 might be (see Figure 7.1).

7.6 Adopt the following notation

$$y=\overline{\omega}_c/(\overline{\omega}_c)_{max}; \quad y_a=(\overline{\omega}_c)_a/(\overline{\omega}_c)_{max}.$$

a) How does the ratio y_a vary with θ_0?
b) Explain why the curve $(\overline{\omega}_c)_{max}$ in Figure 6.19 tends to become indistinguishable from the straight line $(\overline{\omega}_c)_{max} = \theta_0$, when $\theta_0 \to 0$.

7.7 Consider the stream determined by $Q = 1500m^3/s$ and $D = 0.7mm$ ($\gamma_s/\gamma = 1.65$, $\rho = 10^3 Kg/m^3$, $\nu = 10^{-6} m^2/s$). The regime development of S is totally by meandering (no degradation). Use for the total friction factor $c = 13.0$ at all stages.
a) Determine the regime values B_R, h_R, S_R and $(\theta_0)_R$, assuming that the slope S_0 of the initial channel is four times larger than S_R ($S_0 = 4S_R$).
b) Consider now the development stage when S is twice larger than S_R ($S = 2S_R$).
  1 What is the value of the angles θ_0 and $(\overline{\omega}_c)_a$ at this stage?
  2 What is the transport rate $(q_s)_{av}$ at this stage? (Take $B = 0.95B_R$).
  3 What is the channel expansion velocity W'_a at this stage? (Take $\alpha_W \beta_W = 7.5$).

7.8 Prove that the expansion velocity of the angle θ_0, viz $d\theta_0/dt$, is related to the apex-expansion velocity W'_a as follows

$$\theta_0 \frac{d\theta_0}{dt} = \frac{W'_a}{R_a}.$$

REFERENCES

Binns, A.D. and da Silva, A.M.F. (2015). Meandering bed development time: formulation and related experimental testing. *Advances in Water Resources*, 81, 152-160.
Binns, A.D. and da Silva, A.M.F. (2011). Rate of growth and other features of the temporal development of pool-bar complexes in meandering streams. *Journal of Hydraulic Engineering*, 137(12), 1565-1575.
Binns, A.D. and da Silva, A.M.F. (2009). On the quantification of the bed development time of alluvial meandering streams. *Journal of Hydraulic Engineering*, 135(5), 350-360.
Blanckaert, K. (2010). Topographic steering, flow recirculation, velocity redistribution, and bed topography in sharp meander bends. *Water Resources Research*, 46, W09506, doi: 10.1029/2009WR008303.
Blanckaert, K. (2009). Saturation of curvature-induced secondary flow, energy losses, and turbulence in sharp open-channel bends: laboratory experiments, analysis, and modeling. *Journal of Geophysical Research*, 114, F03015, doi: 10.1029/2008J F001137.
Blanckaert, K. and de Vriend, H.J. (2004). Secondary flow in sharp open-channel bends. *Journal of Fluid Mechanics*, Vol. 498, 353-380.
Chen, D. and Duan, J.D. (2006). Simulating sine-generated meandering channel evolution with an analytical model. *Journal of Hydraulic Research*, 44(3), 363-373.
Constantinescu, G., Kashyap, S., Tokyay, T., Rennie, C.D. and Townsend, R.D. (2013). Hydrodynamic processes and sediment erosion mechanisms in an open channel bend of strong curvature with deformed bathymetry. *Journal of Geophysical Research – Earth*, 118(2), 480-496.
Crosato, A. (2008). *Analysis and modelling of river meandering*. IOS Press, Amsterdam, The Netherlands.

Dallali, M. (2016). *Large eddy simulation of suspended sediment in turbulent open-channel flow*. Ph.D. Thesis, Università Degli Studi Di Trieste, Trieste, Italy.

Darby, S.E., Alabyan, A.M. and Van de Wiel, M.J. (2002). Numerical simulation of bank erosion and channel migration in meandering rivers. *Water Resources Research*, 38(9), 1163, doi:10.1029/2001WR000602.

da Silva, A.M.F., El-Tahawy, T. and Tape, W.D. (2006). Variation of flow pattern with sinuosity in sine-generated meandering channels. *Journal of Hydraulic Engineering*, 132(10), 1003-1014.

da Silva, A.M.F. and Bahar, S.M.H. (2003). Migration and expansion of meander loops: a simulation model. *Proceedings of the 30th IAHR Congress*, Thessaloniki, Greece.

da Silva, A.M.F. (1999). Friction factor of meandering flows. *Journal of Hydraulic Engineering*, 125(7), 779-783.

da Silva, A.M.F. (1995). *Turbulent flow in sine-generated meandering channels*. Ph.D. Thesis, Department of Civil Engineering, Queen's University, Kingston, Canada.

Demuren, A.O. (1993). A numerical model for flow in meandering channels with natural bed topography. *Water Resources Research*, 29(4), 1269-1277.

Demuren, A.O. and Rodi, W. (1986). Calculation of flow and pollutant dispersion in meandering channels. *Journal of Fluid Mechanics*, Vol. 172, 63-92.

Duc, B., Wenka, T. and Rodi, W. (2004). Numerical modelling of bed deformation in laboratory channels. *Journal of Hydraulic Engineering*, 130(9), 894-904.

Eke, E.C. (2013). *Numerical modelling of river migration incorporating erosional and depositional bank processes*. Ph.D. Thesis, University of Illinois at Urbana-Champaign, USA.

El-Tahawy, T. (2004). *Patterns of flow and bed deformation in meandering streams*. Ph.D. Thesis, Department of Civil Engineering, Queen's University, Kingston, Canada.

Friedkin, J.F. (1945). *A laboratory study of the meandering of alluvial rivers*. U.S. Waterways Experiment Station, Vicksburg, Mississippi.

Ikeda, S. (1982). Lateral bed load transport on side slopes. *Journal of Hydraulic Engineering*, 108(11), 1369-1373.

Jansen, P.Ph., van Bendegom, L., van den Berg, J., de Vries, M. and Zanen, A. (1979). *Principles of river engineering*. Pitman Publishing Ltd., London.

Jia, Y. and Wang, S.S.Y. (1999). Numerical model for channel flow and morphological change studies. *Journal of Hydraulic Engineering*, 125(9), 924-933.

Kalkwijk, J.P.Th. and de Vriend, H.J. (1980). Computation of the flow in shallow river bends. *Journal of Hydraulic Research*, 18(4), 327-342.

Kang, S. and Sotiropoulos, F. (2011). Flow phenomena and mechanisms in a field-scale experimental meandering channel with a pool-riffle sequence: insights gained via numerical simulation. *Journal of Geophysical Research – Earth*, 116(F3), 1-22.

Kassem, A.K. and Chaudhry, M.H. (2002). Numerical modeling of bed evolution in channel bends. *Journal of Hydraulic Engineering*, 128(5), 507-514.

Kondratiev, N., Popov, I. and Snishchenko, B. (1982). *Foundations of hydromorphological theory of fluvial processes*. (In Russian) Gidrometeoizdat, Leningrad.

Levi, I.I. (1957). *Dynamics of alluvial streams*. State Energy Publishing, Leningrad.

Lien, H.C., Hsieh, T. and Yang, J.C. (1999). Bend-flow simulation using 2D depth-averaged model. *Journal of Hydraulic Engineering*, 125(10), 1097-1108.

Matthes, G.H. (1948). Mississippi River cutoffs. *Transactions of the American Society of Civil Engineers*, 113, 16-39.
Matthes, G.H. (1941). Basic aspects of stream meanders. *Transactions of the American Geophysical Union*, 632-636.
Mosselman, E. (1995). A review of mathematical models of river planform changes. *Earth Surface Processes and Landforms*, 20, 661-670.
Motta, D., Langendoen, E.J., Abad, J.D. and García, M.H. (2014). Modification of meander migration by bank failures. *Journal of Geophysical Research*, 119(5), 1026-1042.
Motta, D., Abad, J.D., Langendoen, E.J. and García, M.H. (2012). A simplified 2D model for meander migration with physically-based bank evolution. *Geomorphology*, Vols. 163-164, 10-25.
Nagata, N., Hosoda, T. and Muramoto, Y. (2000). Numerical analysis of river channel processes with bank erosion. *Journal of Hydraulic Engineering*, 126(4), 243-252.
Nakagawa, H., Tsujimoto, T. and Murakami, S. (1986). Non-equilibrium bed load transport along side slope of an alluvial stream. *Proceedings of the 3rd International Symposium on River Sedimentation*, edited by S.Y. Wang, H.W. Shen and L.Z. Ding, The University of Mississippi, Mississippi, 885-893.
Nasermoaddeli, M.H. (2012). Bank erosion in alluvial rivers with non-cohesive soil in unsteady flow. *Hamburger Wasserbauschriften*, 14, TuTech Verlag, Hamburg, Germany.
Nelson, J.M. and Smith, J.D. (1989a). Evolution and stability of erodible channel beds. In *River Meandering: Water Resources Monograph*, 12, edited by S. Ikeda and G. Parker, American Geophysical Union, 321-337.
Nelson, J.M. and Smith, J.D. (1989b). Flow in meandering channels with natural topography. In *River Meandering: Water Resources Monograph*, 12, edited by S. Ikeda and G. Parker, American Geophysical Union, 69-102.
Okoye, J.K. (1970). *Characteristics of transverse mixing in open channel flows*. Report KH-R-23, W.M. Keck Laboratory for Hydraulics and Water Resources, California Institute of Technology, USA.
Olsen, N.R.B. (2003). Three-dimensional CFD modeling of self-forming meandering channel. *Journal of Hydraulic Engineering*, 129(5), 366-372.
Ottevanger, W., Blanckaert, K. and Uijttewaal, W.S.J. (2012). Processes governing the flow redistribution in sharp river bends. *Geomorphology*, Vols. 163-164, 45-55.
Parker, G. (1984). Discussion of *Lateral bed load transport on side slopes* by S. Ikeda. *Journal of Hydraulic Engineering*, 110(2), 197-199.
Rodi, W. (1980). *Turbulence models and their applications in hydraulics*. IAHR Monograph, Delft, The Netherlands.
Rousseau, Y.Y., Biron, P.M. and Van de Wiel, M.J. (2016). Sensitivity of simulated flow fields and bathymetries in meandering channels to the choice of a morphodynamic model. *Earth Surface Processes and Landforms*, 41, 1169-1184.
Rüther, N. and Olsen, N.R.B. (2007). Modelling free-forming meander evolution in a laboratory channel using three-dimensional computational fluid dynamics. *Geomorphology*, 89, 308-319.
Shimizu, Y. (1991). *A study on prediction of flows and bed deformation in alluvial streams*. (In Japanese), Report, Civil Engineering Research Institute, Hokkaido Development Bureau, Sapporo, Japan.

Shimizu, Y. and Itakura, T. (1989). Calculation of bed variation in alluvial channels. *Journal of Hydraulic Engineering*, 115(3), 367-384.

Smith, J.D. and McLean, S.R. (1984). A model for flow in meandering streams. *Water Resources Research*, 20(9), 1301-1315.

Stoesser, T., Ruether, N. and Olsen, N.R.B. (2010). Calculation of primary and secondary flow and boundary shear stresses in a meandering channel. *Advances in Water Resources*, 33(2), 158-170.

Struiksma, N. and Crosato, A. (1989). *Analysis of a 2-D bed topography model for rivers*. In *River Meandering: Water Resources Monograph*, 12, edited by S. Ikeda and G. Parker, American Geophysical Union, 153-180.

Struiksma, N., Olesen, K.W., Flokstra, C. and de Vriend, H.J. (1985). Bed deformation in curved alluvial channels. *Journal of Hydraulic Research*, 23(1), 57-79.

Struiksma, N. (1985). Prediction of 2-D bed topography in rivers. *Journal of Hydraulic Engineering*, 111(8), 1169-1182.

Termini, D. (1996). *Evoluzione di un canale meandriforme a fondo inizialmente piano: studio teorico-sperimentale del fondo e le caratteristiche cinematiche iniziali della corrente*. Ph.D. Thesis, Department of Hydraulic Engineering and Environmental Applications, University of Palermo, Italy.

van Balen, W., Blanckaert, K. and Uijttewaal, W.S.J. (2010). Analysis of the role of turbulence in curved open-channel flow at different water depths by means of experiments, LES and RANS. *Journal of Turbulence*, 11(12), 1-34.

Vasquez, J., Millar, R. and Steffler, P. (2011). Vertically-averaged and moment model for meandering river morphology. *Canadian Journal of Civil Engineering*, 38(8), 921-931.

Vasquez, J.A., Steffler, P.M. and Millar, R.G. (2008). Modeling bed changes in meandering rivers using triangular finite elements. *Journal of Hydraulic Engineering*, 134(9), 1348-1352.

Velikanov, M.A. (1955). *Dynamics of alluvial streams. Vol. II. Sediment and bed flow*. State Publishing House for Theoretical and Technical Literature, Moscow.

Wu, W., Rodi, W. and Wenka, T. (2000). 3D numerical modeling of flow and sediment transport in open channels. *Journal of Hydraulic Engineering*, 126(1), 4-15.

Yalin, M.S. (1992). *River mechanics*. Pergamon Press, Oxford.

Zolezzi, G., Luchi, R. and Tubino, M. (2012). Modeling morphodynamic processes in meandering rivers with spatial width variations. *Reviews of Geophysics*, 50, RG4005, doi: 10.1029/2012RG000392.

Appendix A

Sources of dune and ripple data

1a. Adriaanse, M. (1986). *De ruwheid van de Dergsche Maas bij hoje afvoeren.* (In Dutch) Rijkswaterstaat, RIZA, Nota 86.19.

2a. Annambhotla, V.S., Sayre, W.W. and Livesey, R.H. (1972). Statistic properties of Missouri river bed forms. *Journal of Waterways, Harbours and Coastal Engineering Division*, ASCE, 98(4), 489-510.

3a. Ashida, K. and Tanaka, Y. (1967). A statistical study of sand waves. *Proceedings XII IAHR Congress*, Fort Collins, Colorado, 2, 103-110.

4a. Banks, N.L. and Collinson, J.D. (1975). The size and shape of small scale ripples: an experimental study using medium sand. *Sedimentology*, 12, 583-599.

5a. Barton, J.R. and Lin, P.N. (1955). *Sediment transport in alluvial channels.* Rept. No. 55JRB2, Civil Engineering Department, Colorado AM College, Fort Collins, Colorado.

6a. Bishop, C.T. (1977). *On the time-growth of dunes.* M.Sc. Thesis, Department of of Civil Engineering, Queen's University, Kingston, Canada.

7a. Casey, H.J. (1935). *Bed load movement.* Ph.D. Dissertation, Technische Hochschula, Berlin.

8a. da Cunha, L.V. (1969). *River Mondego, Portugal.* Report, Laboratório Nacional de Engenharia Civil, LNEC, Lisbon.

9a. East Pakistan Water and Power Development Authority (1969). *Flume studies of roughness and sediment transport of movable bed of sand.* Annual Report of Hydraulic Research Laboratory, Dacca, 1966-69.

10a. Fok, A.T. (1975). *On the development of ripples by an open channel flow.* M.Sc. Thesis, Department of Civil Engineering, Queen's University, Kingston, Canada.

11a. Fredsoe, T. (1981). Unsteady flow in straight alluvial streams. *Journal of Fluid Mechanics*, Vol. 102, Part 2, 431-453.

12a. Grigg, N.S. (1970). Motion of single particle in alluvial channels. *Journal of the Hydraulics Division*, ASCE, 96(12), 2501-2518.

13a. Guy, H.P., Simons, D.B. and Richardson, E.V. (1966). Summary of alluvial channel data from flume experiments 1956-1961. *U.S. Geological Survey Professional Papers*, 462-I, 1-96.

14a. Haque, M.I. and Mahmood, K.M. (1983). Analytical determination of form friction factor. *Journal of Hydraulic Engineering*, 109(4), 590-610.

15a. Hubbell, D.W. and Sayre, W.H. (1964). Sand transport studies with radioactive tracers. *Journal of the Hydraulics Division*, ASCE, 90(3), 39-68.

16a. Hung, C.S. and Shen, H.W. (1979). Statistical analysis of sediment motions of dunes. *Journal of the Hydraulics Division*, ASCE, 105(3), 213-227.

17a. Hwang, L.S. (1965). *Flow resistance of dunes in alluvial streams*. Ph.D. Thesis, California Institute of Technology, Pasadena, California.

18a. Jain, S.C. and Kennedy, J.F. (1971). The growth of sand waves. *Proceedings I International Symposium on Stochastic Hydraulics*, University of Pittsburgh Press, Pittsburgh, 449-471.

19a. Julien, P.Y. (1992). *Study of bed form geometry in large rivers*. Report Q1386, Delft Hydraulics, Emmerlood, The Netherlands.

20a. Lane, E.W. and Eden, E.W. (1940). Sand waves in Lower Mississippi River. *Journal of Western Society of Engineers*, 45(6), 281-291.

21a. Lau, Y.L. and Krishnappan, B. (1985). Sediment transport under ice cover. *Journal of Hydraulic Engineering*, 111(6), 934-950.

22a. Mahmood, K. and Amadi-Karvigh, H. (1976). Analysis of bed profile in sand bed canals. *Proceedings of Rivers'76*, Annual Symposium of the Waterways, Harbors and Coastal Engineering Division of ASCE, Colorado State University, Fort Collins, Colorado.

23a. Mantz, P.A. (1983). Semi-empirical correlations for fine and coarse sediment transport. *Proceedings of the Institution of Civil Engineers*, Vol. 75, Part 2, 1-33.

24a. Mantz, P.A. (1980). Laboratory flume experiment on the transport of cohesionless silica silts by water streams. *Proceedings of the Institution of Civil Engineers*, Vol. 69, Part 2, 977-994.

25a. Martinec, J. (1967). The effect of sand ripples on the increase of river bed roughness. *Proceedings XII IAHR Congress*, Fort Collins, Colorado, A21:1-9.

26a. Matsunashi, J. (1967). On a solution of bed fluctuation in an open channel with a movable bed and steep slopes. *Proceedings XII IAHR Congress*, Fort Collins, Colorado.

27a. NEDECO (1973). *Rio Magdalena and Canal del Dique Project, Mission Tecnica Colombo-Holandesa*. NEDECO Report, NEDECO, The Hague.

28a. Nordin, C.F. (1976). *Flume studies with fine and coarse sand*. Report 76-762, U.S. Geological Survey, Denver, Colorado.

29a. Nordin, C.F. (1971). Statistical properties of dune profiles. *U.S. Geological Survey Professional Papers*, 562-F, 1-41.

30a. Nordin, C.F. and Algert, J.H. (1966). Spectral analysis of sand waves. *Journal of the Hydraulics Division*, ASCE, 92(5), 95-114.

31a. Nordin C.F. and Beverage, J.P. (1965). Sediment transport in the Rio Grande, New Mexico. *U.S. Geological Survey Professional Papers*, 462-F, 1-35.

32a. Nordin, C.F. (1964). Aspects of flow resistance and sediment transport – Rio Grande near Bernalillo, New Mexico. *U.S. Geological Survey Water Supply Papers*, 1498-H, 1-41.

33a. Onishi, Y., Jain, S.C. and Kennedy, J.F. (1976). Effects of meandering in alluvial channels. *Journal of the Hydraulics Division*, ASCE, 102(7), 899-918.

34a. Peters, J.J. (1978). Discharge and sand transport in the braided zone of the Zaire estuary. *Netherlands Journal of Sea Research*, 12, Issues 3-4, 273-292.

35a. Pratt, C.J. (1970). *Summary of experimental data for flume tests over 0.49mm sand*. Department of Civil Engineering, University of Southampton.

36a. Raichlan, F. and Kennedy, J.F. (1965). The growth of sediment bed forms from an initially flattened bed. *Proc. XI IAHR Congress*, Leningrad, Vol. 3, Paper 3.7.

37a. Seitz, H.R. (1976). *Suspended and bedload sediment transport in the Snake and Clearwater Rivers in the vicinity of Lewiston, Idaho*. Report 76-886, U.S. Geological Survey, Boise, Idaho.

38a. Shen, H.W. and Cheong, H.F. (1977). Statistical properties of sediment bed profiles. *Journal of the Hydraulics Division*, ASCE, 103(11), 1303-1321.

39a. Shinohara, K. and Tsubaki, T. (1959). *On the characteristics of sand waves formed upon the beds of open channels and rivers*. Report of the Research Institute for Applied Mechanics, Kyushu University, Japan, Vol. VII, No. 25.

40a. Simons, D.B., Richardson, E.V. and Hausbild, W.L. (1963). Some effects of fine sediment on flow phenomena. *U.S. Geological Survey Water Supply Papers*, 1498-G, Washington.

41a. Simons, D.B., Richardson, E.V. and Albertson, M.L. (1961). Flume studies of alluvial streams using medium sand. *U.S. Geological Survey Water Supply Papers*, 1498-A, 76 pp.

42a. Singh, B. (1960). *Transport of bed load in channels with special reference to gradient and form*. Ph.D. Thesis, London University, England.

43a. Stein, R.A. (1965). Laboratory studies of total load and bed load. *Journal of Geophysical Research*, 70(8), 1831-1842.

44a. Termes, A.P.P. (1986). *Dimensies van beddindvormen onder permanente stromingsom standigheden bij hoog sedimenttransport*. (In Dutch) Verslag onderzoek, M2130/Q232.

45a. Toffaleti, F.B. (1968). *A procedure for computation of the total river sand discharge and detailed distribution, bed to surface*. Technical Report No. 5, Committee of Channel Stabilization, U.S. Army Corps of Engineers.

46a. Williams, G.P. (1970). Flume width and water depth effects in sediment transport experiments. *U.S. Geological Survey Professional Papers*, 562-H, 1-37.

47a. Yalin, M.S. and Karahan, E. (1979). Steepness of sedimentary dunes. *Journal of the Hydraulics Division*, ASCE, 105(4), 381-392.

48a. Znamenskaya, N.S. (1963). Experimental study of the dune movement of sediment. *Soviet Hydrology: Selected papers*. American Geophysical Union, No. 3, 253-275.

Appendix B

Sources of bar data

1b. Ahmari, H. and da Silva (2011). Regions of bars, meandering and braiding in da Silva and Yalin's plan. *Journal of Hydraulic Research*, 49(6), 718-727.

2b. Ashida, K. and Shiomi,Y. (1966). *On the hydraulic of dunes in alluvial channels*. Annual Report No. 9., Disaster Prevention Research Institute, Kyoto University, Japan.

3b. Boraey, A.A. (2014). *Alternate bars under steady state flows: time of development and geometric characteristics*. Ph.D. Thesis, Department of Civil Engineering, Queen's University, Kingston, Canada.

4b. Chang, H.Y., Simons, D. and Woolhiser, D. (1971). Flume experiments on alternate bar formation. *Journal of the Waterways, Harbors and Coastal Engineering Division*, ASCE, 97(1), 155-165.

5b. da Silva, A.M.F. (1991). *Alternate bars and related alluvial processes*. M.Sc. Thesis, Department of Civil Engineering, Queen's University, Kingston, Canada.

6b. Fujita, Y., Nagata, N. and Muramoto, Y. (1989). Experiments on multiple bar formation and stream braiding on fine sand bed. *Disaster Prevention Research Institute Annuals*, Kyoto University, Japan, No. 32, Part B-2, 595-618.

7b. Fujita, Y. and Muramoto, Y. (1989). Multiple bars and stream braiding. *Proceedings of the International Conference on River Regime*, edited by W.R. White, published on behalf of Hydraulics Research Ltd., Wallingford, John Wiley and Sons, 289-300.

8b. Fujita, Y., Akamatsu, H. and Muramoto, Y. (1986). Experiments on formative process of double row bars and braided streams. *Disaster Prevention Research Institute Annuals*, Kyoto University, Japan, No. 29, Part B-2, 451-472.

9b. Fujita, Y. and Muramoto, Y. (1985). Study on the process of development of alternate bars. *Disaster Prevention Research Institute Bulletin*, Kyoto University, Japan, 35(3), 55-86.

10b. Fujita, Y., Muramoto, Y. and Furukawa, T. (1982). Formative conditions of meso-scale river bed configuration. *Disaster Prevention Research Institute Annuals*, Kyoto University, Japan, No. 25, Part B-2, 429-449.

11b. Fujita, Y., Muramoto, Y. and Horiike, S. (1981). Study on the process of alternate bar development. *Disaster Prevention Research Institute Annuals*, Kyoto University, Japan, No. 24, Part B-2, 411-431.

12b. Fujita, Y. (1980). *Fundamental study on channel changes in alluvial rivers*. Thesis presented to Kyoto University, Kyoto, Japan, in partial fulfillment of the requirements for the degree of Doctor of Engineering.

13b. García, M. and Niño, Y. (1993). Dynamics of sediment bars in straight and meandering channels: experiments on the resonance phenomenon. *Journal of Hydraulic Research*, 31(6), 739-761.
14b. Iguchi, M. (1980). *Tests for fine gravel transport in a large laboratory flume.* (In Japanese) Report for National Science Foundation, School of Earth Science, Tsukuba University, Ibaragi, Japan.
15b. Ikeda, H. (1983). *Experiments on bed load transport, bed forms, and sedimentary structures using fine gravel in the 4-meter-wide flume.* Environmental Research Center Papers, No. 2, The University of Tsukuba, Japan.
16b. Ikeda, S. (1984). Prediction of alternate bar wavelength and height. *Journal of Hydraulic Engineering*, 110(4), 371-386.
17b. Jaeggi, M. (1983). *Alternierende Kiesbanke.* Mitteilungen der Versuchsanstal fur Wasserbau, Hydrologie und Glaziologie, Zurich, No. 62.
18b. Kinoshita, R. (1980). *Model experiments based on the dynamic similarity of alternate bars.* (In Japanese) Research Report, Ministry of Construction, Aug.
19b. Kinoshita, R. (1961). *Investigation of the channel deformation of the Ishikari River.* (In Japanese) Memorandum No. 36, Science and Technology Agency, Bureau of Resources.
20b. Kuroki, M., Kishi, T. and Itakura, T. (1975). *Hydraulic characteristics of alternate bars.* (In Japanese) Report for National Science Foundation, Department of Civil Engineering, Hokkaido University, Hokkaido, Japan.
21b. Lanzoni, S. (2000a). Experiments on bar formation in a straight flume. 1. Uniform sediment. *Water Resources Research*, 36(11), 3337-3349.
22b. Lanzoni, S. (2000b). Experiments on bar formation in a straight Flume 2. Graded sediment. *Water Resources Research*, 36(11), 3351-3363.
23b. Muramoto, Y. and Fujita, Y. (1978). The classification of meso-scale bed configuration and the criteria of its formation. Proceedings of the 22nd Japanese Conference on Hydraulics, Japan Society of Civil Engineers, 275-282.
24b. Muramoto, Y. and Fujita, Y. (1977). Study on meso-scale river bed configuration. *Disaster Prevention Research Institute Annuals*, Kyoto University, Japan, No. 20, Part B-2, 243-258.
25b. Tubino, M. (1991). Flume experiments on alternate bars in unsteady flow. In *Fluvial hydraulics of mountain regions: Lecture Notes in Earth Sciences*, Vol. 37, edited by A. Armanini and G. Di Silvio, Springer-Verlag, Germany, 103-117.
26b. Valentine, E.M., Benson, I.A., Nalluri, C. and Bathurst, J.C. (2001). Regime theory and the stability of straight channels with bankfull and overbank flow. *Journal of Hydraulic Research*, 39(3), 259-268.
27b. Villard, P.V. and Church, M. (2005). Bar and dune development during a freshet: Fraser River Estuary, British Columbia, Canada. *Sedimentology*, 52, 737-755.
28b. Watanabe, Y., Sato, K., Hoshi, K. and Oyama, F. (2002). Experimental study on bar formation under unsteady flow conditions. *Proceedings of River Flow 2002, 1st International Conference on Fluvial Hydraulics*, Louvain-La-Neuve, Belgium, edited by D. Bousmar and Y. Zech, A.A. Balkema, The Netherlands, Vol. 2, 813-823.
29b. Yoshino, F. (1967). Study on bed forms. *Collected Papers*, Vol. 4, Department of Civil Engineering, University of Tokyo, Japan, 165-176.

Appendix C

Sources of friction factor data

1c. Bogardi, J. and Yen, C.H. (1939). *Traction of pebbles by flowing water*. Ph.D. Thesis, State University of Iowa.

2c. Brownlie, W.R. (1981). *Compilation of alluvial channel data: laboratory and field*. Report No. KH-R-43B, W.M. Kech Laboratory of Hydraulics and Water Resources, California Institute of Technology, Pasadena, California.

3c. Chyn, S.D. (1935). *An experimental study of the sand transporting capacity of the flowing water on sandy bed and the effect on the composition of the sand*. Thesis presented to the Massachussetts Institute of Technology, Cambridge, Massachussetts.

4c. Einstein, H.A. (1944). *Bed load transportation in Mountain Creek*. Technical Paper 55, U.S. Department of Agriculture, Soil Conservation Service.

5c. Gibbs, C.H. and Neill, C.R. (1972). *Interim report on laboratory study of basket-type bed-load samplers*. Research Council of Alberta in association with Department of Civil Engineering, University of Alberta, Canada.

6c. Ho, Pang-Yung (1939). *Abhangigkeit der Geschiebebewegung von der Kornform und der Temperature*. Preuss. Versuchsanst. fur Wasserbau and Schiffbau, Berlin, Mitt., Vol. 37.

7c. Jorissen, A.L. (1938). Étude expérimentale du transport solide des cours d'eau. *Revue Universelle des Mines*, Belgium, 14(3), 269-282.

8c. Knott, J.M. (1974). *Sediment discharge in the Trinity River basin, California*. Water Resource Investigations Report 73-49, U.S. Geological Survey.

9c. MacDougall, C.H. (1933). Bed-sediment transportation in open-channels. *Transactions of the Annual Meeting 14*, American Geophysical Union, 491-495.

10c. Mahmood, K. et al. (1979). *Selected equilibrium-state data from ACOP canals*. Report No. EWR-79-2, Civil, Mechanical and Environmental Engineering Department, George Washington University, Washington D.C.

11c. Mavis, F.T., Liu, T. and Soucek, E. (1937). *The transportation of detritus by flowing water*. Bulletin 11, Iowa University Studies in Engineering.

12c. Meyer-Peter, E. and Müller, R. (1948). Formulas for bed load transport. *Proceedings of the Second Meeting of International Association for Hydraulic Structures Research*, IAHR, Stockholm, 39-64.

13c. Samide, G.W. (1971). *Sediment transport measurements*. M.Sc. Thesis, University of Alberta, Canada.

14c. Sato, S., Kikkawa, H. and Ashida, K. (1958). Research on the bed load transportation. *Journal of Research*, Public Works Research Institute, Vol. 3, Research Paper 3, Construction Ministry, Tokyo, Japan, 21 pp.

15c. United States Army Corps of Engineers, U.S. Waterways Experiment Station (1936). *Studies of light-weight materials, with special reference to their movement and use as model bed material.* Technical Memorandum 103-1, Vicksburg, Mississippi.

Appendix D

Sources of regime data

1d. Ackers, P. and Charlton, F.G. (1970). The geometry of small meandering streams. *Proceedings of the Institution of Civil Engineers*, Paper 7328S, London.

2d. Ackers, P. (1964). Experiments on small streams in alluvium. *Journal of the Hydraulics Division*, ASCE, 90(4), 1-37.

3d. Bray, D. (1979). Estimating average velocity in gravel-bed rivers. *Journal of the Hydraulics Division*, ASCE, 105(9), 1103-1122.

4d. Brownlie, W.R. (1981). *Compilation of alluvial channel data: laboratory and field*. Report No. KH-R-43B, W.M. Keck Laboratory of Hydraulics and Water Resources, California Institute of Technology, Pasadena, California.

5d. Center Board of Irrigation and Power (1976). *Library of canal data (Punjab and Sind; Upper Ganga, U.S. canals)*. Technical Report No. 15, June.

6d. Chitale, S.V. (1970). River channel patterns. *Journal of the Hydraulics Division*, ASCE, 96(1), 201-221.

7d. Church, M. and Rood, R. (1983). *Catalogue of alluvial river channel regime data*. Department of Geography, University of British Columbia, Vancouver, Canada.

8d. Cinotto, P.J. (2003). *Development of regional curves of bankfull-channel geometry and discharge for streams in non-urban, Piedmont physiograpic province, Pennsylvania and Maryland*. Geological Survey Water Resources Investigations, Report 2003-4014.

9d. Colosimo, C., Coppertino, V.A. and Veltri, M. (1988). Friction factor evaluation in gravel-bed rivers. *Journal of Hydraulic Engineering*, 114(8), 861-876.

10d. Hey, R.R. and Thorne, C.R. (1986). Stable channels with mobile beds. *Journal of Hydraulic Engineering*, 112(8), 671-689.

11d. Higginson, N.N.J. and Johnston, H.T. (1988). Estimation of friction factor in natural streams. In *Proceedings of the International Conference on River Regime*, edited by W.R. White, published on behalf of Hydraulics Research Ltd., Wallingford, John Wiley and Sons, 251-266.

12d. Kellerhals, R., Neill, C.R. and Bray, D.I. (1972). *Hydraulic and geomorphic characteristics of rivers in Alberta*. Alberta Cooperative Research Program in Highway and River Engineering, Alberta, Canada.

13d. Khan, H.R. (1971). *Laboratory study of alluvial river morphology*. Ph.D. Dissertation, Colorado State University, Fort Collins, Colorado.
14d. Lan, Y.Q. (1990). *Dynamic modeling of meandering alluvial channels*. Ph.D. Dissertation, Colorado State University, Fort Collins, Colorado.
15d. Lawlor, S.M. (2004). *Determination of channel-morphology characteristcs, bankfull dischare, and various design-peak discharges in Western Montana*. U.S. Geological Survey Scientific Investigations, Report 2004-5263.
16d. Odgaard, A.J. (1987). Stream bank erosion along two rivers in Iowa. *Water Resources Research*, 23(7), 1225-1236.
17d. Schumm, S.A. (1968). River adjustments to altered hydrologic regime – Murrumbidgee River and Paleochannels, Australia. *U.S. Geological Survey Professional Papers*, 598, 1-63.
18d. Sherwood, M. and Huitger, A.C. (2005). *Bankfull characteristics of Ohio streams and their relation to peak streamflows*. U.S. Geological Survey Scientific Investigations, Report 2005-5153.
19d. Simons, D.B. and Albertson, L. (1960). Uniform water conveyance channels in alluvial material. *Journal of the Hydraulics Division*, ASCE, 86(5), 33-71.
20d. Westergaard, B.E., Mulvihill, C.I., Ernst, A.G. and Baldigo, B.P (2005). *Regionalized equations for bankfull-discharge and channel characteristics of streams in New York state – Hydrologic region in central New York*. U.S. Geological Survey Scientific Investigations, Report 2004-5247.

Appendix E

Sources of meandering and braiding data

1e. Abbado, D., Slingerland, R. and Smith, N.D. (2005). Origin of anatamosis in the upper Columbia River, Canada. *Proceedings of the 7th International Conference on Fluvial Sedimentology*, Special Publication of the International Association of Sedimentologists, edited by N.E. Lincoln, M.D. Blum, S.B. Marriott and S.F. Leclair, Blackwell, Oxford, 35, 3-15.

2e. Ackers, P. and Charlton, F.G. (1970). The geometry of small meandering streams. *Proceedings of the Institution of Civil Engineers*, Paper 7328S, London.

3e. Annable, W.K. (1994). *Morphologic characteristic of water courses in southern Ontario*. Report, Credit Valley Conservation Authority, Ontario, Canada.

4e. Ashmore, P. (1991). Channel morphology and bed-load pulses in braided, gravel-bed streams. *Geografiska Annaler*, Series A, Physical Geomorphology, 73(1), 37-52.

5e. Bray, D.I. (1979). Estimating average velocity in gravel-bed rivers. *Journal of the Hydraulics Division*, ASCE, 105(9), 1103-1122.

6e. Burge, L.M. (2005). Wandering Miramichi Rivers, New Brunswick, Canada. *Geomorphology*, 69, 253-274.

7e. Chitale, S.V. (1970). River channel patterns. *Journal of the Hydraulics Division*, ASCE, 96(1), 201-221.

8e. Church, M. and Rood, K. (1983). *Catalogue of alluvial river channel regime data*. Report, Natural Sciences and Engineering Research Council of Canada, Department of Geology, University of British Columbia, Vancouver, BC, Canada.

9e. Crosato, A. and Mosselman, E. (2009). Simple physics-based predictor for the number of river bars and the transition between meandering and braiding. *Water Resources Research*, 45, W03424, doi:10.1029/2008WR007242.

10e. Dietrich, W.E. and Smith, J.D. (1983). Influence of the point bar on flow through curved channels. *Water Resources Research*, 19(5), 1173-1192.

11e. Fujita, Y., Nagata, N. and Muramoto, Y. (1989). Experiments on multiple bar formation and stream braiding on fine sand bed. *Disaster Prevention Research Institute Annuals*, Kyoto University, Japan, No. 32, Part B-2, 595-618.

12e. Fujita, Y. and Muramoto, Y. (1989). Multiple bars and stream braiding. *Proceedings of the International Conference on River Regime*, edited by W.R. White, published on behalf of Hydraulics Research Ltd., Wallingford, John Wiley and Sons, 289-300.

13e. Fujita, Y. and Muramoto, Y. (1982). Experimental study on stream channel processes in alluvial rivers. Disaster Prevention Research Institute Bulletin, Kyoto University, Japan, Vol. 32, part I, No. 288.

14e. Hey, R.D. and Thorne, C.R. (1986). Stable channel with mobile gravel beds. *Journal of Hydraulic Engineering*, 12(8), 671-689.

15e. Jansen, J.D. and Nanson, G.C. (2004). Anabranching and maximum flow efficiency in Magela Creek, northern Australia. *Water Resources Research*, 40, W04503, doi:10.1029/2003WR002408.

16e. Kellerhals, R., Neill, C.R. and Bray, D.I. (1972). *Hydraulic and geometric characteristics of rivers in Alberta*. Alberta Cooperative Research Program in Highway and River Engineering.

17e. Kinoshita, R. (1980). *Model experiments based on the dynamic similarity of alternate bars*. (In Japanese) Research Report, Ministry of Construction of Japan, Aug.

18e. Lapointe, M.F. and Carson, M.A. (1986). Migration pattern of an asymmetric meandering river: the Rouge River, Quebec. *Water Resources Research*, 22(5), 731-743.

19e. Leopold, L.B. and Wolman, M.G. (1957). River channel patterns: braided, meandering and straight. *U.S. Geological Survey Professional Papers*, 282-B, 39-84.

20e. Neill, C.R. (1973). Hydraulic geometry of sand rivers in Alberta. *Proceedings of the Symposium on Fluvial Processes and Sedimentation*, Subcommittee on Hydrology, Associate Committee on Geodesy and Geophysics, National Research Council of Canada, 9, 341-380.

21e. Odgaard, A.J. (1987). Streambank erosion along two rivers in Iowa. *Water Resources Research*, 23(7), 1225-1236.

22e. Odgaard, A.J. (1981). Transverse bed slope in alluvial channel bends. *Journal of the Hydraulics Division*, ASCE, 107(12), 1677-1694.

23e. Pizzuto, J.E. and Meckelnburg, T.S. (1989). Evaluation of a linear bank erosion equation. *Water Resources Research*, 25(5), 1005-1013.

24e. Prestegaard, K.L. (1983). Bar resistance in gravel bed streams at bankfull stage. *Water Resources Research*, 19(2), 472-476.

25e. Schumm, S.A. and Khan, H.R. (1972). Experimental study of channel patterns. *Geological Society of America Bulletin*, 83, 1755-1770.

26e. Schumm, S.A. (1968). River adjustments to altered hydrologic regime – Murrumbidgee River and paleochannels, Australia. *U.S. Geological Survey Professional Papers*, 598, 1-62.

27e. Smith, N.D. (1971). Transverse bars and braiding in the Lower Platte River, Nebraska. *Geological Society of America Bulletin*, 82(12), 3407-3420.

28e. Struiksma, N. and Klaasen, G.J. (1988). On the threshold between meandering and braiding. *Proceedings of the International Conference on River Regime*, edited by W.R. White, published on behalf of Hydraulics Research Ltd., Wallingford, John Wiley and Sons, 107-120.

29e. Tooth, S. and McCarthy, T.S. (2004). Controls on the transition from meandering to straight channels in the wetlands of the Okavango delta, Botswana. *Earth Surface Processes and Landforms*, 29, 1627-1649.

30e. Tooth, S. and Nanson, G.C. (2004). Forms and processes of two highly contrasting rivers in arid central Australia, and the implications for channel-pattern discrimination and prediction. *Geological Society of America Bulletin*, 116(7-8), 802-816.
31e. Valentine, E.M., Benson, I.A., Nalluri, C., and Bathurst J.C. (2001). Regime theory and the stability of straight channels with bankfull and overbank flow. *Journal of Hydraulic Research*, 39(3), 259-268.
32e. van den Berg, J.H. (1995). Prediction of alluvial channel pattern of perennial rivers. *Geomorphology*, 12, 259-279.

Subject index

Printed and bound by CPI Group (UK) Ltd, Croydon, CR0 4YY
24/10/2024
01778290-0001